Climate Change Law and Policy in the Middle East and North Africa Region

Climate Change Law and Policy in the Middle East and North Africa Region provides an in-depth and authoritative examination of the guiding principles of climate change law and policy in the MENA region.

This volume introduces readers to the latest developments in the regulation of climate change across the region, including the applicable legislation, institutions, and key legal innovations in climate change financing, infrastructure development, and education. It outlines participatory and bottom-up legal strategies—focusing on transparency, accountability, gender justice, and other human rights safeguards—needed to achieve greater coherence and coordination in the design, approval, financing, and implementation of climate response projects across the region. With contributions from a range of experts in the field, the collection reflects on how MENA countries can advance existing national strategies around climate change, green economy, and low carbon futures through clear and comprehensive legislation.

Taking an international and comparative approach, this book will be of great interest to students, scholars, and practitioners who work in the areas of climate change, environmental law and policy, and sustainable development, particularly in relation to the MENA region.

Damilola S. Olawuyi, SAN, FCIArb. is an associate professor of petroleum, energy, and environmental law at Hamad Bin Khalifa University (HBKU) College of Law, Doha, Qatar. He is also Chancellor's Fellow and Director of the Institute for Oil, Gas, Energy, Environment and Sustainable Development (OGEES Institute), Afe Babalola University, Ado Ekiti, Nigeria. He is an Independent Expert of the Working Group on Extractive Industries, Environment, and Human Rights Violations in Africa formed by the African Commission on Human and Peoples' Rights.

Routledge Studies in Environmental Policy

For more information about this series, please visit <https://www.routledge.com/
Routledge-Studies-in-Environmental-Policy/book-series/RSEP>

Climate Change Law and Policy in the Middle East and North Africa Region

Edited by Damilola S. Olawuyi

Routledge
Taylor & Francis Group
LONDON AND NEW YORK

earthscan
from Routledge

First published 2022
by Routledge
2 Park Square, Milton Park, Abingdon, Oxon OX14 4RN

and by Routledge
605 Third Avenue, New York, NY 10158

Routledge is an imprint of the Taylor & Francis Group, an informa business

British Library Cataloguing-in-Publication Data
A catalogue record for this book is available from the British Library

Library of Congress Cataloging-in-Publication Data
Names: Olawuyi, Damilola S. (Damilola Sunday), 1983- editor.
Title: Climate change law and policy in the Middle East and
North Africa region/[edited by] Damilola S. Olawuyi.
Other titles: Routledge studies in environmental policy.
Description: Abingdon, Oxon; New York, NY: Routledge, 2021. |
Series: Routledge studies in environmental policy |
Includes bibliographical references and index.
Identifiers: LCCN 2021006919 (print) | LCCN 2021006920 (ebook) |
ISBN 9780367490324 (hardback) | ISBN 9781032052441 (paperback) |
ISBN 9781003044109 (ebook)
Subjects: LCSH: Climatic changes–Law and legislation–Middle East. |
Climatic changes–Law and legislation–Africa, North. |
Climatic changes–Government policy–Middle East. |
Climatic changes–Government policy–Africa, North.
Classification: LCC KMC706.5 .C58 2021 (print) | LCC KMC706.5
(ebook) | DDC 344.5604/634–dc23
LC record available at https://lccn.loc.gov/2021006919
LC ebook record available at https://lccn.loc.gov/2021006920

ISBN: 978-0-367-49032-4 (hbk)
ISBN: 978-1-032-05244-1 (pbk)
ISBN: 978-1-003-04410-9 (ebk)

Typeset in Goudy
by Deanta Global Publishing Services, Chennai, India

Contents

Contributors

Dalal Aassouli, Assistant Professor of Islamic Finance, College of Islamic Studies, Hamad Bin Khalifa University, Doha, Qatar.

Mohsen Abdollahi, Associate Professor of International and Environmental Law, Head of Human Rights and Environmental Law department, Shahid Beheshti University, Tehran, Iran.

Rasha Abu-El-Ruz, Teaching Assistant, Department of Public Health, College of Health Sciences, Qatar University, Doha, Qatar.

Karam Turk-Adawi, Assistant Professor of Public Health, Department of Public Health, College of Health Sciences, Qatar University, Doha, Qatar.

Saad Belkasmi, PhD Candidate in Law, Hassan first University, Morocco.

Hilary Christina Bell, Assistant Professor, College of Law, Hamad Bin Khalifa University, Doha, Qatar.

Laila Dalaa, PhD Candidate in Law, Hassan first University, Morocco.

Ali O. Diriöz, Assistant Professor, International Entrepreneurship Department, TOBB Economics and Technology University, Ankara, Turkey.

Georgios Dimitropoulos, Associate Professor of Law at Hamad Bin Khalifa University (HBKU) College of Law, Doha, Qatar.

Alexander Ezenagu, Assistant Professor, College of Law, Hamad Bin Khalifa University, Doha, Qatar.

Riyad Fakhri, Professor of Law and Director of the Research Laboratory in Business Law at Hassan first University, Morocco.

Roxana A. Mastor, Project Manager, The United Nations Office for Project Services (UNOPS), Jerusalem.

Robert Home, Emeritus Professor of Land Management, Anglia Ruskin University, United Kingdom.

Samira Idllalène, Cadi Ayyad University, Polydisciplinary Faculty, Law Department (Safi, Morocco), International Development Cooperation Research Laboratory (LRCID), Faculty of Legal, Economic and Social Sciences, Marrakesh, Morocco.

Cameron Kelly, General Counsel, Australian Renewable Energy Agency, Sydney, Australia.

Almas Lokhandwala, LL.M. Candidate at Harvard Law School, Cambridge, Massachusetts, United States of America.

Elizabeth Maruma Mrema, United Nations Assistant Secretary General & Executive Secretary, Secretariat of the Convention on Biological Diversity.

Damilola S. Olawuyi, SAN, FCIArb, Associate Professor of Petroleum, Energy, and Environment Law at Hamad Bin Khalifa University (HBKU) College of Law, Doha, Qatar. He is also Chancellor's Fellow and Director of the Institute for Oil, Gas, Energy, Environment, and Sustainable Development (OGEES Institute), Afe Babalola University, Ado Ekiti, Nigeria.

Ioannis Papageorgiou, Energy Economic Development Associate, The United Nations Office for Project Services (*UNOPS*), Jerusalem.

Andreas Rechkemmer, Senior Professor at the College of Public Policy, Hamad Bin Khalifa University, Doha, Qatar.

S. Duygu Sever, PhD, Department of International Relations, Koç University, İstanbul, Turkey.

Aphrodite Smagadi (PhD EUI), Legal Officer, United Nations Environment Programme.

Foreword

The book *Climate Change Law and Policy in the Middle East and North Africa Region* is a key effort of the legal community to raise awareness, educate, and provide solutions through strategies, policies, and legislation for one of the regions of the world most critically affected by climate change. It also provides insights on opportunities that can help build resilience, such as renewable energy patterns or Islamic climate finance as a catalyst to adaptation and mitigation and to working towards Sustainable Development Goal 13 on Climate Action. This book aims to support Middle East and North Africa (MENA) countries by promoting an exchange of information and best practices among countries that share common climate challenges and legal traditions.

Environmental law can be an effective tool for countries in the MENA region to address the negative effects of climate change, such as rising temperatures, reduced land for residential purposes and agriculture, and water scarcity. Such efforts will also go a long way in supporting efforts to "building back better" after the COVID-19 pandemic and in turning the tide globally when it comes to meeting the biggest planetary crisis of our time, climate change. Actions that countries take now will be critical towards achieving the objectives of the United Nations Framework Convention on Climate Change and the Paris Agreement. These actions will also further promote and advance environmental rule of law and justice.

As the recently published United Nations Environment Programme (UNEP) Global Climate Change Litigation Report, 2020 Status Review shows, there is need for more coherent and effective laws as a means to reduce negative impacts on the environment and the health of present and future generations. UNEP has a long history of supporting countries in the MENA region in building capacity in the area of environmental law. We remain committed to providing such support and hope that this book, which details best practices and underlines the potential of harnessing the growing body of climate change law and policy within the region, will be a useful contribution towards this goal.

Arnold Kreilhuber
Acting Director, Law Division
United Nations Environment Programme

Preface and acknowledgements

Climate change poses complex economic, social, and environmental threats to the Middle East and North Africa (MENA) region—arguably, more so than to any other region of the world. Apart from climate-induced fatal heat waves and debilitating sea level rise, climate change could have wide-ranging effects on extant and future energy infrastructure in the region. This could include potential failure of facilities, reduced life expectancy of buildings, and increased operational and maintenance expenditure of electricity, water, and aviation infrastructure, amongst others. Even without climate change, MENA countries are currently subjected to tough arid conditions and extreme heat, which typically affect the structural integrity, operation, and life span of water, energy, coastal, and transportation infrastructure. Climate change will only escalate these pre-existing conditions. The problem of climate change raises the need for greater development and enforcement of laws in order to tackle the climate change challenge on time in the region.

However, despite the rise in public awareness and policy formulation on climate change across the MENA region, only a few countries have developed clear and comprehensive legal and institutional frameworks on climate change. Furthermore, climate change education is still at an alarming stage of infancy in the region. One key reason for the slow pace of the expansion of legal education on climate change in the region is the absence of an authoritative text that unpacks the nature and guiding principles of climate change law and policy in the region. For example, while there are several cutting-edge books on climate change law and policy in other regions, the law, governance, cultural and ethical dimensions of climate change mitigation, and adaptation in the MENA region have yet to receive detailed, authoritative, and book-length exposition. This book fills that gap.

This book provides an in-depth and authoritative examination of the guiding principles of climate change law and policy in the MENA region. It introduces readers to the latest developments in the regulation of climate change across the MENA region, including the applicable legislation, institutions, and key legal innovations in climate change financing, infrastructure development, and disruptive legal education. Yet this book is not simply a stock-taking exercise. It also explores larger questions on participatory and bottom-up legal strategies—focusing on transparency, accountability, gender justice, and other human rights safeguards—needed to achieve greater coherence and coordination in the design, approval, financing, and implementation of climate response projects across the region. The book also provides legal assessments and reflections on how MENA countries can advance existing national strategies and visions on climate change, green economy, and a low carbon future amongst others, through clear and comprehensive legislation.

Overall, the book aims to enhance an understanding of how the terms "climate change", "climate change law", "climate responsibility", and "climate governance" are conceptualized in the MENA region, with special focus on the influence of Islamic principles on climate change law in the region. It appraises how the guiding principles of international and regional instruments on climate change are applied, implemented, and enforced at national and municipal levels across the MENA region. It also explores how legal innovations relating to blockchain technology, artificial intelligence, disruptive legal education, climate smart infrastructure, climate disaster response, and Islamic climate financing, amongst others, are shaping climate regulation and governance across the MENA region.

The book is prepared in a user-friendly style to enhance its utility among its primary audience, namely students, corporations, environment departments and ministries, law firms, courts, arbitrators, notably international and regional committees, and tribunals before whom arguments over climate change impacts and projects often come for resolution. The book analyzes the key roles that international institutions such as the United Nations Environment Programme (UNEP), the International Renewable Energy Agency, the International Maritime Organization, and the World Trade Organization; regional institutions such as the League of Arab Nations, the Gulf Cooperation Council, the Asian Development Bank, and the African Development Bank; as well as national institutions such as ministries and departments, play in climate change mitigation and adaptation.

It remains for me to register my profound gratitude to many helping hands, without whom the publication of this book would not have been possible. First and foremost, my thanks and appreciation go to Almighty God for the successful publication of this book. Secondly, I appreciate the kind motivation and support of the Founder and Chancellor of Afe Babalola University—Aare Afe Babalola, SAN, OFR, CON, FNIALS, FCIArb., LL.D—for his unflinching support to me and my family. Aare Afe Babalola's selfless commitment and dedication to education, justice, sustainable development, and the rule of law have greatly inspired my interest in this field.

Furthermore, I am especially indebted to everyone at Qatar Foundation and Hamad Bin Khalifa University (HBKU), Doha, Qatar, for their exceptional love and support over the years. This book was initially conceived at the inaugural conference of the Association of Environmental Law Lecturers in MENA Universities (ASSELLMU) hosted by HBKU in November, 2018. Special thanks to Dean Susan L. Karamanian at the HBKU College of Law for her exceptional contributions to the success of that conference and for her ongoing dedication and support for ASSELLMU's events and activities. Thanks are also due to the UNEP and the Qatar National Research Fund for the research funding and technical support without which this book, as well as all other ASSELLMU initiatives and programmes, would not have been possible. I am deeply grateful to Elizabeth Mrema and Aphrodite Smagadi at UNEP for their support and contributions to the success of this book.

Special thanks are also due to the editorial staff of Routledge for the smooth and professional review process which guided this book through to its timely completion. A thank you also goes to Sarah L. Macleod (Schulich School of Law, Dalhousie University, Canada) as well as Umair Dogar (Qatar Foundation) for providing remarkable and thoughtful research assistance. They assisted greatly in collating chapters at submission, review, and editing stages. I also acknowledge the dedication and collegiality of all the contributing authors, whose substantial research and commitment to leading-edge scholarship in this field have been pivotal to the production of this book.

Finally, I would like to thank my dear wife Oluwabunmi and our wonderful children, Titilayo and Oluwatoni, for their love, support, and inspiration. Thanks are also due to my exceptional students whose class contributions and ideas greatly shaped the conceptualization and development of this book.

This book has endeavoured to state the position of the law as of 30 January 2021, although authors have been able to take into account subsequent developments in one or two instances.

Damilola S. Olawuyi
Doha, Qatar

About the editor

Damilola S. Olawuyi, SAN, FCIArb, is an associate professor of petroleum, energy, and environment law at Hamad Bin Khalifa University (HBKU) College of Law, Doha, Qatar. He is also Chancellor's Fellow and Director of the Institute for Oil, Gas, Energy, Environment, and Sustainable Development (OGEES Institute), Afe Babalola University, Nigeria.

A prolific and highly regarded scholar, Professor Olawuyi has practiced and taught energy law in Europe, North America, Asia, Africa, and the Middle East. He has served as a visiting professor at Columbia Law School, New York, China University of Political Science and Law, and senior visiting research fellow at the Oxford Institute for Energy Studies. In 2019, he was a Herbert Smith Freehills visiting professor at Cambridge University. He was formerly an international energy lawyer at Norton Rose Fulbright Canada LLP where he served on the firm's global committee on energy investments in Africa. He has delivered lectures on energy law in over 40 countries.

Professor Olawuyi has published close to a hundred articles, book chapters, and books on petroleum law, energy, and international environmental law. His most recent book publications include *Local Content and Sustainable Development in Global Energy Markets* (Cambridge University Press, 2021), *The Human Rights-Based Approach to Carbon Finance* (Cambridge University Press, 2016) and *Extractives Industry Law in Africa* (Springer, 2018). Professor Olawuyi serves on the executive committees and boards of several organizations. He is Vice Chair of the International Law Association; co-chair of the Africa Interest Group of the American Society of International Law (2016–2019); and member of the Academic Advisory Group of the International Bar Association's Section on Energy, Environment, Natural Resources, and Infrastructure Law (SEERIL). He is the Editor-in-Chief of the *Journal of Sustainable Development Law and Policy*. He is an Independent Expert of the Working Group on Extractive Industries, Environment, and Human Rights Violations in Africa, formed by the African Commission on Human and Peoples' Rights.

Professor Olawuyi holds a doctorate (DPhil) in energy and environmental law from the University of Oxford; a master of laws (LL.M.) from Harvard University; and another LL.M. from the University of Calgary. He has been admitted as Barrister and Solicitor in Alberta, Canada; Ontario, Canada; and

Nigeria. In 2020, he was elevated to the rank of Senior Advocate of Nigeria (Queen's Counsel equivalent), the highest mark of distinction and excellence in the Nigerian legal profession.

Professor Olawuyi is a regular media commentator on all aspects of natural resources, energy and environmental law. Further information about his profile and publications can be found at <www.damilolaolawuyi.com>.

Abbreviations

AAOIFI	Accounting and Auditing Organization for Islamic Financial Institutions
ACRI	Arab Climate Resilience Initiative
AEEP	Africa-EU Energy Partnership
AFAPCC	Arab Framework Action Plan on Climate Change
AOAD	Arab Organization for Agricultural Development
ASDRR	Arab Strategy for Disaster Risk Reduction
ASSELLMU	Association of Environmental Law Lecturers in Middle East and North African Universities
AWC	Arab Water Council
AWMC	Arab Ministerial Water Council
AYCM	Arab Youth Climate Movement
CAMRE	Council of Arab Ministers Responsible for the Environment
CBI	Climate Bonds Initiative
CDM	Clean Development Mechanism
CFC	Chlorofluorocarbon
COVID-19	Coronavirus Disease 2019
DFI	Development Finance Institutions
ESD	Education for Sustainable Development
ETS	Emissions Trading Systems
EU	European Union
EVD	*Ebola* Virus Disease
FDE	Energy Development Fund
FIT	Feed-in Tariffs
GCC	Gulf Cooperation Council
GDP	Gross Domestic Product
GHG	Greenhouse Gases
H$_2$O	Water Vapour
HFC	Hydrofluorocarbons
IEA	International Energy Agency
IFAAS	Islamic Finance Advisory and Assurance Services
INDC	Intended Nationally Determined Contributions
InforMEA	Information Portal on Multilateral Environmental Agreements

IPCC	Intergovernmental Panel on Climate Change
IsDB	Islamic Development Bank
JCPOA	Joint Comprehensive Plan of Action
JI	Joint Implementation
KACARE	King Abdullah City for Atomic and Renewable Energy
KSA	Kingdom of Saudi Arabia
LAS	League of Arab States
LEED	Leadership in Energy and Environmental Design
LEPAP	Lebanon Environmental Pollution Abatement Program
MASEN	Moroccan Agency for Solar Energy
MBA	Market-Based Approaches
MDB	Multilateral Development Bank
MEA	Multilateral Environmental Agreements
MENA	Middle East and North Africa
MERS	Middle East Respiratory Syndrome
N_2O	Nitrous Oxide
NCF	National Climate Funds
NDC	National Determined Contributions
NGO	Non-Governmental Organizations
NGPB	National Green Participative Banks
NOx	Nitrogen Oxide
O_3	Tropospheric Ozone
OECD	Organisation for Economic Co-operation and Development
OIC	Organization of Islamic Cooperation
OPT	Occupied Palestine Territories
PADELIA	Programme on Capacity Building for the Development of Environmental Law and Institutions in Africa
PDT	Public Trust Doctrine
PFC	Perfluorocarbons
PPP	Public Private Partnership
RECREEE	Regional Centre for Renewable Energy and Energy Efficiency
REDD+	Reducing Emissions from Deforestation and Forest Degradation
RES	Renewable Energy Sources
SAGIA	Saudi Arabia General Investment Authority
SARS	Severe Acute Respiratory Syndrome
SDG	Sustainable Development Goal
SE4All	Sustainable Energy for All
SF_6	Sulphur Hexafluoride
SIDS	Small Island Developing States
SLCP	Short-lived Climate Pollutants
SO_2	Sulphur Dioxide
SSA	Sub-Saharan Africa
SWF	Sovereign Wealth Funds
UN	United Nations

UNCCD	United Nations Convention to Combat Desertification
UNCED	United Nations Conference on Environment and Development
UNDP	United Nations Development Programme
UNEP	United Nations Environment Programme
UNFCCC	United Nations Framework Convention on Climate Change
UNOCHA	United Nations Office for the Coordination of Humanitarian Affairs
UNOPS	United Nations Office for Project Services
UNSC	United Nations Security Council
VCLT	Vienna Convention of the Law of Treaties
WFP	World Food Programme
WTO	World Trade Organization
ZCL	Zoonotic Cutaneous Leishmaniasis

Part I

Introductory context and principles

1 Nature and sources of climate change law and policy in the MENA region

Damilola S. Olawuyi

Introduction

The aim of this book is to examine the guiding principles of climate change law and policy in the Middle East and North Africa (MENA) region.[1] It explores the applicable legislation, institutions, as well as lessons learned from emerging innovative and bottom-up approaches to climate regulation across the region. It also provides holistic assessments and reflections on how MENA countries can advance existing national strategies and visions on climate change, through clear and comprehensive legislation.

Climate change is one of the greatest concerns facing our present generation.[2] Simply defined, climate change is an increase in the average temperature of the atmosphere at an abnormal rate, caused mainly by the anthropogenic (human-induced) emission of gases known as greenhouse gases (GHGs) that trap the sun's heat in the atmosphere.[3] GHGs that contribute to climate change include carbon dioxide (CO_2); nitrous oxides (N_2O); water vapour (H_2O); short-lived climate pollutants (SLCPs),[4] such as black carbon, methane (CH_4), tropospheric ozone (O_3); and fluorinated gases, such as chlorofluorocarbons (CFCs), hydro-fluorocarbons (HFCs), perfluorocarbons (PFCs), and sulfur hexafluoride (SF_6).[5] These GHGs, which are mainly by-products of industrial activities, absorb heat radiation that should normally escape to space, thereby heating the atmosphere at an abnormal rate. Due to increasing human dependence on the combustion of fossil fuels, such as coal, oil, and gas, that emit GHGs, the concentrations of GHGs in the atmosphere are increasing at exponential and alarming rates.[6] For example, in 2013, scientific studies concluded that atmospheric concentrations of CO_2 (one of the potent GHGs) exceeded 400 parts per million for the first time in human history, while the emissions of other GHGs, such as methane and nitrous oxide, have continued to increase.[7] With the increased manifestation of extreme weather events across the world, such as devastating floods, droughts, and tropical cyclones across the MENA region,[8] there is now increased consensus that human activity is disrupting the global climate system; and that if left unchecked, the direct and indirect effects of climate change may threaten lives, livelihoods, and public infrastructure all over the world.[9]

Climate change poses even more serious economic, social, and environmental threats to the MENA region—arguably, more so than any other region in the world. Although the MENA region is not one homogeneous geographical unit, MENA countries have historical similarities in culture, language (Arabic), religion (Islam), geography, as well as in their contributions and vulnerabilities to climate change.[10] The MENA region, especially the Gulf countries—Kuwait, Iran, Bahrain, Oman, Qatar, Saudi Arabia, and the United Arab Emirates (UAE)—is home to some of the world's highest exporters of oil and natural gas.[11] Also, for many years, the oil sector has accounted for over 80 per cent of export earnings in North African countries such as Egypt, Libya, and Algeria.[12] Due to the fossil fuel-dependent nature of the economies of several MENA countries, the MENA region has the fastest growth rate in GHG emissions globally.[13] For example, in 2018, the MENA region was responsible for 3.2 billion tonnes of CO_2 and 8.7 per cent of global GHGs.[14] Efforts to address global climate change, and the growing transition away from carbon-intensive fossil fuels across the world, could, therefore, result in fundamental changes in the economies, energy mix, lifestyles, and overall development of several MENA countries.[15]

More importantly, climate change raises complex existential threats to several MENA countries.[16] Apart from the unique geographical vulnerabilities of MENA countries, which contribute to their low adaptive capacity, the region is home to a number of small and low lying islands, such as Qatar and Bahrain, countries that have a dual vulnerability to climate change, both as arid countries and as small island developing states.[17] Even without climate change, several MENA countries are currently subjected to arid and desert-like conditions which typically make farming and agriculture difficult and near impossible. As such, climate change would only escalate these pre-existing social, economic, and environmental conditions, and could intensify the cycle of food shortage, desertification, water scarcity, depleting fish stocks, and the spread of diseases in the MENA region.[18] Recent World Bank studies indicate that the MENA region is expected to have the greatest economic losses from climate-related water scarcity, estimated at 6 to 14 per cent of the gross domestic product (GDP) by 2050.[19] For example, the MENA region is home to 12 out of the world's 17 most water-stressed countries.[20]

Apart from climate-induced water scarcity, flooding, and fatal heat waves, climate change could have wide-ranging effects on extant and future infrastructure in MENA countries.[21] This includes the potential failure of facilities, reduced life expectancy of buildings, and an increased operational and maintenance expenditure of electricity, water, and aviation infrastructure, amongst others.[22] To effectively address the impacts of climate change on critical infrastructure in the MENA region, there is a need for increased investment in climate-smart infrastructure, i.e. buildings, structures, and systems that reduce GHG emissions, and improve society's ability to adapt to, and cope with, the risks posed by climate change.[23]

Despite the grim reality of climate change, the MENA region remains one of the least prepared regions facing it.[24] Clear and comprehensive legal frameworks on climate change have not been so easily forthcoming in many parts of the region. Furthermore, climate change education is still at an alarming stage of infancy in the region.[25] Currently, only very few universities in the MENA region have dedicated law courses that advance knowledge and capacity on climate change law. According to the findings of recent regional conferences on environmental law, one key reason for the slow pace of environment legal education in the region is the absence of an authoritative text that unpacks the nature and guiding principles of climate change law and policy in the region.[26] While earlier scholarships have provided country-focused analyses on climate change in select countries and sectors, a detailed, authoritative, and book-length exposition of the multi-dimensional, region-wide, and intersectional nature of climate change law and policy in the MENA region has remained absent. This book fills that gap.

This book provides an in-depth and authoritative exposition of the nature, scope, and content of law, policy, and institutional frameworks for advancing climate change action in MENA countries. It also explores practical challenges and social justice issues that stifle the design, approval, financing, and implementation of climate response projects across the region, and how they can be holistically addressed.

This chapter provides a conceptual overview of the nature and scope of climate change law in the MENA region. The second section of this chapter provides an overview of the different sources of law that underpin climate change law and policy in the region. It discusses the character, status, and force of the different sources, including the interrelationships between them. The following section outlines the overall aim and structure of the book.

Sources of climate change law in the MENA region

Climate change law is that branch of law that provides the general legal framework for the regulation and protection of human and natural elements of the environment from the risk of climate change and its potential environmental and socio-economic consequences.[27] It includes the body of laws, rules, regulations, and statutes that regulate the emission of GHGs that cause climate change, as well as national plans, strategies, and policies aimed at ameliorating the anticipated adverse consequences of climate change.

Climate change law and policy have two equally important and interrelated objectives: mitigation and adaptation.[28] Climate mitigation refers to efforts that are designed to reduce the emission of GHGs that cause climate change. Mitigation includes the reduced use of carbon-intensive fuels across all key sectors and industries.[29] Climate change adaptation, on the other hand, refers to efforts aimed at coping with the actual or anticipated impacts of climate change.[30] It includes policy adjustments, physical projects, and infrastructural changes that are proposed as a means of ameliorating the anticipated adverse consequences of climate change. For example, increased investment in climate-smart infrastructure that can deal with more frequent rainfall, flooding, bushfires, heatwaves, and rising sea levels.

The discipline of climate change law has developed incrementally over the course of the last three decades, drawing largely from principles and norms in different traditional areas of law such as torts law, criminal law, and international law. Generally, there are three key sources of climate change law and policy in the MENA region: international law; domestic legislation; and judicial decisions and scholarly publications.

International law on climate change

International law governs relations between sovereign nations.[31] International climate change law is therefore a field of international law that regulates the behaviour of states and international organizations with respect to the protection and management of the global climate. It establishes the obligations of countries to prevent dangerous anthropogenic interference with the climate systems.

As recognized in Article 38 of the Statute of the International Court of Justice, the major sources of international law, and by extension, international climate law are international conventions and treaties; international customs; general principles of law; and judicial decisions and teachings of highly qualified publicists.[32] According to Article 26 of the 1969 Vienna Convention of the Law of Treaties (VCLT), which sets out the laws that govern treaties in international law, every responsible state is expected to carry out the terms of its obligations under an international agreement.[33] Over the last decade, MENA countries have played active roles in the formation of international law on climate change.[34] However, for an international treaty to have a force of law domestically, in addition to signature, ratification, and accession, the treaty or convention has to be domesticated.[35] Domestication is the process of making an international treaty part of domestic national laws. Provisions on the domestication of treaties are often set out in the constitution. For example, Article 68 of Qatar's Constitution provides that a "treaty or agreement shall have the power of law after ratification and publication in the official Gazette".[36] This provision, as seen in essentially all constitutions across the region, requires that, for an international obligation to take effect domestically, a law accepting that obligation as part of domestic law must first be made.

As discussed in Chapter 2 of this book, the publication of the Brundtland Commission report in 1987, which elaborated the concept of sustainable development; the establishment in 1988, of the Intergovernmental Panel on Climate Change (IPCC), a body of over two thousand scientists, to provide comprehensive scientific assessments on the risk of climate change; and the adoption of the Rio Declaration on Environment and Development at the 1992 Conference on Environment and Development held in Rio de Janeiro (Earth Summit) are key landmark events in the history and development of climate change law and policy in the MENA region.

THE UNITED NATIONS FRAMEWORK CONVENTION ON CLIMATE CHANGE (UNFCCC)

Adopted in 1992 at the Rio Earth Summit, the ultimate aim of the Convention, as set out in Article 2, is to reduce the emission of GHGs that cause climate change. The Convention is very important for being the first legal instrument to formally recognize climate change as a global concern and to espouse aspirational goals on the need to collectively tackle climate change. The Convention also established two key institutional bodies, namely the Conference of the Parties (COP), which is generally charged with regular review of the implementation of the Convention and any related legal instruments,[37] and the Secretariat, which is tasked with providing institutional coordination and support for the COP and other relevant international bodies under the Convention.[38]

However, given that the UNFCCC does not establish legally binding emission reduction targets, it soon became apparent that there was a need to do more if climate change was to be effectively tackled. By 1995, parties to the UNFCCC met in Berlin (the first COP to the UNFCCC) to discuss the need for legally binding emission reduction commitments as a stronger response to the problem of climate change. This led to negotiations to develop an instrument that could strengthen the global response to climate change. Two years later, the Kyoto Protocol was adopted in Kyoto, Japan, and became the first international agreement that establishes legally binding emission reduction targets on climate change. It was adopted in December 1997 and it entered into force in February 2005.

THE KYOTO PROTOCOL

The Kyoto Protocol is recognized as one of the most important global agreements of the late 20th century, firstly because it *commits* industrialized countries to stabilize GHG emissions by setting legally binding emission reduction targets and obligations.[39] These targets add up to an average of five per cent emissions reduction, compared to 1990 levels, over the first five-year period from 2008 to 2012.[40] Secondly, it establishes three flexible, project-based mechanisms: emissions trading (ET), joint implementation (JI), and the clean development mechanism (CDM).[41] These mechanisms allow industrialized countries to achieve their emission reduction objectives at the lowest cost possible by investing in projects that lead to emission reduction and sustainable development.

At the 17th COP to the UNFCCC in Durban, South Africa, Parties agreed to establish an Ad Hoc Working Group on a Durban Platform for Enhanced Action (AWG-DP). The AWG-DP had the mandate to develop "a protocol, another legal instrument or an agreed outcome with legal force under the Convention applicable to all Parties" that would be ready for signing in Paris in 2015.[42] At the 21st COP in Paris, the Paris Agreement was adopted.

THE PARIS AGREEMENT

The Paris Agreement builds upon the goals of the UNFCCC by setting a clear goal to reduce emissions and keep global temperature under 2°C (3.6°F), and limit it to 1.5°C above pre-industrial levels starting in the year 2020.[43] Article 4(1) also emphasizes the need to achieve the long-term temperature goal "on the basis of equity, and in the context of sustainable development and efforts to eradicate poverty". Article 4 grants Parties the discretion to determine and communicate their nationally determined contributions (NDCs) that they intend to achieve in contribution to the overall objectives of the agreement. Furthermore, Article 6 encourages flexibility in implementation by encouraging Parties to use voluntary approaches and internationally transferred mitigation outcomes towards achieving the NDCs.[44] Furthermore, Article 7 encourages Parties to enhance their adaptive capacity, strengthen resilience, and reduce vulnerability to climate change, with a view to contributing to sustainable development. Article 7(5) specifically encourages the need for "a country-driven, gender-responsive, participatory and fully transparent approach" to climate adaptation, taking into consideration vulnerable groups, communities, and ecosystems.[45]

As signatories to the Paris Agreement and the UNFCCC, MENA countries have increasingly adopted national plans, policies, and strategies that recognize the need to lower current levels of GHG emissions.[46] Furthermore, MENA countries have already committed, in their NDCs, to investing in climate-smart energy systems, that is structures and systems that lower GHG emissions, and improve the ability to adapt to, and cope with, the risks posed by climate change.[47] As the future of international climate change governance continues to evolve, MENA countries will need to intensify the scale and ambition of their NDCs, as well as domestic policies and laws, to accelerate climate change mitigation and adaptation.

SOFT LAW INSTRUMENTS

In addition to the above binding international climate law instruments, a number of non-binding and aspirational documents and declarations have emerged at international and regional levels that provide foundations for the development and application of climate change law and policy in the MENA region. A good example at the international level is the United Nations Sustainable Development Goals (SDGs), where SDG 13 encourages all countries to "take urgent action to combat climate change and its impacts", and to "integrate climate change measures into national policies, strategies and planning".[48]

At the regional level, the 1991 Arab Declaration on Environment and Development and Future Prospects called on Arab countries to limit the degradation of the environment and natural resources, and manage them "in a sustainable manner that ensures Arab water and food security, the preservation of ecosystems and biodiversity and the control of desertification".[49] Although non-binding, this Declaration provides normative guidance on how MENA countries could address significant environmental problems facing the region, including climate change. Similarly, in 2001, the Council of Arab Ministers Responsible for the Environment (CAMRE) adopted the Arab Declaration to the World Summit on Sustainable Development, which renewed the commitment of Arab countries to work together to address the vulnerabilities of Arab countries to water scarcity, food insecurity, and climate change, amongst others.[50] Furthermore, the 2007 Arab Ministerial Declaration on Climate Change contains regional aspirations by all Ministers responsible for environment and climate change to include "policies to deal with climate change issues in all sectors within national and regional policies for sustainable development".[51] It calls on all MENA countries to prioritize climate change mitigation and adaptation through the production and use of cleaner fuels, improving the efficiency of energy use in all sectors, diversifying energy sources, and investing in cleaner production techniques and environmentally friendly technologies, amongst others.[52] Likewise, the 2017 Rabat Islamic Declaration on Environment Protection and Achieving Sustainable Development Goals adopted by Ministers for Environment and Climate Change in member states of the Organization of Islamic Cooperation (OIC), called on Islamic countries to advance climate change mitigation and adaptation in all aspects of planning at national levels.[53] Furthermore, the League of Arab States issued the Arab Framework Action Plan on Climate Change 2010–2020, which provides a framework for Arab countries to address climate change challenges while achieving sustainable development targets.[54]

Though not legally binding, these soft law instruments provide strong normative guidance on how governments, national authorities, investors, and other stakeholders in the MENA region can adopt domestic laws and policies to address the adverse impacts of climate change.

Domestic legislation

Although the nature and scope of domestic legislation relating to climate change varies from one MENA country to another, climate change laws and policies at domestic levels can be divided into three main categories.

THE CONSTITUTION

The constitution is the supreme law and it has binding force on all authorities and persons in a country.[55] The primary source of law across the MENA region is the constitution as it sets the basic principles and norms that define other legislation enacted by the legislature or the Al-Shoura Council. While climate change is not specifically mentioned, the constitutions of several MENA countries expressly codify provisions on environmental protection and sustainable development.[56] For example, Article 33 of Qatar's Constitution of 2003 provides that the state "endeavours to protect the environment and its natural balance, to achieve comprehensive and sustainable development for all generations".[57] Similarly, Article 9(h) of Bahrain's Constitution of 2002 provides that "the State shall take the necessary measures for the protection of the environment and the conversation of wildlife".[58] These constitutional provisions directly incorporate the protection of all aspects of the environment, including the climate, from degradation and pollution, and provide strong foundational basis for subsequent policy and legislative action to address climate change.

Also given that Islam is the dominant religion in the MENA region, Sharia law is constitutionally recognized as the principal source of law in essentially all the countries of the region.[59] For example, Article 1 of Qatar's 2005 Constitution

Table 1.1 Key declarations and instruments relating to climate change by regional bodies in the MENA region

Instrument	Highlights
Arab Declaration on Environment and Development and Future Perspectives, Cairo, 1991	The Declaration called on all Arab countries to limit the degradation of the environment and natural resources, and manage both in a sustainable manner.
Arab Declaration to the World Summit on Sustainable Development, 2001	It renewed the commitment of Arab countries to work together to advance environmental protection and sustainable development, especially by addressing the vulnerabilities of Arab countries to water scarcity, food insecurity, and climate change.
Abu Dhabi Declaration on the Future of Environmental Action in the Arab World, 2001	It identifies the priority of environmental problems in the Arab world as: "the acute shortage and deteriorating quality of sources of water; the paucity and deteriorating quality of exploitable land; the imprudent consumption of natural resources; urban sprawl and its associated problems; the degradation of marine, coastal and watered areas". It also identifies the need for clean production methods and technologies to reduce emissions.
Arab Charter on Human Rights, 2004	Article 38 states that "every person has the right to an adequate standard of living for himself and his family, which ensures their well-being and a decent life, including food, clothing, housing, services and the right to a healthy environment. The States parties shall take the necessary measures commensurate with their resources to guarantee these rights".
The Arab Ministerial Declaration on Climate Change of December 6, Cairo, 2007	It contains regional aspirations by all Ministers responsible for environment and climate change to include "policies to deal with climate change issues in all sectors within national and regional policies for sustainable development". The Declaration also provides that: "Adaptation to measures that address climate change shall be fully consistent with the economic and social development and in such a way so as to achieve sustainable economic growth and eradication of poverty".
Cairo Declaration on Development Challenges and Population Dynamics in a Changing Arab World, 2013, (paras 66, 85, and 87)	It called on Arab countries to develop regional and local climate change response measures that take into account the distribution, vulnerability, and resilience of the targeted populations.
Rabat Islamic Declaration on Environment Protection and Achieving Sustainable Development Goals, 2017	Called on Islamic countries to advance climate change mitigation and adaptation in all aspects of planning at national levels. It also affirms the importance of a green economy transition.

(*Continued*)

Table 1.1 Continued

Instrument	Highlights
Arab Framework Action Plan on Climate Change, 2010–2020	It creates a ten-year master plan for climate mitigation and adaptation in the Arab region.
Pan-Arab Renewable Energy Strategy, 2030 (League of Arab States)	It identifies the integration of renewable energy as a tool for addressing climate change vulnerabilities in the Arab region. It sets a target of increasing installed renewable energy power generation capacity across the region by the year 2030.

clearly states that Qatar is "an independent sovereign Arab State. Its religion is Islam and the Sharia law shall be the principal source of its legislation".[60] Consequently, the four sources of Sharia law: the noble *Quran*; the *Sunnah* or teachings of the Prophet Muhammad (peace and blessings be upon him); *Ijma* or teachings of consensus of various scholars of Islam; and *Qiyas* or analogical reasoning both deductive and inductive—all provide pivotal foundations for the development and practice of environmental law and policy, as well as climate change regulation across the region.[61] Islamic principles emphasize that humankind is a mere steward or trustee (*khalīfah*) of the earth and not a proprietor nor a disposer or one who ordains; and that all Muslims have the solemn duty to maintain and preserve the natural environment from disequilibrium or damage.[62] Consequently, as discussed in Chapters 3 and 12 of this book, Islamic principles have widely influenced various aspects of climate regulation such as climate financing, resilience planning, and green endowments (*Awqafs*).[63]

PRIMARY LEGISLATION

All MENA countries have enacted different environment-related legislation (also known as acts, statutes, decrees, or laws) that seeks to limit environmental pollution and ensure a safe and healthy environment for the public.[64] Such legislation provides the basis for national authorities to penalize and sanction all forms of air pollution, i.e. any chemical, physical, or biological change or modification of the natural characteristics of the atmosphere, in proportions that could be harmful to human life and nature, including contributions to climate change.

However, as earlier noted, while climate awareness and action are increasing across the MENA region, almost all of the countries in the region do not have specific legislation on climate change, a situation which has not fostered a coherent development of climate change principles, norms, and standards across the region.[65] With more available knowledge on the scope and urgency of legislative action required to address climate change, it has become important for countries across the world to move beyond general environmental laws and provide clear and comprehensive law and policy frameworks on climate change.[66] In addition to enacting general environmental protection laws, it is imperative for MENA countries to establish clear and comprehensive laws and policies for domestic regulation of climate change planning in order to provide a coherent framework for monitoring implementation and enforcement.[67]

The importance of developing specific domestic legislation to advance climate change mitigation and resilience planning cannot be overemphasized.[68] A clear legal framework on climate change can provide a legal basis and obligation for project planners and stakeholders to integrate climate resilience into the design, operation, and maintenance of energy infrastructure.[69] For example, Article 7(5) of the Paris Agreement specifically encourages countries to integrate climate adaptation into relevant socio-economic and environmental policies and actions, where appropriate. This will include designing climate-specific laws and policies, as well as mainstreaming climate change considerations into national planning and development processes in all key ministries and sectors.[70]

SECONDARY LEGISLATION

The third main source of domestic climate change regulation flows from the second. Beneath the tier of primary environmental or climate change legislation is a wide range of detailed regulations, by-laws, national visions and plans, ministerial decisions, and guidelines that are used to flesh out much of the detail of the primary legislation. Ministries or departments that oversee environment and climate change regulation often release regulations, rules, notices, warrants, and guidelines pursuant to their enabling primary legislation, which provide clarity and further information on the national strategies, visions, and policies that investors and operators in key sectors are expected to comply with across their entire operations. Given that they are legal instruments of the state released by national authorities with enabling powers under an Act of Parliament, such documents significantly shape policy and practice. Failure to comply with such secondary instruments may also result in loss of business opportunities and in some cases penalties and sanctions, depending on the enforcement powers that have been granted by the primary statute.

For example, while direct climate change laws are generally few across the MENA region, some MENA countries have put in place regulations, action plans, directives, and national visions that address various aspects of climate change mitigation and adaptation.[71] For example, in Qatar, Article 64 of the Executive By-Law for the Environment Protection Law contains elaborate provisions on the reduction of all sources of air pollution, especially the emission of several GHGs that cause climate change.[72]

Furthermore, secondary legislation includes instruments that establish special committees on climate change and the CDM. For example, Bahrain's Resolution No. 51 of 2007 establishes a Joint Committee on Climate Change, with wide powers to monitor and control all activities, projects, and programmes related to climate change in the country.[73] Qatar's Resolution of the Council of Ministers also establishes a Committee on Climate Change and CDM that plays similar roles in laying down strategies, policies, and communications on climate change in the country.[74] Similarly, Oman and the UAE have released national action plans on climate change with detailed climate mitigation and adaptation goals.[75] Also, Qatar's National Vision (QNV) 2030 expressly outlines Qatar's plans to play a "proactive and significant regional role in assessing the impact of climate change and mitigating its negative impacts", and to actively support international efforts to address climate change.[76] Furthermore, in 2016, Oman issued a Ministerial Decision for the management of climate affairs, which requires "greenhouse gas emitting projects" to obtain a climate affairs permit.[77]

While the wide range of action plans and strategies on climate change show increased political commitment across the MENA region to address the problem of climate change, a lot more can be done to increase the coherence and effectiveness of such plans. There is a need to develop holistic climate response strategies that could enhance climate change mitigation and adaptation across diverse sectors and industries. Contextualizing the unique threats of climate change at the domestic level, within a broader climate risk reduction and response strategy, will not only help to clearly identify priority areas in terms of vulnerabilities, physical damages, and interruptions to critical energy infrastructure, but could also yield additional benefits that can promote the full realization of extant national visions on decarbonization, energy diversification, energy efficiency, and climate resilience.[78]

Judicial decisions and scholarly publications

In addition to primary sources of law, such as the constitution, Acts and codes, and international treaties, the development of climate change law and policy in the MENA region is highly influenced by secondary sources of law, such as judicial decisions of courts and tribunals, as well as scholarly works of highly qualified jurists. However, contrary to the *stare decisis* rule applied in the common law system, case law does not constitute a binding source of law in civil law countries in the MENA region.[79] Also, decisions of courts, such as the Court of Cassation and Court of Appeal, do not normally bind either themselves or lower courts.[80] Previously decided cases can, however, be consulted persuasively to guide the court, especially in emerging areas of law.

Furthermore, courts may consult the teachings, commentaries, and opinions of highly qualified jurists and scholars (e.g., law professors, attorneys, judges, etc.), especially in shaping the development of new, emerging, and developing areas of law, such as climate change law. These include the reports of international bodies, such as the International Law Commission, the Academic Advisory Group constituted by the International Bar Association's Section on Energy, Environment, Natural Resources and Infrastructure Law, environmental law academy of the International Union for Conservation of Nature (IUCN), and the International Law Association, that offer authoritative analyses of a wide variety of issues pertaining to climate change law. A good example is the seminal publication by the IUCN in 1983 on the Islamic principles on environmental law.[81]

Similarly, a number of non-binding documents have been released by eminent scholars and clerics to spur climate change action in the MENA region. For example, in 2015, the Islamic Declaration on Global Climate Change was adopted by a coalition of eminent Islamic clerics and scholars.[82] The Declaration notes that climate change is a result of human disruption of the perfect ecological equilibrium (mīzān) bestowed by the Almighty creator, and called upon all Islamic nations to drastically reduce their GHG emissions in order to address the "corruption (fasād) that humans have caused on Earth in our relentless pursuit of economic growth and consumption".[83] The Bogor Declaration on Muslim Action on Climate Change 2010 equally stresses the need for all Muslim countries to address climate change through education, resilience planning, development of green Islamic urban cities, and Islamic labelling for environmentally friendly goods and services.[84] These non-binding declarations provide faith-based frameworks and guidance on how governments, national authorities, investors, and other stakeholders across the MENA region can contribute to global climate action.

Furthermore, a number of advocacy movements are emerging across the region that provide grassroots education, resources, and local awareness on climate change law and policy. For example, the Arab Youth Climate Movement (AYCM) has emerged as one of the largest youth-led climate education, capacity development, and advocacy groups in the MENA region with branches in more than 15 countries, including Egypt, Qatar, Jordan, Syria, Iraq, Palestine, Lebanon, Algeria, and Morocco, amongst others.[85] Likewise, Green Building Councils across the MENA region continue to provide reports, information, and best practices that support informed policy making on low-carbon, energy-efficient, and environmentally sustainable practices in building design and construction.[86] The expert reports and educational resources provided by these organizations have greatly shaped the rapid evolution and development of climate change law and policy across the MENA region.

Climate change law and policy is evidently a rapidly evolving field in the MENA region. While the codification of specific climate change legislation remains in progress in many countries of the region, a wide range of environmental legislation and regulations, including constitutional principles on environmental protection and sustainable development, widely outlaw the emission of pollutants that may degrade human and natural environments. Furthermore, as discussed in Chapter 6 of this book, all MENA countries have designated or established focal ministries, departments, or committees that oversee climate change mitigation and adaptation, including CDM implementation. The chapters of this book review and evaluate how the plethora of policies, instruments, and scientific studies are shaping responses to climate change across the region.

Aim, scope, and structure of the book

The extreme vulnerability of MENA countries to the problem of climate change raises the need for greater development and enforcement of laws in order to tackle the climate change challenge on time in the MENA region. As noted earlier, while earlier scholarships have provided country-focused analyses and surveys of existing laws and policies on climate change in select countries and sectors, the multi-dimensional, region-wide, and intersectional nature of climate change law and policy in the MENA region has yet to receive a detailed, book-length exposition. This analytical gap and the ensuing regulatory disconnect have prevented systemic and comparative evaluation of the normative underpinnings, functionality, and best practices on climate change law and policy in the region. With case studies from across the region, this book provides in-depth analysis and exposition of how climate change laws and policies are evolving in the MENA region, as well as the challenges that remain. It also offers recommendations on the guiding principles of a sustainable and rights-based approach to the design, application, and implementation of climate change law and policy.

While this book offers a scan of the sources and underpinning principles of climate change law and policy in the region, it is clearly acknowledged that the substantive chapters cannot unpack and analyze every applicable piece of legislation and instrument in all MENA countries. The central organizing principle and aim of this book, therefore, is to enhance an understanding of the guiding principles, legislation, policies, and overarching institutions on climate change across the MENA region. It also fosters an understanding of the unique legal, policy, and governance challenges faced in the region in seeking to achieve the SDGs, and other international instruments and documents relating to climate action.

Structure

Each of the chapters of the book provides a detailed and rigorous background of the regulatory context on climate change, the barriers that remain, and innovative approaches for advancing various aspects of climate change regulation, such as financing, infrastructure development, resilience planning, urban development, and carbon taxation, amongst others.

This book is organized into three parts to reflect a transition from theoretical concepts to key practical discussions. Part I of the book, comprising five chapters, introduces the values, principles, and instruments that underpin climate change law and practice across the region. It considers the key regulatory approaches to environmental regulation in the MENA region, most especially the influence of Sharia law, and the use of command and control and incentives-based regulation (especially through government subsidies) to enforce climate change mitigation and adaptation.

Part II of the book consists of a series of geographical case studies that apply these frameworks to countries in the MENA region. The case studies identify concerns of weak legal structures, social exclusion, environmental tradeoffs, lack of transparency, and incoherent application of mitigation standards that tend to limit climate action in some MENA countries. This part also highlights positive and high leverage climate change laws, policies, and strategies in the region, the contexts in which they are being implemented, barriers to their effective implementation, and innovative legal approaches to promote such strategies.

Part III of the book offers reflections on the case studies and addresses how lessons from the diverse jurisdictions may inform thoughts on how to effectively design, apply, and implement sustainable and rights-based climate change laws in the MENA region.

The 17 chapters of this book provide multi-jurisdictional and systematic exposition of how MENA countries can improve law, policy, and institutional capacity on climate change mitigation, adaptation, urban development, climate disaster risk reduction, and resilience planning.

Conclusion

Addressing the widescale adverse impacts of climate change on lives and livelihoods in the MENA region will require clear and coherent laws and policies that place climate change squarely at the heart of national planning and decision making. A comprehensive analysis and study of the structural and non-structural challenges that arise in the design, application, and implementation of climate law and policy across the region can help us to plot a comprehensive path for achieving policy coherence and reform.

The systemic and multi-jurisdictional survey of the unique and underlying features of climate change regulation in diverse jurisdictions offered by this book can simplify the task of advancing low-carbon energy transition and climate change action in key sectors and industries in MENA countries.

Notes

1 Twenty countries are typically included as part of the MENA region: Algeria, Bahrain, Djibouti, Egypt, Iran, Iraq, Israel, Jordan, Kuwait, Lebanon, Libya, Malta, Morocco, Oman, Qatar, Saudi Arabia, Syria, Tunisia, United Arab Emirates, and Yemen. Similarly, some classifications include Sudan, Turkey, and the State of Palestine, a *de jure* sovereign state. See The World Bank, "Middle East and North Africa", <https://data.worldbank.org/region/middle-east-and-north-africa> accessed 5 January 2021.
2 E. Chan, "Climate Change is the World's Greatest Threat – In Celsius or Fahrenheit?" (2018) 60 *Journal of Environmental Psychology* 21–26.
3 United Nations General Assembly, *United Nations Framework Convention on Climate Change (UNFCCC)*, 20 January 1994, UN Doc. A/RES/48/189; D. Olawuyi, *Principles of Nigerian Environmental Law* (Afe Babalola University Press 2015) 108–113.
4 SLCPs are gases that remain in the atmosphere for a much shorter period of time than carbon dioxide (CO_2), but have a higher global warming potential compared to CO_2. Reducing SLCPs has been identified in scientific studies as one of the most effective strategies for constraining global climate change. See D. Zaelke, N. Borgford-Parnell, and others, "Primer on Short-Lived Climate Pollutants" (2013) Institute for Governance and Sustainable Development, *IGSD Working Paper* 5–10.
5 See Intergovernmental Panel on Climate Change (IPCC), "Climate Change 2007", the Fourth Assessment Report, in which the IPCC stated that it is "very likely" that human activity is the main driver of the rise in temperatures since 1950 [AR4]. (All IPCC reports are available at www.ipcc.ch/reports/.)

6 A. M. Haywood and others, "Large-Scale Features of Pliocene Climate: Results from the Pliocene Model Intercomparison Project" (2013) 9 *Climate of the Past* 191, 192.

7 Ibid.

8 For example, Oman has been hit by at least four cyclones over the last 12 years. Similarly, an unprecedented number of devastating floods have hit Qatar, Kuwait, Bahrain, and UAE over the last years. In 2018, the Kuwait's minister of public works resigned due to damaging floods. See "Kuwait public works minister resigns after damaging flood", 19 November 2018 <www.reuters.com/article/us-kuwait-floods/kuwa it-public-works-minister-resigns-after-damaging-flood-idUSKCN1NE2EN>; H. M. Fritz and others, "Cyclone Gonu Storm Surge in Oman" (2010) 86 *Estuarine, Coastal and Shelf Science* 102–106.

9 As of 24 July 2020, 1755 jurisdictions and local governments had declared a climate emergency. See <https://climateemergencydeclaration.org/climateemergency-dec larations-cover-15-million-citizens/>; D. Olawuyi, "Financing Low-Emission and Climate-Resilient Infrastructure in the Arab Region: Potentials and Limitations of Public-Private Partnership Contracts" in W. L. Filho and A. A. Meguid (eds), *Climate Change Adaptation in the Arab Region: Case Studies and Best Practice* (Berlin: Springer 2017) 533–547.

10 The MENA region is considered one of the most vulnerable regions to climate change impacts. See IPCC, Synthesis Report, 2014, available at IPCC, n 5, 151–152.

11 BP, "Statistical Review of World Energy, 67th Edition" (BP 2018) 12.

12 Ibid. Also see Organization of Petroleum Exporting Countries (OPEC), "Annual Statistical Bulletin 2017" (OPEC 2017) 30.

13 R. A. Abbass, P. Kumar, and A. El-Gendy, "An Overview of Monitoring and Reduction Strategies for Health and Climate Change Related Emissions in the Middle East and North Africa Region" (2018) 175 *Atmospheric Environment* 33–43.

14 Ibid. Also see Global Carbon Atlas, "Carbon Emissions" <www.globalcarbonatlas.org /en/CO2-emissions>.

15 S. Tagliapietra, "The Impact of the Global Energy Transition on MENA Oil and Gas Producers" (2019) 26 *Energy Strategy Review* 26.

16 IPCC, Synthesis Report, 2014, n 5, 151–152. See also Economic and Social Commission for Western Asia (ESCWA), "Regional Initiative for the Assessment of the Impact of Climate Change on Water Resources and Socio-Economic Vulnerability in the Arab Region (RICCAR): Climate Projections and Extreme Climate Indices for the Arab Region" United Nations, League of Arab States, 2015, E/ESCWA/ SDPD/2015/Booklet.2.

17 See D. Verner, *Adaptation to a Changing Climate in the Arab Countries: A Case for Adaptation Governance and Leadership in Building Climate Resilience* (Washington, DC: World Bank 2012), 197; M. Luomi, *The Gulf Monarchies and Climate Change: Abu Dhabi and Qatar in an Era of Natural Unsustainability* (Oxford: Oxford University Press 2014) 2–5.

18 Olawuyi, n 9.

19 World Bank, *Beyond Scarcity: Water Security in the Middle East and North Africa. MENA Development Report* (Washington, DC: World Bank 2018).

20 United Nations Development Programme, Regional Bureau for Arab States, *Water Governance in the Arab Region: Managing Scarcity and Securing the Future* (United Nations Publications 2013) 11–28.

21 Olawuyi, n 9; H. Assaf, "Impact of Climate Change: Vulnerability and Adaptation Infrastructure" in M. K. Tolba and N. W. Saab (eds), *Arab Environment Climate Change – Impact of Climate Change on Arab Countries* (Arab Forum for Environment and Development 2009) 113–120.

22 J. Sieber, "Impacts of, and Adaptation Options to, Extreme Weather Events and Climate Change Concerning Thermal Power Plants" (2013) 121 *Climatic Change* 55–66.

23 Olawuyi, n 9.

24 UNEP Environment, *Outlook for the Arab Region: Environment for Development and Human Well-Being* (2010) 394–395. <https://www.unenvironment.org/resources/repor t/environment-outlook-arab-region-environment-development-and-human-well-be ing>. See also United Nations Economic and Social Commission for Western Asia (ESCWA) et al. (2017) Arab Climate Change Assessment Report—Main Report. E/ ESCWA/SDPD/2017/RICCAR/Report.

25 On the importance of environmental education, see P. Molthan-Hill and others, "Climate Change Education for Universities: A Conceptual Framework from an International Study" (2019) 226 *Journal of Cleaner Production* 1092, 1093.

26 D. Olawuyi, "Conference Highlights Need to Introduce Environmental Law to Higher Education Curricula in the Middle East" (International Union for Conservation of Nature 2018) <www.iucn.org/news/world-commission-environmental-law/201811/ conference-highlights-need-introduce-environmental-law-higher-education-curricul a-middle-east> accessed 5 January 2021.

27 J. Peel, "Climate Change Law: The Emergence of a New Legal Discipline" (2012) 32 *Melbourne University Law Review* 922–925.

28 These two objectives are captured in Article 2 of the UNFCCC, n 3.

29 Ibid.

30 Ibid.

31 Olawuyi, n 3.

32 United Nations, *Statute of the International Court of Justice*, 18 April 1946, 33 UNTS 993.

33 United Nations, *Vienna Convention on the Law of Treaties*, 23 May 1969, 1155 UNTS 331.

34 *The Permanent Constitution of the State of Qatar 0/2004*, article 68 [Qatar Constitution].

35 Olawuyi, n 3, 80–83.

36 Qatar Constitution, n 34.

37 UNFCCC, n 3, Article 7.

38 Ibid, Article 8.

39 D. Olawuyi, "Rethinking the Place of Flexible Mechanisms in Kyoto's Post 2012 Commitments" (2010) 6 *Journal of Law, Environment and Development* 23–35.

40 Ibid.

41 Ibid.

42 Ibid.

43 UNFCCC, *Paris Agreement under the United Nations Framework Convention on Climate Change, Conference of the Parties Twenty-First Session Paris*, 30 November to 11 December 2015, FCCC/CP/2015/L.9, Articles 2(1), 4, 6, and 7.

44 Ibid.

45 Ibid.

46 As of 2020, all MENA countries have signed the Paris Agreement, 15 countries ratified it and 13 submitted their first Nationally Determined Contributions (NDC) (Paris Agreement—Status of Ratification, https://unfccc.int/process/the-paris-agreement/sta tus-of-ratification).

47 D. Olawuyi, "Advancing Innovations in Renewable Energy Technologies as Alternatives to Fossil Fuel Use in the Middle East: Trends, Limitations, and Ways Forward" in D. Zillman, M. Roggenkamp, L. Paddock, and L. Godden (eds), *Innovation in Energy Law and Technology: Dynamic Solutions for Energy Transitions* (Oxford: Oxford University Press 2018) 354–370.

48 United Nations, *Transforming Our World: The 2030 Agenda for Sustainable Development*, Resolution adopted by the General Assembly on 25 September 2015, UN Doc. A/ RES/70/1.

49 Arab Declaration on Environment and Development and Future Prospects, adopted by the Arab Ministerial Conference on Environment and Development (10 September 1991) E/ESCWA/ENVHS/1992/1.

50 Arab Declaration to the World Summit on Sustainable Development, 2002, available at: www.hlrn.org/img/documents/Arab_Declaration_Sustainable_Dev.pdf.
51 Arab Ministerial Declaration on Climate Change adopted by the Council of Arab Ministers responsible for the Environment, 19th session (6 December 2007) E/ESCWA/SDPD/2012/Technical Paper.2.
52 Ibid.
53 Rabat Islamic Declaration on Environment Protection and Achieving Sustainable Development Goals adopted by 7th Islamic Conference of Ministers, Organization of Islamic States (25 October 2017).
54 The Arab Framework Action Plan on Climate Change (AFAP-CC; 2010–2020). See also the Arab Strategy for Water Security in the Arab Region to Meet the Challenges and Future Needs for Sustainable Development 2010–2030 and its associated action plan; and the Arab Strategy for Disaster Reduction 2020.
55 Olawuyi, n 3.
56 See D. Olawuyi, "Human Rights and the Environment in Middle East and North African (MENA) Region: Trends, Limitations and Opportunities" in J. May and E. Daly (eds), *Encyclopedia of Human Rights and the Environment, Indivisibility, Dignity and Legality* (Cheltenham: Edward Elgar 2018).
57 Ibid.
58 Ibid.
59 K. S. Vikør, "Sharīʿah" in E. El-Din Shahin (ed), *The Oxford Encyclopedia of Islam and Politics* (Oxford: Oxford University Press 2014).
60 Qatar Constitution, n 34.
61 Israel, Lebanon, Malta, and Turkey are key exceptions to this. Article 1A of Israel's Basic Law: Human Dignity and Liberty (1992) recognizes Israel as "a Jewish and democratic state"; while Article 9 of the Constitution of the Republic of Lebanon (rev. 2004) provides for freedom of all religions. Also, Article 2 of Turkey's Constitution of 1982 states that Turkey is a "democratic, secular and social state".
62 See IUCN, "Islamic Principles for the Conservation of the Natural Environment", IUCN Environmental Policy and Law Paper 20, 1983.
63 See also I. Ozdemir, "What Does Islam Say about Climate Change and Climate Action?" *Al Jazeera* (Doha, 12 August 2020).
64 See, for example, Qatar's Law No. 30 of 2002 on Environmental Protection; Algeria's Law 83-03 05 1983 on environmental protection; Egypt's Law No. 4 of 1994 on Environment and Morocco's Framework Law No. 99-12 of 6 March 2014 on the national charter for the environment and sustainable development. See also Jordan's Law No 13 of 2012 concerning Renewable Energy and Energy Efficiency Law (Adopted on 16 April 2012, published in Al Jarida Al Rasmiyya, 2012- 04- 16, No 5153) 1610–1618.
65 Malta is one exception as it has enacted the Climate Action Act, 2015 No. XVII of 2015 (as amended by Act No. XXXVI of 2020).
66 See D. Olawuyi, "Energy Poverty in the Middle East and North African (MENA) Region: Divergent Tales and Future Prospects" in I. Del Guayo, L. Godden, D. N. Zillman, M. F. Montoya, and J. J. Gonzalez (eds), *Energy Law and Energy Justice* (Oxford University Press 2020).
67 See Government of Qatar, "State of Qatar Second Voluntary National Review 2018 Submitted to the High-Level Political Forum", 2018, 42–433, identifying the need for comprehensive legislation on climate change risk management and disaster risk reduction in Qatar.
68 Olawuyi, n 66.
69 See Chapter 17 of this book.
70 See United Nations Development Programme, "Paving the Way for Climate-Resilient Infrastructure: Guidance for Practitioners and Planners", September 2011.
71 See, for example, Jordan's Climate Change Regulation No.79 of 2019, issued pursuant to Article 30 of the Environmental Protection Law No.6 of 2017, which designates

the Ministry of the Environment as the National Committee for Climate Change with powers to propose climate change policies and strategies for climate change mitigation and adaptation. See also Iran's Regulation of the Climate Change Convention and Kyoto Protocol Official Gazette No. 18775, 17 August 2009; also The National Climate Change Plan of the United Arab Emirates 2017–2050; and Morocco's Politique du Changement Climatique au Maroc 01 January 2014, LEX-FAOC152565.

72 Decision No. (30) of 2002 of the President of the Supreme Council of Environment and Natural Protection concerning the issuance of the Executive By-Law for The Environment protection Law.

73 Bahrain's Resolution No. 51 of 2007 establishing the Joint Committee on Climate Change.

74 See Qatar's Resolution of the Council of Ministers No. 15 of 2011 establishing the Committee on Climate Change and Clean Development at the Ministry of Environment.

75 See Ministry of Municipality and Environment of Qatar, "Qatar's Climate Change Strategy (CCS) for the Urban Planning and Development Sector" (MME Qatar) <www.mme.gov.qa/QatarMasterPlan/English/strategicplans.aspx?panel=ccs> accessed 6 January 2021; United Arab Emirates Ministry of Climate Change and Environment, "UAE's National Climate Action Plan of the United Arab Emirates 2017–2050" (MCCE UAE) <www.moccae.gov.ae/assets/30e58e2e/national-climate -change-plan-for-the-united-arab-emirates-2017-2050.aspx> accessed 6 January 2021; Ministry of Environment and Climate Affairs, Sultanate of Oman, "Second National Communication" (UNFCCC December 2019).

76 See Qatar National Vision 2030 <www.gco.gov.qa/en/about-qatar/national-vision 2030/>.

  See also UAE National Vision 2021 <www.vision2021.ae/en>; Bahrain National Vision 2030 <www.bahrainedb.com/en/about/Pages/economic%20vision%202030 .aspx>; Saudi Arabia Vision 2030 <http://vision2030.gov.sa/en>; Kuwait National Development Plan <www.newkuwait.gov.kw/en/plan/>; Oman Vision 2040.

77 Sultanate of Oman, Ministerial Decision No. 20 of 2016 Regulations for the Management of Climate Affairs.

78 Olawuyi, n 66.

79 J. Dainow, "The Civil Law and the Common Law: Some Points of Comparison" (1966–1967) 15 *The American Journal of Comparative Law* 419.

80 For example, Article 7 of Qatar's Law 10/2003 of the Judiciary Act.

81 IUCN, n 62.

82 Islamic Declaration on Global Climate Change, August 18, 2015 <https://unfccc.int/ news/islamic-declaration-on-climate-change>.

83 Ibid, paras 1.3 and 2.3.

84 See also British Earth Mates Dialogue Center and the Kuwait Ministry of Awqaf and Islamic Affairs, "Muslim Seven Year Action Plan on Climate Change 2010–2017" (International Environmental Forum) <https://iefworld.org/fl/WindsorARCMuslim _summary091020.pdf > accessed 6 January 2021.

85 N. Shaf, "The Arab World's Best Weapon against Climate Change? Its Young People", 18 January 2019 <www.weforum.org/agenda/2019/01/the-arab-worlds-best-weapon-a gainst-climate-change-its-youth/>.

86 See, for example, the Qatar Green Building Council (QGBC). See also the World Green Building Council, "MENA Regional Network" <https://www.worldgbc.org/our -regional-networks/middle-east-north-africa> accessed 6 January 2021.

2 The United Nations Environment Programme – promoting climate law education in the Middle East and North Africa

Elizabeth Maruma Mrema and
Aphrodite Smagadi

Introduction

In today's reality, we deal with several interconnected and transboundary environmental problems that require different environmental solutions. Over the last two decades, environmental law teaching has been developed in many parts of the world as a way to equip young legal professionals with the knowledge they would need to deal with this wide range of arising legal issues. Given the rapid development of this branch of law, it could be expected that environmental law is an obligatory subject for every student in her legal education curriculum. But alas! Despite its recognized importance, environmental law is yet not taught in all educational law/legal institutions. In private conversations with young law students during the second conference of the Association for Environmental Law Lecturers in Middle East and North Africa (MENA) Universities,[1] several students stated that environmental law was either not taught at their universities or an optional subject. Nonetheless, they were interested in learning more and thus their interest and commitment to attend the 2019 Environmental Law Lecturers conference held in Settat, Morocco.

In the meantime, climate law has developed as a subsection of environmental law and has grown so much that it claims its own space as a separate legal discipline. However, there are still areas for further growth and development.

The aim of this chapter is to present an overview of the work of the United Nations Environment Programme (UNEP) throughout its existence over almost 50 years in the promotion of environmental law, especially in the area of climate change. It also discusses ongoing initiatives and programmes that seek to nurture and promote the teaching of climate change law in MENA universities.

This chapter is divided into seven sections. After this introduction, the second section discusses the origins of environmental law. The third section will dive into the development of climate change and examine the history of different international legal efforts. The fourth section addresses the impacts of climate change in everyday life. The fifth section quickly addresses the need for climate change law in the MENA region, and the sixth section discusses the activities of

UNEP in encouraging environmental law and climate law teaching. This is followed by the concluding section.

The genesis of environmental law

The precursors of environmental law existed already in the first half of the 20th century, but environmental law really evolved only during the past 50 years. Its evolution was so rapid that today we speak about a solid body of law in the area of the protection of the environment at the national, regional, and international levels.

To understand the development of environmental law, it is noteworthy to discuss briefly the context that contributed to this,[2] starting with treaty law and case law.

The forerunners in treaty and case law

Around the end of the 19th century and the start of the 20th century, the international community adopted treaties that touch upon some aspects of the protection of the environment. Those treaties regulated the capture, hunting, and fishing of certain species, such as birds, fur seals, and whales.[3] But, at this stage, the level of protection of the environment was not yet known or understood as grasped since the 1970s. The approach of these treaties is utilitarian as they did not aim at the protection of fur seals, birds, or whales as such, but at the regulation of their exploitation so as to maximize derived economic, trade, and commercial benefits, for example, from whales' fat, fur, and hunting for food.[4]

Then, international case law was an important factor in the development of the environmental law. In the 1941 *Trail Smelter* case,[5] a smelter plant operator in British Columbia, Canada, was found to emit hazardous gas (sulphur dioxide) that caused harm to agriculture and forests across the border in Washington State in the United States. The arbitral award in that case articulated the principle that a state should exercise due diligence and prevention to make sure that the activities carried out in its territory, whether by a public or private entity, do not lead to any significant harm on another state.

During the same decade, the 1949 *Corfu Channel* case[6] before the ICJ was significant in the development of international environmental law. Although not directly related to the protection of the environment but to state sovereignty and the law of the sea, this case provided insight into the rights and obligations of states for activities that take place in one's territory with possible impacts on other states. The United Kingdom had alleged that due to Albania's failure to warn about mines in its territorial waters, two British navy ships had been damaged and naval personnel had lost their lives. The Court found that Albania was obliged to make known the presence of mines in its territorial waters.

A decade later, the 1956 *Lac Lanoux* case[7] between France and Spain tackled the use of waters of the Lake Lanoux in the Pyrenees between Spain and France. France planned to use the waters for electricity generation and Spain expressed concerns that the quality of that water, then coming to its territory, would not be good anymore. The arbitral tribunal stressed that if one state conducts an activity in its territory that has a potential impact on the territory of another state, the decision-making authority for the activity has to notify and provide all supporting information to the other state, which should be given an opportunity to express its views on the design of the project.

In the landmark *Gabčikovo-Nagymaros* case in 1997,[8] Hungary claimed that Czechoslovakia, by unilaterally deciding to control the shared waters of Danube River, had violated the provisions of the 1977 treaty relating to projects undertaken by watercourse states along the river. In its judgement, the ICJ introduced the notion of evolving provisions, namely that the interpretation and application of older treaties is changing in view of the development of new norms of international environmental law, including the precautionary principle, intergenerational equity, or environmental impact assessment.[9]

In the above cases, there were only two states involved. But the judgements were very important as the principles they set out, such as transboundary cooperation, good faith, notification, and consultation, laid the seeds for the further development of international environmental law.

The turning point

Other than law, discussions in ethics, politics, and economics in the 1960s and 1970s shaped the thinking around environmental protection. This is the time when Rachel Carson's *Silent Spring* (1962)[10] raised awareness about the use of chemicals and pesticides (especially DDT) in agriculture and their negative effects on the environment, the time when economists voiced their doubts about the concept of economic growth and the need to consider social factors to define development,[11] and the time when ecological political parties made their appearance in the political scene of Europe and the United States.

Under those circumstances, the need for the development of law to protect the environment and the interests of the states and of the planet was acknowledged. The discussions were then relayed to the United Nations, and in 1969 the General Assembly called for a conference[12] aimed "to serve as a practical means to encourage, and to provide guidelines [...] to protect and improve the human environment and to remedy and prevent its impairment".[13]

As a result, the United Nations Conference on the Human Environment[14] was held in Stockholm, from 5 to 16 June 1972.[15] It brought together states, international organizations, civil society (non-governmental organizations and the scientific community), and the private sector, and led to a number of documents of political, legal, ethical, financial, and institutional value and significance.[16] It also led to the decision for the establishment of the United Nations Environment Programme (UNEP) in the same year[17] as a global institutional framework body to assess and support actions on environmental matters.

Soon after, UNEP was established in 1974. UNEP is an integral part of the United Nations with institutional and budgetary autonomy. At that time, neither the Charter of the United Nations nor the constitutive documents of other organizations established in the middle of the 20th century included any reference to the environment,[18] a clear sign that, until UNEP was created as part of the UN, the international community had not yet conceived that the protection of the environment should be a subject of global importance.

The establishment of modern environmental law

After the 1972 Stockholm conference, environmental law continued to grow steadily and speedily. The UN called for more conferences, which represent major milestones in that evolution: shortly after the cold war era and one year after the fall of the Berlin Wall, the 1992 United Nations Conference on Environment and Development (UNCED) in Rio de Janeiro focused on the theme of the environment and development;[19] at the 2002 World Summit on Sustainable Development (or Earth Summit 2002), the international community committed to multilateralism as a means to achieve sustainable development; and at the 2012 United Nations Conference on Sustainable Development (or Rio+20) in Rio de Janeiro, more thought was given to the state of play and what was aspired to in terms of action for the protection of the environment. The adoption of the 17 ambitious Sustainable Development Goals (SDGs) and 169 targets by the General Assembly in 2015[20] determine the national and international agenda and a lot of hope has been placed in a "better" world by 2030.

An explosive growth of multilateral environmental agreements (MEAs) further refined international environmental law and provided invaluable guidance to states on how to design national environmental law and support institutional frameworks. Slowly, there has been a transition from the first generation of MEAs focusing on sectoral environmental issues, such as protecting specific species and/ or ecosystems and banning some practices related to those species and ecosystems,[21] to the second generation of MEAs emerging in the 1980s and dealing with complex global environmental problems in a multi-sectoral approach.[22]

The global climate change regime

Climate change – an issue of global concern

Climate change discourse focuses on the problem of the greenhouse effect as a cause of global warming. The Earth has the capacity to regulate naturally occurring greenhouse gases (GHGs), so these do not cause a problem by themselves. The problem is the dramatic rise of such gases due to human activity (fossil energy, increased methane emissions in agriculture) since the beginning of the industrialization era at the end of the 19th century. The increased anthropogenic emissions exacerbated the greenhouse effect beyond the Earth's natural capacity to autoregulate itself and led to climate change. Climate change is accompanied by several major consequences, including rising sea levels that threaten the existence of certain island countries currently at zero sea level, desertification, biodiversity loss, habitat fragmentation, and negative effects on ecosystems. Worse still, the impacts continue to threaten human wellbeing and the planet due to an increase in epidemics and pandemics, such as malaria, dengue, Middle East Respiratory Syndrome (MERS), Severe Acute Respiratory Syndrome (SARS), Ebola Virus Disease (EVD), avian and birds influenza, and the Coronavirus Disease 2019 (COVID-19) in areas where they did not exist before.

The impacts of industrialization, urbanization, and population growth on the planet's climate and natural equilibrium and the risks associated with the greenhouse effect were observed by scientists quite early with data on global surface temperature increase recorded at the end of the 19th century.[23] Almost a century later, the international community mobilized in regard to the risks associated with the greenhouse gas effect phenomenon. It had become apparent that humans needed to devise long-term measures and strategies and drastically change their behaviour and lifestyles to adapt to and escape the consequences of climate change. The problem was not national but global and complex and could not be addressed by individual states. States recognized that the climate change response required international cooperation and collaboration in an integrated manner to combat a transboundary and cross-border issue that was greater than ever before.

The climate change treaty regime

The first step for the development of a global legal framework was to create a group of experts, the Intergovernmental Panel on Climate Change (IPCC),[24] a joint UNEP-World Meteorological Organization (WMO) body that, to date, brings together a wide range of sciences, ranging from hard science, such as math, physics, and chemistry, to humanities, such as economics, social anthropology, political science, and law. The treaty text for the United Nations Framework Convention on Climate Change (UNFCCC) was prepared in record time[25] and was tabled for adoption at the 1992 Rio Conference,[26] together with the Convention on Biological Diversity (CBD).[27] On the recommendation of Agenda 21 adopted in Rio,[28] the United Nations Convention to Combat Desertification in Those Countries Experiencing Serious Drought and/or Desertification, particularly in Africa (UNCCD) was adopted in 1994.[29] The three treaties, popularly known as the "Rio Conventions", have different objectives, but the dynamics between climate, biodiversity, and land are interrelated, interconnected, and interdependent and thus an integrated approach is necessary to meet the complex demands and challenges each attempts to regulate.[30]

After its adoption, the UNFCCC was opened for signature in 1992 and entered into force in March 1994. A main feature of the Convention is that it is a framework treaty that puts in place normative and institutional elements and a long-term vision for a global climate change regime based on the principles of common concern of humankind, the precautionary principle, and the principle of sustainable development.[31] All Parties make general commitments[32] to address climate change through climate change mitigation and adaptation to eventual impacts of climate change. The Convention acknowledges that the greenhouse effect is primarily due to the activities of developed countries and proclaims the key principle of common but differentiated responsibilities, whereby a certain group of countries (those listed in Annex I to the Convention) bear additional responsibilities and are called to provide financial and technical assistance to the developing countries.[33] To this end, financial mechanisms were put in place to support projects that would promote the implementation of the Convention, such as emissions tracking and strategies on the decarbonization of the atmosphere or renewables.[34]

The Convention does not stipulate firm commitments. The need for a stronger framework had already been discussed at the first session of the Conference of the Parties in 1995[35] and soon led to the adoption of the Kyoto Protocol in 1997.[36] The Kyoto Protocol includes more stringent rules to achieve the Convention objectives. The main goal is for Parties included in Annex I (mostly Organization for Economic Co-operation and Development (OECD) countries) to reduce their overall emissions of certain greenhouse gases[37] by at least five per cent below 1990 levels in the commitment period 2008 to 2012.[38] Parties included in Annex I agreed to differentiated responsibilities amongst themselves. Tensions on the degree of implementation of the commitments continued after the Kyoto Protocol's adoption in 1997. The 2001 Marrakesh Accords aimed to clarify how the targets set in the Protocol were to be met.

The Kyoto Protocol is the first international legal instrument on the environment that employs tools and mechanisms of the market economy, namely, joint implementation, the clean development mechanism, and international emissions trading,[39] and attempts to harness the muscle of the private sector to achieve environmental objectives.[40]

The implementation of the Protocol inaugurated a whole new international carbon trading market.[41] The Protocol set targets from 2008 to 2012, with the Doha Amendment setting targets from 2013 to 2020.[42]

In 2015, after acrimonious negotiations, Parties to the UNFCCC adopted the Paris Agreement[43] aiming to keep a global temperature rise this century below 2°C above pre-industrial levels and to pursue efforts to limit the temperature increase even further to 1.5°C.[44] The Agreement also calls for a new technology and capacity building framework to support countries to deal with climate change impacts.[45] Parties are required to undertake and communicate their efforts to meet the Agreement's objective through Nationally Determined Contributions (NDCs).[46]

The transversal nature of climate change law

At the international level

Climate change is inevitably interrelated and interdependent on other specialized areas of environmental law, such as biodiversity loss, and desertification and drought concerns. At the international level, climate change law has also penetrated other areas of international law, not immediately related to the environment, and its understanding is critical.[47]

Firstly, because it employs trade and finance techniques, climate change law has a complex relationship with international trade and finance law, especially the agreements under the World Trade Organization (WTO) or the European Union directives. The relationship between international environmental law and trade law has been subjected to extensive discussions in legal scholarship,[48] and climate change exemplifies this relationship. Subsidies for new technologies, tax relief, or other fiscal instruments that promote climate-friendly practices, or the question of whether trade barriers may be justified for certain production methods (GATT XX (b) and (g)) have caused a lot of debate amongst trade lawyers representing different, and at times, divergent state interests.[49]

Secondly, notions of equity and human rights, in regard to climate change, although not immediately obvious, are fundamental. Climate change causes desertification and affects the quantities of drinking water, the quality of farmed land for food production, and agriculture. These, in turn, impact on the rights of populations to water and food security and thus reduce peoples' opportunity to live at a certain standard of life with dignity. The mortality rate increases and the rights of vulnerable populations, children, and women are especially affected. Because of the great impact climate change may have on the quality of life in some locations, forced emigration and refugees are inevitable and may cause tensions amongst sovereign states.[50]

Thirdly, climate change is receiving growing attention regarding peace and security. The effects of climate change may be so devastating that conflict for resources may be inevitable. Already, climate change is recognized as one of the main drivers of population displacement. The Security Council has long denied the relevance of the climate change discussions within its forum, but there has been a shift in the last couple of years,[51] and the North Atlantic Treaty Organization has in recent years been the host of debates on climate change and related security risks.[52]

Lastly, but not least, the interrelationship between climate change and the law of the sea has now been established.[53] Historically, international climate change law has focused on land-based mitigation and adaptation, but the health of the oceans, the effects of climate warming on marine life and the coral reefs, the welfare of the people whose livelihoods depend on the oceans, and the changing baselines affecting the rights of states have rendered the review of this relationship necessary. "[T]he importance of ensuring the integrity of all ecosystems, including oceans" was first acknowledged in 2015,[54] and the IPCC concluded a large report on the subject in 2019.[55]

At the domestic level

At the national level, environmental law practitioners are naturally expected to comprehend the complexity of climate change law. Countries' efforts to transpose international treaty law relating to climate change into the national level need local legal experts to assist them. The experts need prerequisite knowledge of the specific jurisprudence when they advise on the most appropriate mitigation and adaptation measures, the organization of the domestic carbon market, or on environmental impact assessment and environmental monitoring measures and cases. In addition, climate change law has occupied quite a few areas of legal practice as examined below, which makes its understanding essential for legal professionals.

We see that climate change concerns infiltrate a wide range of legal practice.[56] As the first environmental law area incorporating market economy instruments, society needs more corporate lawyers understanding the specifics of climate change law to serve investors and consumers. Especially after the adoption of the Paris Agreement, the contribution of the private sector to help countries meet their obligations is significant. Countries are crafting policies, laws, and regulations for climate finance to encourage and build investor confidence.[57]

Moreover, climate change law leads to increased corporate responsibility for GHG emissions. Following the adoption of the Paris Agreement, climate change legislation has proliferated as countries strive to meet their emission reduction targets.[58] As a result, large corporations seek advice from lawyers on the new frameworks and corporate liabilities, risks, and opportunities.[59]

The use of conventional energy sources for electricity, heating, and transport is one of the main causes of GHGs. To reduce climate change challenges, countries shape their energy legislation to regulate new sources of no- or low-carbon generation and energy efficiency, transforming the national energy sector.[60] The private and public sectors require legal practitioners who understand the energy and climate interests.

Like energy law, land use planning law is a critical part of an overall national climate response portfolio. Due to the rising sea levels and coastal erosion, the greater frequency of cyclones, floods and avalanches, drought, water scarcity, and heat stress, governments are introducing land use planning policies and related mitigation and adaptation measures to respond to climate change. According to the IPCC "[l]and-use zoning, spatial planning, integrated landscape planning, regulations, incentives (such as payment for ecosystem services), and voluntary or persuasive instruments (such as environmental farm planning, standards and certification for sustainable production, use of scientific, local and indigenous knowledge and collective action), can achieve positive adaptation and mitigation outcomes".[61] Parties to the UNFCCC have included land use activities in their intended INDCs and NDCs.[62]

The potential impact of climate change on property, health, crops, and life, in general, has attracted the attention of the insurance industry during the past decade.[63] As global warming has been associated with extreme weather phenomena and the potential impact of climate change on global economy, ecology, human health, and human welfare, insurers are called to assume increased property and liability risks. Climate change is at the top of the insurance industry agenda as regards the management of climate-related risk and the calculation of premiums.

In 1990, Finland was the first country to introduce carbon taxes, an important economic incentive to promote climate-friendly activities.[64] Countries may opt to levy carbon taxes on activities deemed to harm the environment.[65] Countries may also impose border taxes on commodities, if they deem that their production method is harmful to the environment, and induce other countries to move towards a low-carbon economy.

Given the expansion of climate change law in a multitude of other areas, courts worldwide have experienced an increased number of lawsuits.[66] In such litigation, some climate-related lawsuits tend to be common cases, such as those challenging the issuance or denial of a permit for solar panels, those relating to compliance with the national emissions trading system, and those pertaining to the enforcement of reporting obligations on gas emissions. However, the class of suits that skyrocketed in the recent past have been those that raise a debate about a state's obligations to reduce GHGs or human and constitutional rights relating to the environment.[67] In addition, to deter environmental crime, such as the illicit trade in substances that may negatively affect the environment,[68] prosecutors would need to understand the specifics of climate change.

The analysis above clearly shows that climate change law cannot be seen and treated separately or in isolation from other disciplines of law as it links and connects with several other branches. It is one of the most important areas of international environmental law and it is closely related and interconnected to other areas of law. As the international climate change regime is transposed and implemented at the domestic level and as the necessity to act locally becomes increasingly relevant, there is inevitably the need for across the board climate change legal professionals (legal advisors, trial lawyers, judges, prosecutors) who comprehend the complexity of the international climate change system.

Climate change lawyers in the MENA region

The MENA region is seriously affected by climate change and the socio-economic projections are gloomy as climate change challenges its natural capital and affects the availability of arable land and freshwater.[69] In Africa,[70] the projections run counter to Africa's vision in its Agenda 2063[71] for inclusive growth, peace, and good governance by means of investment in renewable energy, biodiversity conservation, fair and equitable use of the genetic resources, increased agricultural productivity, clean air and water, as well as better capacity to climate change. Likewise, the Middle East region is vulnerable to climate change impacts which are exacerbated by conflicts, wars, and political instability, as well as unsustainable production and consumption methods.[72]

To move towards climate resilience and sustainable growth, the MENA region would need to harness the wealth of opportunities offered by climate change, build on good governance and human capacities and make climate-friendly choices on energy, food production, and natural resources exploitation. MENA society requires more lawyers conversant in the cross-cutting area of climate change law as it works to meet its goals both internationally and domestically.

The lawyering skills and values identified in the 1992 report commissioned by the American Bar Association,[73] including problem solving, legal analysis and reasoning, legal research, and factual investigation are still applicable in the 21st century. With the interdisciplinary nature of climate change law, the region needs to carry out legal research and analysis, and understand the facts and the underlying issues in cases that are not solely climate change law cases but may also touch upon climate change. Lawyers are expected to think of solutions and thus need to be well trained in the complexities and sophistication of climate change law as a complex and evolving body of law within the environmental law field. To respond to this need, many universities around the world offer climate change specialized graduate and post-graduate degrees.[74] However, universities in the MENA region rarely offer such courses or degrees.

In view of the growing importance of climate change, generally and in law, it is imperative that the fundamentals of climate change law be made a part of legal education, rather than a self-standing, post-graduate legal specialization. In fact, environmental law in undergraduate education has been stressed since the 1970s.[75] Teaching climate change law in law schools will help students to better understand the scientific, as well as socio-economic issues, and the legal process. This is increasingly acknowledged with increased interest in Islamic finance and renewables.[76]

The role of UNEP in promoting environmental law and policy

As mentioned above, UNEP was established in the early 1970s as part of the major turn of international thinking on the protection of the environment. Since its establishment, UNEP, in collaboration with partners, has played a pivotal role in the development of international environmental law[77] in the international sphere and has assisted individual countries in building and strengthening capacities in environmental law. In response to its mandate in Agenda 21,[78] UNEP has provided policy, legal, and technical advisory services to almost all member states of the UN and has been a primary source of information and expertise in that regard. This has mostly been accomplished through the UNEP's Law Division programme[79] which seeks to address the needs of different countries[80] and to support transparency and accountability in legislature, judiciaries, and policy-making institutions regarding environmental law.

UNEP's capacity building activities to promote environmental law and policy cannot be summarized in a few paragraphs. This section only gives a glimpse of the work UNEP undertakes in the MENA region, but it also outlines some other important initiatives that have the potential to stimulate environmental law, including climate law, education, and practice in the MENA region. Even though there is no separate regional office for MENA within UNEP, UNEP has and collaborates with its Africa Office in Nairobi for North Africa and the West Asia Office in Manama for the Middle East.[81]

PADELIA – *ten-year capacity building focused on Africa*

The fact that UNEP is based in Kenya helped the promotion of environmental law in Kenya and Africa. Retired Professor Charles Okidi, then at the University of Nairobi School of Law, inspired and trained many young legal scholars in environmental law since the 1980s. When he joined UNEP, Professor Okidi worked persistently on the promotion of environmental law in African countries through the Programme on Capacity Building for the Development of Environmental Law and Institutions in Africa (PADELIA).

PADELIA[82] aimed to build and strengthen in-country capacity in environmental law throughout two phases spanning a period of about ten years starting in 1994. In Phase I, support was provided to individual countries, whereas in Phase II, there was a shift towards subregional projects to build and enhance cooperation on transboundary and cross-cutting issues between the Phase I countries and other countries in their subregion.[83]

Project activities depended a lot on the endogenous country capacity to deliver. To this end, national legal experts were identified and designated for each country to coordinate the activities at the national level. PADELIA directly benefited selected countries in Africa, but it also supported Africa-wide activities with a lasting impact on the promotion of environmental law, including climate change law,[84] targeting government institutions as well as parliamentarians,[85] judiciary,[86] private sector, and civil society,[87] to mention a few.

PADELIA's legacy continues to empower UNEP to replicate, adapt, and upscale the lessons learned from that project with training and capacity building in environmental law. PADELIA enhances and empowers various stakeholders in Africa, and beyond.[88]

Legal scholars' network in the MENA region

UNEP has supported countries and stakeholders in building and enhancing their capacity in environmental law through the exchange of knowledge and lessons learned. In Africa, the Association of Environmental Law Lecturers in African Universities (ASSELLAU) was established in 2006[89] and, for the past 15 years, it strives to mainstream environmental legal education into university curricula in Africa. ASSELLAU membership continues to grow and its impacts on universities in the region are strong.[90]

Inspired by the work of ASSELLAU, at the initiative of Professor Damilola Olawuyi of Hamad Bin Khalifa University and in partnership with UNEP, the Association of Environmental Law Lecturers in the Middle East and North Africa (ASSELLMU) was established in 2018.[91] Leading environmental law scholars, practitioners, and policy leaders come together to discuss legal innovations and approaches for addressing climate change impacts in the MENA region. ASSELLMU aspires to develop an environmental law curriculum for universities in the region, to strengthen the relationship between academia and enforcement actors (police and the judiciary), and to celebrate the excellence of young researchers from MENA, who will be provided with an opportunity to present their research on environmental law in future ASSELLMU events.

Targeted training

UNEP works together with judicial and prosecution education and training institutes, police academies, and customs training institutes[92] to strengthen implementation and enforcement of environmental law. This type of training provides an opportunity for enforcement officials to understand the main concepts and principles of environmental law and to review their training curricula on environmental law, including on environmental crime.[93] It is hoped that more is done in the MENA region in this regard.

In addition, UNEP collaborates with individual countries[94] or other organizations, such as the Arab Organization for Agricultural Development (AOAD),[95] to enhance the capacity of government officials from the Arab region on how to effectively negotiate and participate in intergovernmental environmental negotiations.[96] With a view to boosting the engagement of the least developed countries in intergovernmental climate change negotiations, UNEP offered a series of workshops in selected countries.[97] Participants in these trainings have recognized that their personal skills and their countries' institutional capacities had been enhanced as a result of the training offered by UNEP, and their negotiation skills and active engagement in climate change processes are clearly seen in practice.

Knowledge products

Studies and publications are important tools for the promotion of environmental law education. Many of these knowledge products may constitute valuable materials in the context of environmental law teaching, including climate law.[98] The curriculum can easily be adapted and used by national bar associations and law societies in their respective jurisdictions. These products promote continuous legal education and their use will enable lawyers to effectively enforce and implement environmental law, at both national and regional levels. Knowledge products ensure continuous legal education on environmental law.

As laws codifying national and international responses to climate change have grown during the past years, so has the litigation regarding their validity and/or application. The Status of Climate Change Litigation provides valuable information in this context.[99] Moreover, UNEP's work on legal readiness for climate finance focusing on opportunities for private sector stimulates collaboration and mutual learning between public and private stakeholders on climate finance.[100]

Electronic tools on environmental law education

In today's digital society,[101] electronic tools play a critical and pivotal role in training and education. UNEP directs the United Nations Information Portal on Multilateral Environmental Agreements (InforMEA), as well as ECOLEX, an online information centre on environmental law.[102] Free online e-learning courses with e-certificates are available in InforMEA and already widely used by lecturers to prepare their environmental law courses, as well as utilized by universities as part of their curriculum.[103] At the time of writing, UNEP was working on a West Asia regional platform for MEAs and a specific and targeted regional forum on InforMEA in the Arabic language to consolidate environmental law knowledge in West Asia.

In close collaboration with partners, a Climate Change Toolkit has been developed[104] to be used by experts assisting countries in crafting climate legislation and for academia and research institutions that are keen to analyze the ever-growing body of climate change-related legislation, including climate-specific and other climate-related sectoral legislation, throughout the world. The Toolkit is organized in modules (urban planning, agriculture, etc.) to help countries see the link between sectoral legislation and effective climate change action, analyze the legislation in a comprehensive manner, identify climate-related gaps and priority areas for climate law review, and potential areas of legislative or regulatory reform.

Conclusion

Climate change complexity still puzzles many scholars with the effects it has on land, biodiversity, human health, and everyday life. It raises scientific, political, sociological, and legal problems that call for scientific, economic, and legal strategies and solutions. Climate change is a cross-cutting issue as it does not relate to the environment only but to several other sectors, including transport, agriculture, forestry, and energy, while the means to tackle the problem touch upon taxes, investments, insurance, and trade, along with environmental and health policies. Climate change touches almost every major field of law and has evolved to claim its own autonomy from the sphere of environmental law.

Legal education is, therefore, a pivotal and prerequisite element of the global response to climate change. To be ready for the decade to come, the region should strengthen its capacity to address climate change considerations and to contribute to the environmental dimension of the 2030 Agenda. To this end, it is important to harness opportunities and invest in human capital, including skilled and knowledgeable lawyers. The basics of environmental law, including climate change law, should be one of the core modules of any typical university law curriculum so that young lawyers who are the future leaders are well equipped for their career development and readiness to take leadership roles on this subject in the future.[105] Continuing education of legal professionals is necessary to keep abreast of developments in the area and to further enhance appreciation and understanding of the ways environmental protection issues have penetrated other areas of legal practice.

The UNEP Law Division, which leads UNEP's global efforts in the field of environmental law, governance, and policy issues, continues to be well positioned to provide this support to countries. UNEP will continue its efforts to promote exchange and collaboration among educational institutions with shared legal tradition, such as ASSELLMU and ASSELLAU, and to develop knowledge products and tools that can be employed by educational institutions and interested legal professionals.

Notes

1 The Association for Environmental Law Lecturers in Middle East Universities was established in 2018 upon the initiative of Professor Damilola Olawuyi of Hamad Bin Khalifa University Law School, Doha, Qatar, and is supported by the United Nations Environment Programme, Law Division, at an inaugural conference held in Doha, from 4 to 5 November 2018. The objective was to promote environmental law teaching in the universities in the region. After that first conference and due to the commonalities among Arab countries in the MENA, the Association's scope has been expanded to include universities in North Africa. The second conference was hosted by Hassan First University of Settat, Morocco, from 4 to 5 November 2019.

2 E. B. Weiss, "The Evolution of International Environmental Law" (2011) 54 *Japanese Yearbook of International Law* 1–27.

3 1900 Convention for the Preservation of Wild Animals, Birds and Fish in Africa (never entered into force and later replaced by the 1933 Convention Relative to the Preservation of Fauna and Flora in their Natural State (London, adopted 8 November 1933; in force 14 January 1936)); the 1946 International Convention for the Regulation of Whaling (Washington, USA, adopted 2 December 1946; in force since 1 July 1948); and the North Pacific Fur Seal Convention of 1911 (Washington, USA; in force since 7 July 1911).

4 Ibid.

5 Trail Smelter Case (*United States v Canada*) (16 April 1938 and 11 March 1941) 3 RIAA 1905–1982.

6 Corfu Channel Case (*UK v Albania*) (Assessment of Compensation) [1949] ICJ Rep 15 XII 49.

7 Lake Lanoux Case (*France v Spain*) (1957) 12 RIAA 281.

8 Gabčikovo-Nagymaros Dam (*Hungary v Slovakia*) (Separate Opinion of Vice-President Weeramantry) [1997] ICJ Rep 115.

9 M. Fitzmaurice, "The International Court of Justice and International Environmental Law" in C. J. Tams and J. Sloan (eds), *The Development of International Law by the International Court of Justice* (Oxford University Press 2013).

10 R. Carson, *Silent Spring* (Boston, MA: Houghton Mifflin Company 1962).

11 D. H. Meadows, D. L. Meadows, J. Randers, and W. W. Behrens III, *The Limits to Growth: A Report for the Club of Rome's Project on the Predicament of Mankind* (Universe Books 1972).

12 UN General Assembly Resolution 2398 (XXIII), *Problems of the Human Environment*, 3 December 1968, UN Doc. A/Res/2398/23; UN General Assembly Resolution 2581 (XXIV), *United Nations Conference on the Human Environment*, 15 December 1969, UN Doc. A/RES/2581(XXIV).

13 UN General Assembly Resolution 2581 (XXIV), *United Nations Conference on the Human Environment*, 15 December 1969, UN Doc. A/RES/2581(XXIV).

14 The conference was dedicated to the *human* environment as an expression of the predominant anthropocentric approach as regards environmental concerns.

15 G. Handl, "Declaration of the United Nations Conference on the Human Environment (Stockholm Declaration), 1972 and The Rio Declaration on Environment and Development, 199" (United Nations Audiovisual Library of International Law 2012) <https://legal.un.org/avl/ha/dunche/dunche.html>.

16 Ibid. The Stockholm Declaration sets out principles and general legal, political, and ethical directions. The UN Conference on the Human Environment also recognized that some countries would like to protect the environment but have no capacity or means. To address these disparities, it decided to establish a financial mechanism for the environment. The Conference also adopted a Programme of Action for the future, while it foresees the creation of the United Nations Environment Programme.

17 UN General Assembly Resolution 2997 (XXVII), *Institutional and Financial Arrange-*

ments for International Environmental Co-operation, 15 December 1972, UN Doc. A/RES/2997(XXVII).

18 Such as UNESCO, FAO, and WMO.

19 Handl, n. 15.

20 UN General Assembly Resolution 70/1, *Transforming Our World: The 2030 Agenda for Sustainable Development*, 21 October 2015, UN Doc. A/RES/70/1, with SDG13 devoted to Climate Action, while climate considerations play an important role for the achievement of other SDGs.

21 For example, UN, *Convention on International Trade in Endangered Species of Wild Fauna and Flora*, 3 March 1973 UNTS 243 (CITES or the Washington Convention); UN, *Convention on the Conservation of Migratory Species of Wild Animals*, 23 June 1979, 1651 UNTS 333 (CMS or the Bonn Convention); UN, *Convention on Wetlands of International Importance Especially as Waterfowl Habitat*, 21 December 1971, 996 UNTS 245; UN Educational, Scientific and Cultural Organisation (UNESCO), *Convention Concerning the Protection of the World Cultural and Natural Heritage*, 16 November 1972.

22 For example, UN, Vienna Convention for the Protection of the Ozone Layer, 22 March 1985, 26 ILM 1516; UN, *United Nations Framework Convention on Climate Change*, 9 May 1992, 1771 UNTS 107; UN, *Convention on Biological Diversity*, 5 June 1992, 1760 UNTS 7 [UNFCCC]; UN, *United Nations Convention to Combat Desertification in Those Countries Experiencing Serious Drought and/or Desertification, Particularly in Africa*, 14 October 1994, 1954 UNTS 3.

23 H. Le Treut and others, "2007: Historical Overview of Climate Change" in S. Solomon and others (eds), *Climate Change 2007: The Physical Science Basis. Contribution of Working Group I to the Fourth Assessment Report of the Intergovernmental Panel on Climate Change* (Cambridge University Press 2007).

24 The Intergovernmental Panel on Climate Change (IPCC) <www.ipcc.ch/> accessed 6 January 2021.

25 The negotiations lasted two years, which is uncommon in the international scene, especially when it comes to such complex and multifaceted issues that call for universal agreement. See UNFCCC, "UNFCCC – 25 Years of Effort and Achievement" (UNFCCC) <https://unfccc.int/timeline/> accessed 6 January 2021.

26 D. Bodansky, "The History of the Global Climate Change Regime" in U. Luterbacher, D. F. Sprinz (eds), *International Relations and Global Climate Change* (MIT Press 2001) 23; K. Hampton, "Understanding the International Climate Regime and Prospects for Future Action" (Green Globe Network UK, The Institute for Progressive Policy Research 2004) <www.ippr.org/research/publications/understanding-the-international-climate-regime-and-prospects-for-future-action> accessed 6 January 2021; R. D. Brunner, "Science and the Climate Change Regime" (2001) 34 *Policy Sciences* 1–33.

27 The CBD was opened for signature on 5 June 1992 at the UNCED in Rio and remained open for signature until 4 June 1993. It entered into force on 29 December 1993. See n. 22.

28 UN, General Assembly Resolution 47/188, *Establishment of an Intergovernmental Negotiating Committee for the Elaboration of an International Convention to Combat Desertification in Those Countries Experiencing Serious Drought and/or Desertification, Particularly in Africa*, 12 March 1993, UN Doc. A/RES/47/188, established an Intergovernmental Negotiating Committee to prepare a Convention to combat desertification, particularly in Africa.

29 Convention to Combat Desertification (adopted in 1994, entered into force 1996)

30 S. Díaz and others (eds), *Summary for Policymakers of the Global Assessment Report on Biodiversity and Ecosystem Services* (Intergovernmental Science-Policy Platform on Biodiversity and Ecosystem Services 2019).

31 UNFCCC, n. 22, Art. 3.

32 Ibid, Art. 4.
33 Ibid, n. 22, Preambular paras 3, 18, 20, and 22 Arts. 3 and 11.
34 Ibid. Global Environmental Facility, Arts. 11 and 21, and later the Adaptation Fund in Marrakesh (COP Decision 10/CP.7 – Annex 2 <https://unfccc.int/decisions?f %5B0%5D=session%3A3615>) and the Green Climate Fund with the Copenhagen Accord at COP-15 in 2009 <www.greenclimate.fund/>.
35 COP decision 1/CP.1 (the "Berlin Mandate") <https://unfccc.int/decisions?f%5B0% 5D=session%3A3615>.
36 All Annex I Parties are Parties to the Protocol, except for the USA (signature withdrawn in 2001) and Canada – the latter was a Party that withdrew in 2012. UNF-CCC, *Kyoto Protocol to the United Nations Framework Convention on Climate Change*, 11 December 1997, 2303 UNTS 162 [Kyoto Protocol].
37 Ibid, Annex A, includes a list of targeted GHGs.
38 Ibid, Art. 3, para. 1.
39 Kyoto Protocol, n. 37, Arts. 6, 12, and 17.
40 Unlike the Convention, the Protocol also put in place a strict compliance procedure to monitor and sanction Parties. The Enforcement Branch of the Compliance Committee of the Kyoto Protocol can suspend from eligibility to participate in the international carbon trade. UNFCCC Conference of the Parties, Decision 27/CMP.1 on Procedures and Mechanisms Relating to Compliance under the Kyoto Protocol (30 March 2006) FCCC/KP/CMP/2005/8/Add.3.
41 J. V. Fenhann, M. C. Schletz, and W. Vergara, *Zero Carbon Latin America – A Pathway for Net Decarbonisation of the Regional Economy by Mid-Century* (UNEP DTU Partnership 2015). OECD, *Aligning Policies for a Low-Carbon Economy* (OECD 2015); B. Stephan and M. Paterson, "The Politics of Carbon Markets: An Introduction" (2012) 21 *Environmental Politics* 545–562; S. Fankhauser, "A Practitioner's Guide to a Low-Carbon Economy: Lessons from the UK" (Centre for Climate Change Economics and Policy, Grantham Research Institute on Climate Change and the Environment 2012) <www.lse.ac.uk/granthaminstitute/publication/a-practitioners-guide-to -a-low-carbon-economy-lessons-from-the-uk/> accessed 6 January 2021.
42 See UNFCCC Conference of the Parties, Decision 1/CMP.8 the Doha Amendment (28 February 2013) FCCC/KP/CMP/2012/13/Add.1. The amendment has not entered into force, but Parties have provisionally applied it pending its entry into force.
43 Adopted through decision 1/CP.21 and in force since 2016. See UNFCCC Conference of the Parties, Decision 1/CP.21 on Adoption of the Paris Agreement (29 January 2016) FCCC/CP/2015/10/Add.1.
44 United Nations Framework Convention on Climate Change, Adoption of the Paris Agreement (12 December 2015) UN Doc. FCCC/CP/2015/L.9, Art. 2 [Paris Agreement].
45 Ibid, Arts. 9–12.
46 Ibid, Art. 3.
47 International Bar Association (IBA) Climate Change Justice and Human Rights Task Force, *Achieving Justice and Human Rights in an Era of Climate Disruption* (IBA 2014).
48 M. Halle and others (eds), *Trade and Environment: Resource Book* (Winnipeg, MB: IISD 2007); A. R. Maggio, *Environmental Policy, Non-Product Related Process and Production Methods and the Law of the World Trade Organization* (New York: Springer 2017); D. Y. Bag, *Legal Issues on Climate Change and International Trade Law* (New York: Springer 2016); K. Holzer, *Carbon-Related Border Adjustment and WTO Law* (Edward Elgar Publishing 2014); E. Vranes, *Trade and the Environment: Fundamental Issues in International Law, WTO Law, and Legal Theory* (Oxford University Press 2009); N. Bernasconi-Osterwalder, J. H. Jackson, and E. Brown Weiss (eds), *Reconciling Environment and Trade* (Brill Nijhoff 2008).

49 For example, United States – Gasoline Appellate Body Report, United States – Standards for Reformulated and Conventional Gasoline, WT/DS2/AB/R, adopted 20 May 1996; United States — Shrimp: Appellate Body Report, United States – Import Prohibition of Certain Shrimp and Shrimp Products, WT/DS58/AB/R, adopted 6 November 1998 (Initial Phase) and Appellate Body Report, United States – Import Prohibition of Certain Shrimp and Shrimp Products – Recourse to Article 21.5 of the DSU by Malaysia, WT/DS58/AB/RW, adopted 21 November 2001 (Implementation Phase under Article 21.5; European Communities – Asbestos Panel Report, European Communities – Measures Affecting Asbestos and Asbestos-Containing Products, WT/DS135/R and Add.1, adopted 5 April 2001, as modified by the Appellate Body Report, WT/DS135/AB/R and Appellate Body Report, European Communities – Measures Affecting Asbestos and Asbestos-Containing Products, WT/DS135/AB/R, adopted 5 April 2001. Canada – Salmon and Herring Panel Report, Canada – Measures Affecting Exports of Unprocessed Herring and Salmon, L/6268, adopted 22/03/1988; United States – Canadian Tuna Panel Report, United States – Prohibition of Imports of Tuna and Tuna Products from Canada, L/5198, adopted 22/02/1982.

50 "Non-U.S. Climate Change Litigation: Suits Against Governments" (Climate Case Chart) <http://climatecasechart.com/non-us-case-category/human-rights/> accessed 6 June 2020; UNEP, *The Status of Climate Change Litigation: A Global Review* (UNEP 2017).

51 K. Eklöw, "A Short History of Climate Change and the UN Security Council" (World Economic Forum, 9 January 2020) <https://www.weforum.org/agenda/2020/01/a-short-history-of-climate-change-and-the-un-security-council>; United Nations News, "Climate Change Recognized as 'Threat Multiplier', UN Security Council Debates Its Impact on Peace" (UN News, 25 January 2019) <https://news.un.org/en/story/2019/01/1031322>.

52 At the Summit Meeting of NATO held in Lisbon, Portugal, 19–20 Nov 2010, Heads of State and Government adopted the Strategic Concept for the Defence and Security of the Members of the North Atlantic Treaty Organisation. Paragraph 15 reads: "Key environmental and resource constraints, including health risks, climate change, water scarcity and increasing energy needs will further shape the future security environment in areas of concern to NATO and have the potential to significantly affect NATO planning and operations". See North Atlantic Treaty Organization, "Active Engagement, Modern Defence: Strategic Concept for the Defence and Security of the Members of the North Atlantic Treaty Organization" (NATO 2010) <https://www.nato.int/nato_static_fl2014/assets/pdf/pdf_publicatio ns/20120214_strategic-concept-2010-eng.pdf>. This initial acknowledgement was later followed by the Wales Declaration (Wales Summit 4–5 September 2014), which states that "[k]ey environmental and resource constraints, including health risks, climate change, water scarcity, and increasing energy needs will further shape the future security environment in areas of concern to NATO and have the potential to significantly affect NATO planning and operations" (NATO, "Wales Summit Declaration" <www.nato.int/cps/en/natohq/official_texts_112964.htm>) and the NATO Parliamentary Assembly Resolution 427 on Climate Change and International Security (193 STC 15 E rev. 1 *bis* <www.actu-environnement.com/media/pdf/news-25462-resolution-otan-2015.pdf>).

53 M. McCreath and A. M. Maggio, "Introduction: Climate Change and the Law of the Sea: Adapting the Law of the Sea to Address the Challenges of Climate Change" (2019) 34 *The International Journal of Marine and Coastal Law* 387, 389.

54 Paris Agreement, n. 44, Preamble.

55 P. R. Shukla and others (eds), *Special Report on the Ocean and Cryosphere in a Changing Climate* (IPCC 2019).

56 J. Peel, "Climate Change Law: The Emergence of a New Legal Discipline" (2012) 32 *Melbourne University Law Review* 923-977.

57 UNEP and King's College London, "Legal Readiness for Climate Finance: Private Sector Opportunities – Report and Findings of Roundtable held at King's College London, 25 January 2019" (UNEP 2019) <https://wedocs.unep.org/bitstream/handle /20.500.11822/28219/2019law_clim_finrprt.pdf?sequence=1&isAllowed=y>; UNEP and King's College London, "Climate Finance Law: Legal Readiness for Climate Finance – Report and Findings of Workshop held at King's College London 9–11 March 2018" (UNEP 2018) <https://wedocs.unep.org/bitstream/handle/20.500.11 822/26378/climate_finance_law.pdf?sequence=1&isAllowed=y>.

58 Taking into account the adoption of the Paris Agreement on 12 December 2015, the Grantham Research Institute on Climate Change of the London School of Economics (LSE) points to 152 legislative acts enacted between 2015 and 2020 (inclusive) <https://climatelaws.org/cclow/legislation_and_policies?law_passed_from=2015 &law_passed_to=2020&type%5B%5D=legislative>. A total of 483 acts, including legislative and executive (policies), have been issued in the same period <https:// climatelaws.org/cclow/legislation_and_policies?law_passed_from=2015&law_passed _to=2020&type%5B%5D=legislative&type%5B%5D=executive>.

59 International Bar Association, "Corporate Lawyers in a Climate of Change" (IBA, 30 September 2019) <https://www.ibanet.org/Article/NewDetail.aspx?ArticleUid =82BD7FB4-C46C-4819-A56F-BE9D7F683FA6>.

60 D. Markell, "Climate Change and the Roles of Land Use and Energy Law: An Introduction" (2012) 27 *Journal of Land Use and Environmental Law* 231–243; O. Edenhofer and others (eds), *Renewable Energy Sources and Climate Change Mitigation* (Cambridge University Press 2011).

61 P. R. Shukla and others (eds), *Climate Change and Land: An IPCC Special Report on Climate Change, Desertification, Land Degradation, Sustainable Land Management, Food Security, and Greenhouse Gas Fluxes in Terrestrial Ecosystems* (IPCC 2019).

62 UNFCCC, *Aggregate Effect of the Intended Nationally Determined Contributions: An Update – Synthesis Report by the Secretariat*, 2 May 2016, UN Doc. FCCC/CP/2016/2.

63 P. Jenkins, "Why Climate Change is the New 9/11 for Insurance Companies" (Financial Times, 8 September 2019) <www.ft.com/content/63c80228-cfee-11e9-99a4-b5 ded7a7fe3f>; D. Jergler, "Climate Change and the Reinsurance Implications" (Insurance Journal, 13 June 2019) <https://www.insurancejournal.com/news/national/2019 /06/13/529201.htm>; The Geneva Association, "Climate Change and the Insurance Industry: Taking Action as Risk Managers and Investors – Perspectives from C-Level Executives in the Insurance Industry" (Geneva Association 2018) <www.genevaassociation.org/sites/default/files/researchtopicsdocumenttype/pdf_public/climate_ch ange_and_the_insurance_industry_-_taking_action_as_risk_managers_and_invest ors.pdf>; "Climate Change and Insurance Law: General Report submitted to the AIDA World Congress, Paris, May 2010 (Internatinoal Insurance Law Association 2010) <http://www.aida.org.uk/pdf/Climate%20Change%20and%20Insurance%20 Law%20General%20Report.pdf> accessed 6 January 2021; National Association of Insurance Commissioners (NAIC), *The Potential Impact of Climate Change on Insurance Regulation* (NAIC 2008); J. MacDougald and P. Kochenburger, "Insurance and Climate Change" (2013) 47 *John Marshall Law Review* 719; A. Dlugolecki, "Climate Change and the Insurance Sector Source" (2008) 33 *Geneva Papers on Risk and Insurance Issues and Practice* 71–90.

64 E. Asen, "Carbon Taxes in Europe" (Tax Foundation, 14 November 2019) <https:// taxfoundation.org/carbon-taxes-in-europe-2019/> accessed 6 January 2021.

65 The World Bank, "Pricing Carbon" (The World Bank, 17 June 2019) <https://www .worldbank.org/en/programs/pricing-carbon> accessed 6 January 2021.

66 E. de Wit, S. Seneviratne, and H. Calford, "Climate Change Litigation Update"

(Norton Rose Fulbright, February 2020) <www.nortonrosefulbright.com/en/know ledge/publications/7d58ae66/climate-change-litigation-update> accessed 6 January 2021; J. Setzer and R. Byrnes, *Global Trends in Climate Change Litigation: 2019 Snapshot* (London: Grantham Research Institute on Climate Change and the Environment and Centre for Climate Change Economics and Policy, London School of Economics and Political Science 2019). According to the Grantham Research Institute on Climate Change of the London School of Economics, there have been 70 cases globally from 2017 to 2020 <https://climate-laws.org/cclow/litigation_cases?case_started _from=2017>.

67 *Urgenda Foundation (on behalf of 886 individuals) v The State of the Netherlands (Ministry of Infrastructure and the Environment)*, First instance decision, HA ZA 13-1396, C/09/456689, ECLI:NL:RBDHA:2015:7145, ILDC 2456 (NL 2015), 24th June 2015, Netherlands; The Hague; *District Court or Juliana v. United States*, Case No. 6:15-cv-01517-TC, U.S. District Court, District of Oregon.

68 UN, *Amendment to the Montreal Protocol on Substances that Deplete the Ozone Layer*, 15 October 2016, UN Doc. C.N.872.2016.TREATIES-XXVII.2.f.

69 D. Broom, "How the Middle East is Suffering on the Front Lines of Climate Change" (World Economic Forum, 5 April 2019) <www.weforum.org/agenda/2019/04/middle -east-front-lines-climate-change-mena/> accessed 6 January 2021.

70 O. C. Ruppel, "Aspects of International Climate Change Law and Policy from an African Perspective" in O. C. Ruppel and E. D. Kam Yogo (eds), *Environmental Law and Policy in Cameroon – Towards Making Africa the Tree of Life | Droit et politique de l'environnement au Cameroun – Afin de faire de l'Afrique l'arbre de vie* (Baden-Baden, Deutschland: Nomos Verlagsgesellschaft mbH 2018); UNEP, "GEO-6 Regional Assessment for Africa" (UNEP 2016) <www.unenvironment.org/resources/assessment /geo6regionalassessmentafrica#:~:text=The%20GEO%2D6%20regional%20assessment,well%2Dbeing%20are%20continuously%20enhanced>.

71 "Agenda 2063: The Africa We Want" (African Union) <https://au.int/en/agenda2063/overview> accessed 6 January 2021.

72 UNEP, n. 73.

73 American Bar Association, *Legal Education and Professional Development – An Educational Continuum: Report of the Task Force on Law Schools and the Profession: Narrowing the Gap* (American Bar Association 1992).

74 M. Mehling, H. van Asselt, K. Kulovesi, and E. Morgera, "Teaching Climate Law: Trends, Methods and Outlook" (2020) 32 (3) *Journal of Environmental Law* 417–440.

75 L. L. Laska, "Environmental Law for Undergraduates: Concepts and Curricula" (1973) 51 *Peabody Journal of Education* 33–41.

76 See also a collection of contributions in: P. Kameri-Mbote and C. Odote (eds), *Blazing the Trail: Professor Charles Okidi's Enduring Legacy in the Development of Environmental Law* (University of Nairobi, School of Law 2019) 1-25.

77 I. R. Bulska, "Contribution of the United Nations Environment Programme (UNEP) to the Development of International Environmental Law" in Kameri-Mbote and Odote, ibid, 258.

78 UN Department of Public Information, *Agenda 21: Programme of Action for Sustainable Development, Rio Declaration on Environment and Development, Statement of Forest Principles: The Final Text of Agreements Negotiated by Governments at the United Nations Conference on Environment and Development (UNCED), 3–14 June 1992, Rio de Janeiro, Brazil*, 1993, UN Doc. [ST/]DPI/1344.

79 Before 2016, known as the Division of Environmental Law and Conventions (DELC).

80 The Law Division is headquartered in Nairobi and closely coordinates its work with individual countries through UNEP's regional offices in Geneva (for Europe, including Caucasus and Central Asia), Panama (for Latin America and the Caribbean),

Bangkok (for East Asia and the Pacific), Manama (for West Asia), and Nairobi (for Africa).

81 UN Environment's Africa Office <www.unenvironment.org/regions/africa/our-work -africa/un-environments-africa-office and West Asia, www.unenvironment.org/regions/west-asia>.

82 Phase I countries included Burkina Faso, Malawi, Mozambique, Sao Tome and Principe, Kenya, Tanzania, and Uganda. Phase II countries focused on the subregional projects for Sahel (Burkina Faso, Mali, Niger, and Senegal) and for Southern Africa (Malawi, Botswana, Lesotho, and Swaziland), E. M. Mrema, "PADELIA – A Review" (2003) 33 *Environmental Policy and Law* 204.

83 The project was initially known as the UNEP/UNDP/Dutch Joint Project on Environmental Law and Institutions in Africa. E. M. Mrema, "Away from Traditional Project Management: Lessons from the Programme for the Development of Environmental Law and Institutions in Africa (PADELIA)" in Kameri-Mbote and Odote, n. 79, p. 68.

84 Mrema, ibid, 80.

85 Ibid, 81. The Inter-Parliamentary Union (IPU) and UNEP have been assisting national parliaments to build their capacity to formulate and review relevant legislation and to provide effective oversight on the negotiation and implementation of internationally agreed environmental agreements and climate change goals. To this end, workshops for Members of Parliament and parliamentary staff took place in Harare, Zimbabwe, on 8 March 2019 and Bujumbura, Burundi, on 22 March 2019.

86 Ibid.

87 Ibid, 83.

88 Ibid, 92.

89 The International Union for Conservation of Nature Academy of Environmental Law (IUCNAEL) and the International Commission of Environmental Law (ICEL) provided fora for the exchange of views and networking of scholars with special interest in environmental law. The 2nd [IUCNL] Academy Annual Colloquium took place in Kenya in 2004 and gave the opportunity to many African legal scholars and practitioners to participate. On that occasion, in partnership with UNEP, the first Symposium of African Environmental Law Lecturers (Nakuru, Kenya, 29 September – 2 October 2004) was held back-to-back with the Academy symposium, which laid the foundation for the establishment of ASSELLAU (P. Kameri-Mbote, "Building an Army of Environmental Law Scholars: Professor Charles Odidi Okidi's Legacy" in Kameri-Mbote and Odote, n. 79, 95).

90 North Africa's partaking is not as intense as engagement had been from Sub-Saharan Africa. This may be attributed to the unique social, cultural, and legal traditions in the North African countries that blend in more with West Asia rather than with Sub-Saharan countries.

91 The first ASSELLMU conference took place in Doha, from 4 to 5 November 2018, the second conference in Settat from 4 to 5 November 2019, and the third conference is scheduled to take place in Doha, Qatar, in November 2020. See also the "2019 Middle East and North Africa (MENA) Environmental Law Scholars' Conference" (Hamad Bin Khalifa University College of Law), <https://www.hbku.edu.qa/en/academic-events/mena-law-conference> accessed 6 January 2021.

92 The UNEP Law Division also runs the secretariat of the Green Customs Initiative (GCI), a partnership of international entities cooperating to prevent the illegal trade in environmentally sensitive commodities and substances and to facilitate their legal trade. The GCI was launched in 2004 and its partners are: Secretariats of several multilateral environmental agreements (Basel, Rotterdam, Stockholm, Minamata, Cartagena Protocol to the CBD, CITES and Montreal Protocol to the Vienna Ozone Convention), UNEP, OzonAction (discharging UNEP's function as an Implement-

ing Agency of the Multilateral Fund of the Montreal Protocol on Substances that Deplete the Ozone Layer), UNODC, INTERPOL, OPCW, and WCO <www.green-customs.org>.

93 Training workshops with the Zambia Police Service and the Judiciary of Zambia in Lusaka, from 2 to 3 and from 5 to 6 December 2019, respectively (UNEP Law Division 2019 Annual Report, <https://spark.adobe.com/page/BiOKQuLrRtm8V/>). Greening the judiciary in Africa is a current and ongoing initiative that aims to build the capacity of judges in applying and enforcing environmental laws, promoting the environmental rule of law. The first colloquium in 2017 in Johannesburg resulted in the adoption of the Johannesburg Action Plan on Environmental Law <www.judiciary.go.ke/wp-content/uploads/2020/07/Johannesburg-Plan-of-Action.pdf> and the second colloquium took place in 2018 in Maputo <www.unenvironment.org/news-and-stories/blogpost/greening-judiciaries-africa-second-regional-symposium>.   In 2018, UNEP in partnership with the Office of the Director of Public Prosecutions in Uganda held an induction training in Entebbe on "Greening the Police and Criminal Prosecutorial Education in Africa". <https://www.unenvironment.org/news-and-stories/story/environmental-crime-curriculum-police-and-prosecutors-africa-launched>.

94 The workshop on International Environmental Law and Environmental Diplomacy (Ramallah, 18–19 December 2018) was organized by UNEP West Asia Office and the Law Division, at the request of and in consultation with the Environmental Quality Authority (EQA) of Palestine, which also discharges the functions of the Ministry of Environment in the country, with a view to building capacity of government officials (especially designated national focal points of Palestine to MEAs (usually from EQA), as well as recently recruited officials at the ministry responsible for foreign affairs) in the field of international environmental law and diplomacy <www.unenvironment.org/news-and-stories/story/strengthening-capacity-palestinian-officials-international-environmental-law>.

95 The first AOAD-UNEP workshop on environmental diplomacy focused on the United Nations Convention to Combat Desertification (Cairo, 8–10 October 2018) <www.aoad.org/eng/enews15-oct-2018.htm> and the second workshop on the Convention on Biological Diversity (Cairo, 7–9 December 2019) <https://leap.informea.org/sites/default/files/inline-files/Report%20by%20the%20secretariat%20-%20Progress%20in%20the%20implementation%20of%20MV5.pdf para. 20>.

96 In the field of environmental law-making and diplomacy, UNEP and the University of Eastern Finland are organizing every year a high-profile two-week event. The Seventeenth Joint Annual UNEP-University of Eastern Finland MEA Course is scheduled to take place in Joensuu, Finland, in early 2021.

97 In Dili, Timor Leste, on 9 and 10 October 2019, and in Dhulikhel, Nepal, on 14 and 15 November 2019 <https://leap.informea.org/sites/default/files/inline-files/Report%20by%20the%20secretariat%20-%20Progress%20in%20the%20implementation%20of%20MV5.pdf para. 21>.

98 For instance, in 2019, UNEP, in collaboration with the International Bar Association, developed an environmental law curriculum for private legal practitioners, which is being used globally: Framework Model Curriculum on Continuing Legal Education In Environmental Law <https://wedocs.unep.org/bitstream/handle/20.500.11822/30989/EnvLaw.pdf?sequence=1&isAllowed=y>.

99 UNEP, n. 52 (At the time of writing, UNEP was working on an updated study on the status of litigation with a view to publishing in 2021).

100 See n. 57.

101 M. R. Poustie, "Engaging Students and Enhancing Skills: Lessons from the Development of a Web-Supported International Environmental Law Conference Simulation" (2001) 15 *International Review of Law, Computers and Technology* 331–344.

102 InforMEA <www.informea.org/en> started as a digital MEAs "dictionary" and has now expanded to include copious information on MEAs decisions, news, events, and Parties, jurisprudence (national, regional and international). It is supported by the European Union. ECOLEX <www.ecolex.org/> operates jointly with the Food and Agriculture Organization of the United Nations and the International Union for the Conservation of Nature, and includes information on treaties, international soft law, and other non-binding policy and technical guidance documents, national legislation, judicial decisions, and law and policy literature. Users have direct access to the abstracts and indexing information about each document, as well as to the full text of most of the information provided.

103 For example, the Macquarie University (Sydney) uses the InforMEA Climate Law course for the Climate Change Law <https://unitguides.mq.edu.au/unit_offerings/8 4801/unit_guide/print>.

104 The toolkit is developed in partnership with the UNFCCC Secretariat and the Commonwealth Secretariat (Law and Climate Change Toolkit <https://lcc.eaudeweb .ro>; UNEP, "New law and climate change toolkit to guide the implementation of the Paris Agreement, July 12, 2018 <www.unenvironment.org/news-and-stories /blogpost/new-law-and-climate-change-toolkit-guide-implementation-paris-agree ment>.

105 E. Scotford and S. Vaughan, "Environmental Law and the Core of Legal Learning: Framing the Future of Environmental Lawyers" (OUPblog, 15 October 2018) <https://blog.oup.com/2018/10/environmental-law-core-legal-learning/> accessed 6 January 2021; D. M. Ong, "Prospects for Integrating an Environmental Sustainability Perspective within the University Law Curriculum in England" (2016) 50 *The Law Teacher* 276–299.

3 The role of environmental *Waqf* in addressing climate change in the MENA region

A comparative law analysis

Samira Idllalène

Introduction

The aim of this chapter is to examine the potential roles of *Waqf* endowments in climate change action in the Middle East and North Africa (MENA) region.[1] It examines the ethical and legal foundations of environmental *Waqf* as well as its theoretical foundations in comparison to the Public Trust Doctrine (PTD) in common law. It then addresses the legal and institutional barriers that can hinder the revival of the environmental *Waqf* and suggests some recommendations as a roadmap for this revival.

Simply, *Waqf* is "a legal act by which a person, in order to be pleasing to God, strips themselves of one or more of their goods, generally immovable, and assigns the goods in perpetuity for a pious, charitable or social purpose".[2] Such assignment can be in an absolute manner, exclusive of any restriction (public *Waqf*), or may be specifically reserved to a person or several specific persons (family *Waqf*). When the beneficiaries of the right of enjoyment cease to exist, the family (*Waqf*) becomes a public (*Waqf*). As an Islamic legal institution, *Waqf* would be based on the *Hadith* (acts and sayings of the Prophet Muhammad which is the second source of Islamic law) and on *Fiqh* (Islamic science) since the 2nd century after *Hijra* (Islamic calendar).[3]

How is *Waqf* relevant or important to environmental protection? Or conversely, what does environmental protection have to do with *Waqf* endowments? Simply, *Waqf* as a tool for charitable giving can offer multiple benefits for the protection of the environment. The linkages between environmental protection and religious principles cannot be overemphasized, especially in MENA countries. Indeed, while scientists report the alarming situation of environmental degradation in the world and more specifically in the MENA region,[4] some authors argue that spirituality is likely to play a paramount role in environmental protection and natural resources conservation.[5] This Spiritual Ecology movement, which emphasizes the linkages between religion, spirituality, and environmental protection, is already gaining momentum in Islamic countries (Eco-Islam), especially in Asian countries.[6]

Sharia is a primary source of law in many parts of the MENA region and already encompasses both religion and the law. In fact, Sharia is "a way of life".[7]

However, its ecological potential is still largely untapped today in framing climate action. Consequently, environmental law in Muslim countries does not fully reference Sharia, and vice versa. Scholarship on the ecological potential of Islamic law in Muslim countries is largely influenced by philosophers and biologists, but rarely by jurists.[8] While the ecological potential of Sharia is experimented within a few Asian Muslim countries (particularly Indonesia[9] and Malaysia[10]), it is still in its theoretical phases in the MENA region.[11] Yet, this region is confronting the most urgent ecological threats today. Indeed, while some countries are facing drought and poverty, others represent big fossil fuel producers.[12] Therefore, the legal paradox is exacerbated by the economic and ecologic contrast.

Drawing from the history and culture of MENA countries, especially with respect to *Waqf*, can provide adequate solutions to the global climate change conundrum. In parallel, scholarship in comparative law can shed light on a new research and political agendas on Islamic Environmental Law (IEL) able to protect the climate system. Efforts to mainstream religious principles into environmental protection have already resulted in the adoption of the Islamic Declaration on Global Climate Change[13] at the international level. Though not legally binding, the Declaration distils core Islamic principles on climate change, and is likely to provide wider normative influence on the development of IEL in MENA countries. In fact, the Declaration called on well-off nations and oil-producing states to *inter alia* "[l]ead the way in phasing out their greenhouse gas emissions as early as possible and no later than the middle of the century" and "[i]nvest in the creation of a green economy".[14] Referring to a related scriptural basis, this Declaration emphasizes the need for moral and ethical understanding of the roles of Muslims and Muslim countries to take all steps necessary to address climate change through financing, emission reductions, as well as responsive legal and policy efforts.[15]

In light of the increasing importance of Islamic principles in climate change response, this chapter examines the potential roles of *Waqf* in climate change regulation in the MENA region. It is divided into six sections. After this introduction, the second section addresses the meaning, nature and elements of the *Waqf*, while the third section focuses on how comparative law can be useful in the expansion of environmental *Waqf*. The fourth section explains the challenges and barriers that hamper the revival of environmental *Waqf* in the MENA region, while the fifth section discusses innovative legal approaches for addressing the challenges. The chapter then concludes in the final section.

Environmental *Waqf*: background, nature, and elements

The importance of *Waqf* as an important tool for the protection of the environment in the MENA region and in Muslim countries stems predominantly from the pivotal role of Islamic law in MENA countries.

Firstly, constitutions in MENA countries include a clause on the supremacy of Sharia as a source of law.[16] This means that the entire legal system is submitted to Islamic law. Consequently, this Islamic supremacy clause renders Islamic law as a body of supra-constitutional standards such as an "Islamic *Grundnorm*" for which all other laws derive validity. As a part of the *grundnorm*, any other law or policy that is inconsistent with Islamic law will be void to the extent of such inconsistency.[17] Environmental laws in MENA countries are thus also implicitly and greatly influenced by Islamic law and principles. For example, several verses from the Qur'an emphasize that humankind has a solemn duty to keep, maintain, and preserve the natural environment such that it does not disrupt or upset the interests of future generations: "There is no animal on the earth, or any bird that wings its flight, but is a community like you"[18] and "Do not strut arrogantly on the earth. You will never split the earth apart nor will you ever rival the mountains' stature".[19] According to Mekouar, these Islamic principles emphasize that humankind is a mere trustee of the earth, not a proprietor, a disposer, or an autonomous owner. Rather humankind must use the environment as a responsible holder who must answer to God for his actions.[20]

This principle of *Khilafa* (trusteeship) is well embedded within Islamic law and is well illustrated through the *Waqf*. It thus constitutes the foundation of the whole legal system in Muslim countries, including in environmental law.

Waqf is at the heart of Islamic law

Islamic endowment or *Waqf* is as old as what is considered "Islamic civilization".[21] As a tool of social welfare, *Waqf* was used for many different purposes. In fact, all over the Islamic world, magnificent works of architecture "as well as wealth of services vitally important to the society have been financed and maintained for centuries through this system. Many *Awqaf* [plural of *Waqf*] had survived for considerably longer than half a millennium and some even for more than a millennium".[22]

Waqf can be private or public. Private or family *Waqf* are dedicated to the family of the donor (*Waqef*), while public *Waqf* are dedicated to the general community.

Waqf had its boom in the Muslim world from the 2nd century until the fall of the Ottoman Empire.[23] In Istanbul, the birdhouses adorning the city walls are the vestige of a prosperous past of the environmental *Waqf*.[24] *Waqf* was also the main legal instrument for the management of water resources, urban planning, and protection of animals.[25] *Agdal* (or *Hima*), land for the conservation of forest and pastoral environments, also reflects social and spiritual values.[26]

The environmental functionality of *Waqf* goes beyond the institution itself. It mirrors the ancient civilization and cultural epoch of donating or giving valuable property and donations for charitable purposes. This cultural value can be reinvigorated to provide a basis for reforming environmental laws in MENA countries. The flexibility of *Waqf* allows it to have infinite environmental applications. Indeed, "[i]t is possible to constitute land, on a perpetual basis, in [*Waqf*], and this for various purposes, such as the development of agro-pastoral research, the increase of wild fauna and the restoration of its habitats, the creation of village woods, the construction of ponds, the digging of wells, the construction of public gardens, etc. [The *Waqf*] can also consist of the allocation of income from real estate or the financing of similar projects".[27] *Waqf* is in favour of the participatory approach and the perpetual inalienability of resources.

A number of key elements are required to create a valid *Waqf*. These are the donor (*Waqef* or *Mohabiss*), the manager of the *Waqf* (*Nadher* or *Moutawali*), the *Waqf* asset or *Mawquf*, the beneficiary (*Mawqouf alaïh*), and the formula (*Sigha*).[28] These elements must respect the ultimate goal of the *Waqf* that is the *Maslaha* (public benefit or interest).[29]

The donor (*Waqef*)

The donor (an individual or an institution) must have the legal capacity to constitute the *Waqf*. The donor may not necessarily be a Muslim. Nevertheless, he or she is required to establish the *Waqf* for a purpose that does not contradict the principles of Islam. The *Awqaf* Ministries or State Departments could, for example, mobilize their land reserves for environmental purposes such as protected areas, environmental schools, animal shelters, etc.[30]

The *Nadher* (*Mutawalli*), the manager of the *Waqf*

The function of the *Nadher* is to manage *Waqf* assets "in confidence".[31] They are bound by the *Waqf* deed instituted by the donor. But under the supervision of the judge, the *Nadher* is free to implement the *Waqf* in all areas likely to ensure the general interest. *Waqf* can include the creation of an institutional body to ensure its maintenance.[32]

The *Nadher* can also dissolve *Waqf* assets when they no longer meet their public utility function. This possibility is consistent with the ecological principle of adaptive management which is the core of the legal climate system.[33] This flexibility also characterizes *Waqf* assets.

Waqf assets (*Mawquf*)

According to *Fiqh*, *Waqf* assets should satisfy four essential conditions: "it must be a possession (*melk*) and this possession must consist of a good (*'ayn*) materially recognizable and identifiable, capable of producing wealth and having a utility for Humans".[34] If the first condition requires a broad interpretation of the concept of possession, inasmuch as biodiversity or the air is mainly a matter of *res nullius*. The second condition *'ayn*, which is quite similar to the Roman concept of *res*, can be broadly applied to biodiversity and to the atmosphere as natural *res*.[35]

This is an unprecedented regime of goods. In fact, the notion "*Waqf* asset" differs from the economic concept of property in civil law.[36] If an asset is put into a public *Waqf*, it then serves an objective of general interest. Ecosystem services could be part of it. In this sense, the *Habous Bellarj* in Marrakech benefited from the rental income from buildings (hotels).[37]

These revenues were dedicated to the protection of storks who were the beneficiaries of the *Waqf*.[38]

The beneficiaries of the *Waqf*

As for *Waqf* assets, the beneficiaries of the *Waqf* can be unlimited.[39] They can even be designated according to vague formulas: the global community.[40] Natural resources can be among the beneficiaries if it is decided in the *Waqf* deed. For example, a *Hima* (or *Agdal*) can be created by a *Waqf* as carbon sinks.[41] Moreover, beneficiaries are not required to exist when the *Waqf* is instituted because it can also be assigned to future beneficiaries.[42] A powerful application of this would be an atmospheric *Waqf* dedicated to future generations.[43] Therefore, if the *Waqef* has planned to create an environmental *Waqf*, the absence of beneficiaries determined at the outset can divert the *Waqf* to other similar areas. This perpetual character of some permanent *Waqf* confirms the interest of future generations that the principle of sustainable development promotes.[44]

The beneficiary of the *Waqf* has only a right of usufruct over the *Waqf* asset. However, benefiting from rights basically means enjoying legal personhood. It is therefore curious that the ancient *Waqf* have designated animals as beneficiaries, while modern law denies them the prerogative of legal personhood.[45] This issue has not been debated by *Ulamas* or *Fuqahas* (scholars specialized in *Fiqh*), who seem to take the practice for granted.[46]

The *Waqf* Statutes can also be interpreted as such. For example, in article 11 the Moroccan *Waqf* Code provides that the *Waqf* can be instituted for the benefit of all who can legitimately benefit from the resources of the *Waqf* asset.[47] The impersonal formula adopted by this text and several other *Waqf* Codes in the MENA countries could be largely interpreted.[48] In this sense, the ecological *Waqf* prefigures modern law. In fact, the *Waqf* must conform to the *Maslaha*.[49]

The objective of the *Waqf*: *Maslaha*

Ulamas have used the principle of the *Qiyas* (analogy) by rationalizing popular practice in the sense of *Maslaha amma* (public utility).[50] The principles of *Maslaha moursala* and *Istihsan* (looking for good) have facilitated this development.[51] The first principle refers to the idea of public utility, an argument to legitimize a practice for which the legal text (here the *Qur'an*) is silent. The second principle avoids resorting to *Qiyas* when fair considerations require it.

The use of *Maslaha* is subject to conditions.[52] Among them is respect for the spirit of the law, in particular the *Qur'an*. In fact, *Maslaha* should conform to the higher objectives of Islamic law (*Maqasid Al Sharia'*) which are: preservation of life, religion, reason, progeny, and property.[53]

Maqasid Al Sharia' "is a system of values that could contribute to a desired and sound application of the [Sharia]".[54]

The Persian philosopher, al-Ghazali, emphasized that the Masaliḥ (plural of Maslaha) "are potentially limitless and change according to time and context".[55]

Maqasid Sharia' can be extended according to the needs and specifics of each society. The use of Ijtihad[56] and its main tool, Qiyas, allows this.[57]

Environmental protection and the right to a healthy environment are no longer a luxury of developed countries but an emergency to consider. Maqasid Sharia' broadly encompasses the protection of the environment as it is intertwined with the protection of health and the ecological balance (Mizan) of our planet.[58] Sharia principles should therefore be further exploited for the protection of natural resources. In fact, scholars also consider the many junctions that exist between the Sustainable Development Goals (SDGs)[59] and Maqasid Sharia that Waqf illustrates. According to Abdullah, "the institution of waqf has a proven track record of social welfare and development, and it is best suited to patronize the causes that can be served through philanthropy".[60] Therefore, "maintaining the sustainability of resources by conservation of sea, ocean, climate, ecosystem, water and energy coupled with the promotion of economic growth, industrialization, safety of cities, cooperation among all and sustainable consumption patterns include among the aims of both maqasid and SDGs".[61]

In order to highlight the ecological potential of Waqf, scholars may study Ijtihad through comparative law as it reveals many similarities between Waqf and the trust.

Influence of the "trust" concept in comparative law

Waqf assets, whether public (assigned to charitable general interests) or mixed (first assigned to specific beneficiaries, then intended to general charitable purposes) "are closer to the English concept of 'Trust' and, more precisely, of 'charitable Trusts'".[62] According to some authors, the origin of the trust, which appeared during the crusade wars between the 12th and the 13th centuries, is the Waqf itself.[63] According to Verbit, "[a] Waqf is a transfer from A to B for the benefit of C. That is to say, it fits into the classic pattern of the Trust where B is a fiduciary who manages the property for the benefit of C. Using common law terminology C can encompass a variety of present and future interests".[64] Elements constituting the Waqf and the trust are very similar.[65] The Waqf and the trust are sui generis institutions, they both draw their foundation from religion and they both operate almost identically.

In fact, they both require a trustor (or Waqef), a trustee (Nadher or Moutawali), and one or more beneficiaries (beneficiaries or cestui que trust). In addition, both appeal to the judge (Cadi), whose origins are often religious (ecclesiastics in England (Chancellor) and Muslim in countries adopting the Waqf). Similarly, the trust, like the Waqf, can be either private (family Waqf, private trust) or public (Waqf Khaïri, charitable trust), temporary or permanent.[66] In both the Waqf and the trust, only the usufruct is appropriated, for the benefit of specific individuals, or for a general charitable purpose; the corpus becomes inalienable in perpetuity in favour of successive beneficiaries and thus outside inheritance law. Continuity is guaranteed by the successive appointment of trustees or *Moutawallis* (*Nadher*).[67] The similarities between common law and Islamic law are such that in some common law states where a minority of Muslims live, there is no hesitation in recognizing the *Waqf* as a trust, and *vice versa*.[68]

Despite these similarities, the trust has undergone an enormous evolution in contrast to the *Waqf*. There is in fact "a striking contrast between the decline that *Habous* [or *Waqf*] has known for a long time in a country of Islam and the influence of the Trust in England".[69] As such, the trust has been particularly successful in protecting the environment. The National Trust for Places of Historic Interest or Natural Beauty (National Trust) founded in 1895 in England is a perfect application of the charitable environmental trusts.[70]

In the United States, environmental trusts expanded within the framework of the Public Trust Doctrine (PTD) which stimulated the development of environmental law. In his seminal article, Joseph Sax proposed to extend the public trust, initially intended to protect shorelines, navigable waters, and the air, by virtue of "an adaptation of both the English trust and the concept of *res communis* of the Justinian Institutes to other elements of the environment such as wild species".[71] By "liberating the Public Trust Doctrine from its Historical Shackles", Joseph Sax[72] has impacted the environmental common law in the United States and several other countries.[73]

The PTD is based on the idea that assets which cannot be appropriated because of their nature must be managed by a public authority, for the benefit of the population. The government, in this sense, is not considered to be the owner of these assets, but quite simply as a trustee responsible for their management.[74] The PTD should be used, according to Sax, as a tool for "effective judicial intervention"[75] with the aim of protecting the environment and natural resources. He considers that the citizens, as the beneficiary of the trust established on natural resources, could claim in courts that the government complies with its duty as a trustee.[76]

Tribunals have contributed to the development of the PTD since 1821 in *Arnold v Mundy*, where the court rejected a claim of the exclusive right of a person on oyster beds, asserting that "the rivers [...] are common to all citizens".[77] Increasingly, the PTD has been considered as the best tool for achieving ecosystem services, for protecting the federal public trust, for combating the harmful effects of climate change, etc.[78] Authors also systematized the notion of ecological public trust or more specifically that of Nature's Trust.[79]

As an overarching principle, the PTD has expanded in common law countries. Its extension in England has not required major modifications of the law. In fact, "it is logical to conclude that there is a cause of action somewhere in English law based on violation of the Public Trust. It must simply be discovered".[80] Therefore, even if its relations with the English law became quite distant, the PTD represents the extension of the English trusts. Furthermore, in *Kinloch v Secretary of State for India*, the court stated that: "[a] 'higher sense' of Trust inhered in the Crown's control of a phosphate-rich island colony which had an Ordinance providing a commissioner would establish the formula for paying mining royalties".[81] In sum, "the

Trust's continuing background presence is felt through the field" in the PTD.[82] Consequently, the PTD can prevail over the legislation.[83] In fact, "[i]ncreasingly modern agencies fail to protect public resources, and citizens look to the PTD to provide more effective action by requiring fiduciary action".[84] Sax's argument for broader recognition of the PTD stemmed from his belief that, looking around at the paucity of options for concerned citizens: "[o]f all the concepts known to American law, only the [PTD] seems to have the breadth and substantive content which might make it useful as a tool of general application for citizens seeking to develop a comprehensive legal approach to resource management problems".[85]

The growing concept of atmospheric trust litigation builds upon the PTD. As noted in *Juliana et al. v United States*, "[t]he complaint alleges defendants violated their duties as trustees by failing to protect the atmosphere, water, seas, seashores, and wildlife".[86]

The PTD concept has also migrated to other countries, notably through colonization.[87] The arguments underlying the PTD reinforce a theoretical understanding of *Waqf*. In MENA countries, the PTD principle will certainly gain momentum if exploited in the framework of an *Atmospheric Waqf Paradigm*. This encompasses a basis for creating large environmental restoration funds in trust. Some Muslim countries in Asia, such as Indonesia and Malaysia, are already experimenting with the environmental *Waqf* notion.[88]

Applying *Waqf* to climate change regulation in the MENA region: legal challenges and barriers

Despite the rising importance of *Waqf* as a potential tool for financing climate action in the MENA region, there are five key legal and institutional barriers that must be overcome to ensure effective application of *Waqf* principles to advance climate change regulation in the area.

Poor environmental law enforcement

Public policies in the field of environmental protection in the MENA region adopt several types of measures: awareness-raising, planning, and legal rules, amongst others.[89] Yet, the extreme climate conditions have created conflicts over scarce natural resources in this region.[90] In Jordan, for example, environmental risks are induced by a "lack of oversight and monitoring, which leads to abuse of the system, weak law enforcement and an unclear system of penalties and fees. This creates conditions that encourage illegal action".[91] In Morocco, "deforestation affects nearly 25,000 ha per year [...] and more than 600 species are currently considered threatened".[92]

Poorly enforced environmental law is not a phenomenon limited to MENA countries; a UNEP report on the environmental rule of law released in 2019 stated that this phenomenon is global.[93] The report also emphasized that in developing countries, it is mostly linked to colonization. Therefore, based on imported laws from other countries, initial frameworks of environmental laws "failed to represent the conditions, needs, and priorities of the countries into which they were imported".[94]

This is what has prompted some scholars to call for a reinvigoration of customary law institutions such as *Agdal* or *Hema*.[95] The revival of ancient customary and religious institutions could be an answer to the non-enforcement of statutory environmental law. However, this revival may be hindered by the general context where environmental law is not largely respected or enforced. In fact, one of the main reasons for the non-enforcement of environmental law is the lack of awareness.[96] These problems of lack of enforcement will need to be addressed by designing realistic and responsive environmental laws, as well as creating effective institutions to monitor the enforcement of such laws.[97]

Incoherent application of Islamic law in environmental instruments

In several MENA countries, Islamic law is proclaimed as a source of the legal system in the Constitution (the "Islamic Supremacy clause").[98] However, this clause is not applied in environmental matters. The linkages between Islamic principles and environmental protection are yet to be fully reflected in environmental planning, policy making, and legislation.

In general, Islamic law is mainly applied in family law and land law, while commercial law, urban planning, and environmental law, for example, are more influenced by the Western legal model.[99] Yet, several Muslim countries have enacted statutes dedicated to *Waqf*.[100] However, there is no link between these two types of laws. Few countries have started to create environmental *Waqf* such as funds for water supply and clean-ups,[101] but there is no legal text specifically dedicated to environmental *Waqf*. Consequently, courts do not refer to Islamic law in environmental litigation. A judge will recognize the symbolic value of the Sharia but will go little further than this simple recognition.[102]

Nevertheless, there are some exceptions. In fact, an Egyptian court stated on the legitimacy (according to Sharia) of the decision taken by an environmental agency which had not carried out the Environment Impact Assessment (EIA) before the construction of a cement factory. The judge retained in this case the *Fiqh* principle *amr bil ma'rūf wa nahi ani l munkar* (prohibiting good and preventing evil) to allege that the said agency failed to protect "the health of citizens when it continued to allow local cement plant to emit harmful air pollutants".[103] In general, examples of the use of Sharia in the environmental sphere come mostly from Asian countries.[104]

There is a need to address the gaps in environmental rulemaking in the MENA region by reflecting Islamic tenets and principles, such as the *Waqf*, that could provide a basis for creating specialized funds dedicated to environmental purposes. Furthermore, climate change policies and law should integrate important Islamic principles that could propel homegrown and Sharia-compliant financing, technology, and legislation on climate change mitigation and adaptation.

The institutional gap

In MENA countries, state departments in charge of governing the environment are not the same as the state departments in charge of *Waqf* or Islamic affairs. These two institutions work separately. Historically, environment departments were created in the 20th century, while the *Waqf* is an old institution that has been used for centuries.[105]

Muslim countries have dedicated state departments for managing the *Awqaf* (generally called Ministry of *Awqaf* and Islamic Affairs).[106] Theoretically, even though their duties may involve the protection of cultural heritage and resources, they do not have supervisory functions over environmental matters, such as climate change funds or environmental restoration funds. They mostly focus on managing mosques and religious schools.[107] However, over the last decade, these ministries started to work on ecological programmes, such as the Green Mosques.[108] Furthermore, environmental ministers of the MENA region now have annual meetings in the framework of the Islamic World Educational, Scientific and Cultural Organization (ICESCO).[109] The increasing interactions between *Waqf* departments and environment ministries may trigger closer cooperation between these institutions in jointly managing environmental restoration funds and endowments.

Another institutional gap that exists is that unlike the practice in Muslim Asian countries, a number of Islamic institutions in the MENA region are yet to fully embrace the use of Sharia for protecting the environment.[110] The committee of *Ulamas* in many Asian countries, such as Indonesia and Malaysia, has played a major role in allowing Sharia to reinvest the field of environmental protection. For example, from 2006 and 2016, Majliss Ulamas Indonesia (MUI) has issued six *Fatwas* (legal opinions) on environmental matters.[111] These *Fatwas* generally ask the government, private companies, religious groups, and citizens to protect the environment. The *Fatwas* are divided into two parts: a part presenting the legal and ethical basis and a part on the decision taken in the form of guidelines and recommendations. The first part contains references to the *Qur'an* verses and to *Hadith* and *Fiqh* principles related to the issue. For example, the *Fatwa* on Wildlife Conservation for the Preservation of Ecosystem Balance issued in 2014 cited the Imam al-Sharbini's book *Mughni al-Muhtaj* (5/527) which explained the "obligation to protect rare animals and the prohibition on causing them to [become] extinct".[112] Even if they do not have a binding legal effect, environmental *Fatwas* issued by the MUI have an important symbolic meaning because they are a great opportunity for collaboration between religious groups, environmental scholars, and local and international environmental NGOs.[113] In contrast, across the MENA region, only very few *Fatwas* related to environmental protection have been issued.[114] Consequently, scientific research on the relevance and role of *Fatwas* in promoting environmental protection in the MENA region is generally limited. To promote the integration of Islamic principles and norms in climate action and programmes, there is a need for Islamic scholars and institutions to develop normative Islamic principles and *Fatwas* that clarify and unpack how important tools, such as *Waqf*, can advance climate action in light of local contexts and realities.

Inadequacy of scientific research programmes

Despite the potential contributions of IEL to addressing global climate change, the analysis of environmental applications of *Waqfs* remains in its infancy in MENA countries.[115] Research in the Asian Muslim countries is relatively more extensive, although also not abundant.[116] Nevertheless, research programmes are ongoing in the framework of Islamic finance and increasingly in the field of green finance.[117] This can have positive effects in promoting Islamic based financing in the design and implementation of climate mitigation and adaptation projects and policies.

The contrasting perception of climate change

MENA countries do not view climate change in the same way. On one hand, the countries most affected by climate change suffer from water scarcity, droughts, and pollution. On the other hand, many countries "continue to contribute to climate change by burning fossil fuels such as coal, gas and oil in copious amounts".[118] In fact, seven of the top 15 oil-producing states in the world are Muslim countries, six of which are MENA countries. These countries account for over 30 million barrels of oil production per day.[119]

The costs of environmental degradation are estimated to be between 2.1 per cent and 7.4 per cent of GDP in each country.[120] Without designing cleaner production methods and technologies, the Middle East may need to leave about 40 per cent of its oil and 60 per cent of its gas underground in order to comply with the objectives of the international climate regime and contribute to limiting global warming to below 1.5 degrees Celsius. However, "leaving the main source of income underground would be – economically and politically – a very difficult decarbonization option".[121] The dilemma between promoting resource use and accelerating low carbon measures to protect the environment from climate change has not fostered concerted dialogue on the importance of *Waqf* and other traditionally relevant tools for addressing climate change. These contrasting situations are not in favour of the environmental *Waqf*, even if a few scattered measures are undertaken for clean-ups and tree planting funded by *Waqf*.

However, "the recent slump in the price of oil spurred stronger political emphasis on diversification" and the use of renewable energies.[122] Yet, it is necessary to have clear laws and policies on climate change, in order to propel a significant shift to the rapid development and application of environmental *Waqf* in MENA countries.

Advancing the integration of *Waqf* into climate governance in the MENA region

The integration of *Waqf* principles and norms into climate change policies and programmes will necessitate major reforms in the legal and institutional systems, such as the amendment of laws related to *Waqf* in order to enable their expansion to the environment. These reforms can focus on raising awareness and capacity building, reinforcing cooperation, strengthening funding mechanisms, and fostering a broad conception of *Waqf* as an overarching principle.

Raising awareness and capacity building

In order to allow *Waqf* to play a more active role in environmental protection generally and climate action specifically, it is first important to raise awareness of religious authorities by providing solid learning programmes in ecology and environmental law (including comparative law). These programmes should be done in conjunction with targeted and in-depth teachings from *Fiqh*, in particular from the *Maqasid Al Sharia'*, but also from Islamic literature and philosophy.[123] These programmes should also include judges and governmental officials working in the fields of environment and *Awqaf* and other environmental NGOs.

It is also important to educate citizens. Indeed, *Waqf* is a voluntary act. Citizens must be convinced that charitable giving towards nature is as valid as building a mosque. The government could set an example by setting itself up as *Waqef* or as manager of environmental *Waqf*. Furthermore, national authorities can provide legal incentives to stimulate and harness interest in environmentally focused *Waqf*. For example, reducing the legal requirements and procedures for charitable spending can provide permanent flexibility and incentives for more people to be involved. This could encompass both public and private *Waqfs* as the *Waqf* deeds can include provisions for protecting the environment imposed by the government. A family *Waqf* may, for example, require the manager (*Mutawalli*) to plan actions in favour of environmental protection such as rules of conduct in the management of the *Waqf*. At government level, Awkaf Public Foundation (KAPF) in Kuwait has played a major role in pulling public attention to the importance of environmental *Waqf*.[124] KAPF created a dedicated *Waqf* company (Kuwaiti Company for the Environmental Services) which is providing cleaning services and an environmental fund.[125]

Reinforcing cooperation

It is important that religious groups and ministries of *Awqaf* work more closely not only with the departments of environment, but also with other religious groups and ministries in other countries, in particular by benefiting from the experience of organizations such as MUI in Indonesia.[126]

If ministries of environment in Muslim countries already meet within the framework of the ICESCO, as mentioned, it is important to also include the ministries of Islamic affairs in these meetings. Besides, as the previous MUI example has shown, close collaboration between *Ulamas*, academia, and non-governmental organizations can be a productive experience.

In this regard, partnership and cooperation with faith-based groups can also help in securing sustainable financing mechanisms and in fundraising.[127] Therefore, building the capacity of these groups is paramount. Besides, it is important to build a partnership between conservation organizations and religious groups "to achieve both social and ecological benefits while also accessing large public funding opportunities that support the priorities of both groups".[128] The new UN "Faith for Earth Initiative" is a good framework for achieving this goal. It aims to "Encourage, Empower and Engage with Faith-Based Organizations as partners, at all levels, toward achieving the Sustainable Development Goals and fulfilling Agenda 2030".[129]

Integrating and strengthening funding mechanisms

There is also a need to integrate the different options for climate financing to ensure coherence and mutually reinforcing outcomes.[130] Environmental *Waqf* and Islamic finance both have convergent roles. Islamic finance is based on three main principles, namely the prohibition of interest, the prohibition of contract ambiguity, and the prohibition of speculation. It also encompasses the obligation of profit and risk sharing.[131] Islamic finance "can both complement and provide an alternative to traditional sources of funding to realize the SDGs".[132]

Muslim countries which are intensively using Islamic finance are also among the main producers of greenhouse gases in the world (Gulf region and MENA region).[133] Therefore, it is critical that these countries implement the growing Islamic green finance. Islamic finance could be mobilized in ecological projects by using instruments that fit within Sharia rules such as the *mudaraba* and the green *Sukuk*.[134]

The *Waqf* can serve as a tool for green finance. This can occur in small villages, bigger cities, and even at the global level. In fact, a global "atmospheric *Waqf*" can be dedicated to the protection of the climate system as a fund for adaptation and mitigation. It can serve as a platform where all initiatives related to the protection of the climate system can converge and create a dynamic both at global and local levels. This must be backed by a strong legislative framework on climate change, which will provide a basis for incorporating and clarifying all forms of available climate financing tools, including *Waqf*. This means that this legislative enabling environment "may include taxation-related amendments as well as regulatory frameworks that can address the risks peculiar to Islamic financing".[135] In fact, the whole legal system must converge towards the integration of green finance, including banking law, insurance law, taxation, commercial law, and so on.

Foster a broad conception of *Waqf* as an overarching principle: the atmospheric *Waqf* principle

The principle of *Waqf* is embedded within Islamic law. It encompasses the idea of *Khilafa* (trusteeship) which is also related to the natural resources of the planet.[136] As mentioned earlier, the *Khalifa* is responsible for governing natural resources for the benefit of the trustees.[137] Therefore, the principle of *Waqf* cannot be understood outside the idea of *Khilafa* which concerns the global climate system. The "green" *Khilafa* means that the beneficiaries have the possibility to ask the *Khilafa*, who is the manager of the trust/*Waqf* assets, to fulfil his duty of care towards natural resources. Extending the principle of the ecological *Waqf* to the legal system means that courts should apply it in climate change litigation. Furthermore, as the *Khilafa* principle is claimed by the governments in Muslim countries, as it is the very spirit of the law, nothing could prevent it from also being applied in environmental matters.

A parallel with the common law where the principle of the trust was extended by courts in the framework of environmental law is enriching.[138] In fact, as mentioned above, even if the trust is at the heart of the PTD, it does not need to be instituted as in the conventional trust. In the PTD, the trust exists rather as the spirit of the law. There is a "common core of principles [...] forming the foundation for how the Doctrine is applied in each state".[139]

The PTD represents the "chalkboard on which the Constitution is written".[140] In fact, the PTD "functions in a quasi-constitutional way: it establishes overarching fiduciary principles regarding Trust resources that may not be overridden by legislative or executive action".[141] It is, therefore, a guiding principle which should reinforce and orient the action of the governments in MENA countries. As mentioned, there are many similarities between the trust and the *Waqf*. An atmospheric *Waqf* paradigm bears the idea of trusteeship over natural resources. Its main legal argument is grounded in the *Khilafa* principle, especially the environmental *Khilafa*. Therefore, its main objective would be to reinforce the fiduciary duty upon natural resources and the climate system. The atmospheric *Waqf* principle would allow civil society to ask the governments to fulfil their fiduciary duties. It considers that the climate system is a *Waqf* asset that should be kept for the benefit of present and future generations. It also means that the governments can reframe their environmental laws according to the principle of the environmental *Khilafa*.

The atmospheric *Waqf* principle should rely on the work of *Ulama* in collaboration with environmental scholars. It should be based on solid scientific arguments and also on legal arguments explored from ancient environmental *Waqf* and also from *Maqasid Sharia*'. In fact, it is important to push the limits of the *Waqf* for the benefit of the climate system.

Conclusion

MENA countries have the legal, ethical, and institutional foundations for the institution of environmental *Waqf* and may use it as a platform towards a strengthened climate change legal system. They still need to agree to overcome the many constraints that hinder this evolution. Indeed, in the absence of an effective environmental law and the absence of institutional coordination between the departments of the environment and Islamic affairs, it is difficult to develop the environmental *Waqf*. Added to these constraints are the lack of dedicated research programmes and the contrasting perception of climate change between oil-producing countries and poor countries in the same region. Therefore, the solution is to raise awareness, strengthen cooperation, and tap into booming Islamic finance.

The atmospheric *Waqf* paradigm is capable of establishing a profound change in climate change legal systems in the MENA region. This paradigm shift will not require a profound reform of the law, as the legal system already has all the seeds for such an evolution. However, this shift will necessitate a qualitative socioeconomic leap which will strengthen both the action of governments and of citizens. In fact, as was the case in common law countries, the PTD based on the idea of trust, which is similar to *Waqf*, served as a platform for environmental activism in the context of the atmospheric trust litigation. The same evolution is possible in MENA countries, especially as the Arab Spring allowed a dynamic of social activism to launch. Indeed, this movement, which is still latent, cannot deny environmental consideration, particularly in the area of climate change, which is undeniably linked with human rights.

Notes

1 *Waqf* is called *Habous* in North Africa.
2 J. Luccioni, *Les fondations pieuses «Habous au Maroc» depuis les origines jusqu'à 1956* (Imprimerie Royale 1982) 15.
3 Ibid.
4 See Clima-Med, "Acting for Climate in South Mediterranean" <www.climamed.eu> accessed on 6 January 2021.
5 See M. E. Tucker and J. Grimm, "Series Forward" in R. C. Foltz, F. M. Denny, and A. H. Baharuddin (eds), *Islam and Ecology, A Bestowed Trust* (Cambridge University Press 2003) xvi.
6 See V. Donatella, "'Green' Islam and Social Movements for Sustainability" (PhD Thesis, Liberta University Internationale Degli Studi Sociali, Luiss Guido Carli, Rome 2017).
7 J. Schacht, *An Introduction to Islamic Law* (Oxford: Clarendon Press 1964).
8 See, for example, Ö. Ibrahim, "Toward an Understanding of Environmental Ethics from a Qur'anic Perspective" in R. C. Foltz, F. Denny, and A. Baharuddin (eds), *Islam and Ecology, A Bestowed Trust* (Cambridge University Press 2003).
9 See M. F. Majeri and others, "Faiths from the Archipelago. Action on the Environment and Climate Change" (2015) 19 (2) *Worldviews: Global Religions, Culture, and Ecology* 103.
10 S. Ramlan, "Religious Law for the Environment: Comparative Islamic Environmental Law in Singapore, Malaysia, and Indonesia" (2019) NUS Centre for Asian Legal Studies Working Paper 19/03.
11 Ibid.
12 D. S. Olawuyi, "Advancing Innovations in Renewable Energy Technologies as Alternatives to Fossil Fuel Use in the Middle East, Trends, Limitations, and Ways Forward" in D. Zillman, M. Roggenkamp, L. Paddock, and L, Godden (eds), *Innovation in Energy Law and Technology, Dynamic Solutions for Energy Transitions* (Oxford University Press 2018) 354, 360.

13 International Islamic Climate Change Symposium, "Islamic Declaration on Global Climate Change" (Australian Religious Response to Climate Change) [Declaration].
14 Ibid, Article 3.2. See also A. A. Bagader and others, *Environmental Protection in Islam* (IUCN 1994).
15 Declaration, n 13.
16 See, for example, *Constitution of the Arab Republic of Egypt*, 22 September 1971, Article 2.
17 N. Bernard-Maugiron and B. Dupret, "Les principes de la sharia sont la source principale de la legislation" (1999) 2 *Égypte/Monde arabe* 107–126; J. M. Otto, *Sharia and National Law in Muslim Countries: Tensions and Opportunities for Dutch and EU Foreign Policy* (Leiden University Press 2008), 8.
18 Qur'an 6: 38.
19 Qur'an 17: 37.
20 M. A. Mekouar, "Islam et environnement: une éthique pour la conservation" in M. A. Mekouar (ed.), *Etudes en droit de l'environnement* (Okad 1988) 41; D. Olawuyi, *Principles of Nigerian Environmental Law* (Nigeria: Afe Babalola University Press 2015) vi.
21 P. C. Hennigan, *The Birth of a Legal Institution: The Formation of the Waqf in Third-Century A. H. Hanafi Legal Discourse* (Brill, 2004) xiii.
22 M. Cizakca, *A History of Philanthropic Foundations: The Islamic World from the Seventh Century to the Present* (Istanbul: Bogazici University Press 2000).
23 T. Khalfoune, "Le Habous, le domaine public et le trust" (2005) 57 *Revue internationale de droit comparé* 441, 446.
24 S. Barnes, "Elaborate Birdhouses Resembling Miniature Palaces Built in Ottoman-Era Turkey" (My Modern Net, 31 July 2107).
25 Luccionni, n 2, 101. S. Idllalène, "Le *habous*, instrument de protection de la biodiversité ? Le cas du Maroc dans une approche de droit comparé" (2013) 4 *Développement durable et territoires* 1-3.
26 L. Auclair and others "Patrimony for Resilience: Evidence from the Forest Agdal in the Moroccan High Atlas Mountains" (2011) 16 *Ecology and Society* 24.
27 A. A. Bagader, n 14, Principle 15.
28 *Waqf* often takes the form of a solemn act, especially in private *Waqf* but this form is not necessary.
29 M. Z. Abbasi, "The Classical Islamic Law of Waqf: A Concise Introduction" (2012) 26(2) *Arab Law Quarterly* 121.
30 Ibid.
31 Ibid.
32 Ibid.
33 S. Idllalène, *Rediscovery and Revival in Islamic Environmental Law: Back to the Future of Nature's Trust* (Cambridge University Press, 2021).
34 F. Bilici, "Les waqfs monétaires à l'époque ottomane : droit hanéfite et pratique" (1996) 79–80 *Revue du monde musulman et de laMéditerranée* 73, 74.
35 M. C. Wood, *Nature's Trust: Environmental Law for a New Ecological Age* (Cambridge University Press 2013).
36 Khalfoune, n 23.
37 M. El Manouni, "The Role of Moroccan Waqfs in the Era of Bani Marin" (1983) 230 *Majallat Daawat Al Haq* 1-10.
38 Idllalène, n 25.
39 Abbasi, n 32. Idllalène, n 25, n 33.
40 Ibid.
41 Idllalène, n 25 and 33. M. Obaidullah, "Managing Climate Change: Role of Islamic Finance" (2018) 26(1) *Islamic Economic Studies* 31–62, 47–48.
42 Idllalène, n 33.

43 Ibid.
44 M. Abdullah, "Waqf, Sustainable Development Goals (SDGs) and maqasidal-shariah" (2018) 45 *International Journal of Social Economics* 158.
45 R. S. Abate, *Climate Change and the Voiceless, Protecting Future Generations, Wildlife and Natural Resources* (Cambridge University Press 2019) 1–25.
46 See Idllalène, n 33.
47 Dahir No. 236.09.1 of 8 Rabia I 1431 (23 February 2010) concerning the *Awqaf* Code, Official Bulletin Number 5847, 14 June 2010, 3154 (in Arabic).
48 For example, the Royal Decree No. (M / 11) of 2/26/1437 AH on General Authority of *Awqaf* in Saudi Arabia, states that "*Al Mawquf Alayh: The Beneficiary of the Waqf According to the Condition of the Waqf*" (article 1). In Turkey, the Foundation Law No 5737, 20 February 2008, states in its Article 21 that private foundation forests may be planted. The Federal Law No 5 of 2018 in the United Arab Emirates states in its article 1 that the *Mawquf Alayh* is "*Who is Entitled to Benefit from the Waqf, Whether He is a Natural or Legal Person, Initiative, Project or Others*" (article 1). For more details, see Idllalène, n 33.
49 Idllalène, n 33.
50 Ibid.
51 Ibid.
52 Ibid.
53 A. Raysuni, *Theory of the Higher Objectives and Intends of Islamic Law* (Herndon, Virginia: The International Institute of Islamic Thought, 2005).
54 A. Duderija, "Contemporary Muslim Reformist Thought and Maqāṣid cum Maṣlaḥa Approaches to Islamic Law: An Introduction" in A. Duderija (ed.), *Maqasid Al Shari'a and Contemporary Reformist Muslim Thought an Examination* (New York: Palgrave McMillan 2014) 3.
55 Ibid.
56 *Ijtihad* means the effort of interpreting Islamic science and sources of Sharia undertaken by *Ulamas*.
57 Ibid.
58 Mohammad Abdullah, "Waqf, Sustainable Development Goals (SDGs) and Maqasidal-Shariah" (2018) 45(1) *International Journal of Social Economics* 158–172.
59 Ibid.
60 Ibid, 158.
61 Ibid, 159.
62 Khalfoune, n 23.
63 M. M. Gaudiosi, "The Influence of the Islamic Law of Waqf on the Development of the Trust in England: The Case of Merton College" (1988) 136 *University of Pennsylvania Law Review* 1231.
64 G. P. Verbit, *The Origins of the Trust* (Bloomington, IN: Xlibris Corporation 2002) 117.
65 P. Stibbard, D. Q. C. Russel, and B. Bromeley, "Understanding the Waqf in the World of the Trust" (2012) 18(8) *Trust and Trustees* 785–810.
66 Ibid.
67 Gaudiosi, n 64, 1246.
68 M. Cizakca, *A History of Philanthropic Foundations: The Islamic World from the Seventh Century to the Present* (Istanbul: Bogazici University Press 2000), 13–14.
69 Khalfoune, n 23.
70 "Our Cause" (UK National Trust) <www.nationaltrust.org.uk/our-cause> accessed 6 January 2021.
71 J. L. Sax, "The Public Trust Doctrine in Natural Resource Law: Effective Judicial Intervention" (1970) 68 *Michigan Law Review* 471.
72 J. L. Sax, "Liberating the Public Trust Doctrine from Its Historical Shackles" (1980) 14 *University of California Davis Law Review* 185.

73 L. V. de Melo Bento, "Searching for Intergenerational Green Solution: The Relevance of the Public *Trust* Doctrine to Environmental Preservation" (2009) 11 *Common Law Review* 7 12.
74 Ibid.
75 Sax, n 72.
76 Ibid.
77 Ibid, 8.
78 B. J. Morehouse, "Heritage, Public *Trust* and Non-market Values in Water Governance" (2011) 30 *Policy and Society* 323, 327.
79 Wood, n 35.
80 B. Freedman and E. Shirley, "England and the Public Trust Doctrine" (2014) 8 *Journal of Planning and Environmental Law* 839, 841.
81 Ibid.
82 Z. J. B. Plater and others, *Environmental Law and Policy: Nature, Law, and Society* (New York: Aspen 2004), 1120 cited by R. Sagarin and M. Turnipseed, "The Public Trust Doctrine: Where Ecology Meets Natural Resources Management" (2012) 37 *Annual Review of Environmental Resources* 473, 490.
83 Wood, n 35, 14.
84 M. C. Blumm and M. C. Wood, *The Public Trust Doctrine in Environmental and Natural Resources Law*, 2nd ed. (Carolina Academic Press 2015) 8.
85 Sax, n 72, 474.
86 *Juliana v United States*, 217 F. Supp. 3d 1224 (D. Or., 2016).
87 M. C. Blumm and R. D. Guthrie, "Internationalizing the Public Trust Doctrine: Natural Law and Constitutional and Statutory Approaches to Fulfilling the Saxion Vision" (2012) 45 *University of California Davis Law Review* 745-807.
88 M. A. Budiman, "The Role of Waqf for Environmental Protection in Indonesia" (Aceh Development International Conference (ADIC), Kuala Lumpur, 26–28 March 2011) 880, 889.
89 See, for example, R. Dibie and M. Hussein, "Environmental Policies and Issues in Some Arab Countries" in R. Dibie (ed.), *Comparative Perspectives on Environmental Policies and Issues* (Routledge-Francis Taylor 2014), 400–423.
90 K. Waha and others, "Climate Change Impacts in the Middle East and Northern Africa (MENA) Region and Their Implications for Vulnerable Population Groups" (2017) 17 *Regional Environmental Change* 1623-1638.
91 M. Al-Alaween and others (eds), *Water Integrity in the Middle East and North Africa Region: Synthesis Report of Water Integrity Risks Assessments in Jordan, Lebanon, Morocco, Palestine and Tunisia* (UNDP Water Governance Facility at SIWI, Stockholm 2016) 20.
92 Royaume du Maroc, Secrétariat d'Etat auprès du Ministère de l'Energie, des Mines, de l'Eau et de l'Environnement, chargé de l'Eau et de l'Environnement, Département de l'Environnement, mars 2009, *Quatrième rapport national sur la biodiversité*.
93 UNEP, *Environmental Rule of Law: First Global Report* (UNEP 2019).
94 Ibid, 3.
95 See, for example, K. Kakish, "Facilitating a Hima Resurgence: Understanding the Links between Land Governance and Tenure Security West Asia-North" (West Asia-North Africa Institute 2017) <http://wanainstitute.org/sites/default/files/publications/Facilitating%20a%20Hima%20Resurgence.pdf> accessed 6 January 2021.
96 UNEP, n 93, 95.
97 D. Olawuyi, "Human Rights and the Environment in Middle East and North African (MENA) Region: Trends, Limitations and Opportunities" in J. May and E. Daly (eds), *Encyclopaedia of Human Rights and the Environment, Indivisibility, Dignity and Legality* (Edward Elgar 2018).

98 See n 17.

99 See, for example, Otto, n 17, 8.

100 Idllalène, n 33.

101 See, for example, Budiman, n 88.

102 N. Bernard-Maugiron and B. Dupret, n 17.

103 Ramlan, n 13.

104 Idllalène, n 33.

105 Ibid.

106 A simple survey on the Internet confirms this trend. In Muslim countries, Islamic affairs, including *Waqf* are managed by specific governmental Departments. Idllalène, n 33.

107 See, for example, the activities of the Ministry of Awaqf and Islamic affairs in Morocco <www.habous.gov.ma>.

108 The Green Mosque concept is applied in countries such as Malaysia and Morocco. See O. S. Syamimi, I. N. Hanim, M. Zulhaili, and B. Ruwaidah, "Green Mosque: A Living Nexus" (2018) 3 *Environment-Behaviour Proceedings Journal* 53–63.

109 See the ISESCO website <www.icesco.org>.

110 In fact, available resources on legal aspects of eco-theology in Muslim countries are concern mainly these two countries.

111 For an assessment of environmental Fatwas in Asian countries, see Ramlan, n 103; M. F. Majeri and P. Gudah, "Fatwas on Boosting Environmental Conservation in Indonesia" (2019) 10 (10) *Religions* 570-575.

112 This *Fatwa* and few others are available at <www.arcworld.org>.

113 For more details on the experience of MUI see M. F. Majeri, I. S. L. Tobing, A. Binawan, E. Pua, and M. Nurbawa, n 10. The OIC Fiqh Academy, "The OIC Fatwa on Environmental Preservation" (Resolution 185, OIC Fiqh Academy 2009).

114 OIC Fiqh Academy, ibid.

115 N. Al-Duaij, "The Environmental Protection in the Islamic Waqf" (bepress, 10 September 2009) <https://works.bepress.com/eisa_al_enizy/2/> accessed 6 January 2021.

116 See, for example, M. A. Budiman, n 88.

117 For example, the Islamic Development Bank (IBD) created a specific research programme on *Waqf* led by its research institute <https://irti.org/>.

118 N. Haghamed, "The Muslim World Has to Take Climate Action" *Al Jazeera News* (Doha, 4 November 2016) <https://www.aljazeera.com/opinions/2016/11/4/the-muslim-world-has-to-take-climate-action> accessed 6 January 2021.

119 J. J. Kaminski, "The OIC and the Paris 2015 Climate Change Agreement: Islam and the Environment" in L. A. Pal and M. E. Tok (eds), *Global Governance and Muslim Organizations* (New York: Springer International Publishing 2019) 190.

120 A. Al-Sarihi, "Implications of Climate Policies for Gulf States Economic Diversification Strategies" (The Arab Gulf States Institute in Washington, 9 July 2018); Olawuyi, n 12, 354, 355.

121 Ibid.

122 Olawuyi, n 12, 354, 358.

123 In this regard, Seyyed Hossein Nasr calls for a rediscovery of Islamic environmental ethics by using the scriptural basis from the *Qur'an* and *Sunnah* but also from the Sufi sources in both poetry and prose (such as works of Ibn Arabi, Jalal al Din Rumi and Mahmud Shabistari) (S. H. Nasr, "God is Absolute Reality and All Creation His Tajalli" in J. Hart (ed), *The Wiley Blackwell Companion to Religion and Ecology* (Hoboken, NJ: John Willey and Sons Ltd, 2017) 11.

124 Al-Duaij, n 115.

125 Ibid.

126 Majeri and others, n 9.
127 E. Mcleod and M. Palmer, "Why Conservation Needs Religion" (2015) 43 *Coastal Management* 239, 249.
128 Ibid.
129 UN Environment Programme, Faith for Earth Dialogue Synthesis Report, 8 May 2019.
130 D. Aassouli and others, "Green Sukuk, Energy Poverty, and Climate Change: A Roadmap for Sub-Saharan Africa" (2018) Policy Research Working Paper 8680.
131 A. Mirakhor and I. Zaidi, "Profit-and-Loss Sharing Contracts in Islamic Finance" in M. K. Hassan and M. K. Lewis (eds), *Handbook of Islamic banking* (Edward Elgar 2007) 49.
132 Aassouli, n 131, 6.
133 Ibid, 119.
134 A. Sekreter, "Green Finance and Islamic Finance" (2017) 4(3) *International Journal of Social Sciences & Educational Studies* 115.
135 *Mudaraba*, is an Islamic financing technic based on profit and loss sharing system. It permits a partnership between a bank (the investor or *Rab Al Mal*) and a contractor (*Mudarib*) in which the bank pledges to fully finance the project while the contactor have to manage the project (Aassouli, n 131).
136 F. Khalid, "Islam and the Environment – Ethics and Practice an Assessment" (2010) 4 (11) *Religion Compass* 707-716.
137 Ibid.
138 Blumm and Wood, n 85.
139 Ibid, 12.
140 See G. Torres and N. Bellinger, "The Public Trust: The Law's DNA" (2014) 4 *Cornell Law Faculty Publications* 218, 294.
141 Blumm and Wood, n 84, 303.

4 Addressing climate change in the MENA region through regulatory design

Instrument choice questions

Georgios Dimitropoulos and
Almas Lokhandwala

Introduction

This chapter analyzes the use of regulatory policies to address climate change concerns in the Middle East and North Africa (MENA) region. It assesses the effectiveness of current climate change regulatory approaches in the context of the complex socio-economic conditions of the region and discusses options for strengthening their effective implementation.

In the past few decades, the MENA region has experienced a great transformation in terms of sweeping urbanization, industrialization, and acute economic growth.[1] In order to fulfil their socio-economic and development goals, countries in the region have followed urbanization and industrialization policies, leading sometimes to inconsiderate energy use, as well as water and waste management.[2] More recently, they have started developing regulatory responses to the climate change challenges posed to them.[3]

The MENA region is an incongruous region to compare in terms of understanding the climate change policies of the states that comprise it; pollution levels and climate change policies vary significantly across the states. The juxtaposed climate change performance ranking of Morocco, which is considered to be a forerunner worldwide on climate change policies, with Iran and Saudi Arabia, for example, that rank at the bottom, clearly exhibits the region's disparity towards climate change policies.[4] The inconsistent approach to climate protection policy coupled with a burgeoning population – comprised of a large proportion of young adults,[5] uneven economic development,[6] sometimes lack of political[7] and social stability,[8] and a host of security issues,[9] makes the MENA region a unique environment to study when it comes to its approach to climate change issues.

States in the MENA region, when striving to fulfil domestic or international obligations of combating climate change, have followed the worldwide trend of first instituting command-and-control (CAC) approaches to climate protection. CAC is still the most widely used form of regulatory instrument. A trend towards market-based approaches (MBA) has slowly evolved, and the region is now also, albeit reluctantly, integrating climate change "nudges" into its climate change policies and regulations.[10] While there is a general willingness to learn about softer and more flexible instruments and to implement them, this takes time and the focus remains on CAC. Given the complementarity of these regulatory approaches, we recommend that states in the MENA region continue employing all three regulatory instruments in a way that satisfies their unique climatic, as well as socio-economic and political conditions, and that they intensify the collaboration between the administrative agencies managing these climate change regulatory interventions.

After this introduction, the second section of this chapter discusses the unique socio-economic, political, and regulatory diversity of the MENA region, alongside the acute environmental issues that the region faces. The third section explores the global trends on the choice of instruments used to address pollution and climate change issues – beginning with movement from CAC approaches to MBAs, and more recently, to climate change nudges. Three countries from the MENA region – Morocco, Lebanon, and the United Arab Emirates (UAE) – are further examined to showcase the utilization of these regulatory tools in the fulfilment of local, regional, and international obligations for combating climate change in specific contexts. The fourth section offers a broad assessment of the endogenous and exogenous challenges to climate change regulation faced by MENA states, while considering how to better address socio-economic and practical challenges that hinder the effective implementation of these regulatory approaches. The final section concludes.

The Middle East and North Africa: addressing climate change in a diverse region

The MENA region stretches from the Black Sea in the North to the Arabian Sea to the South; and from the Atlantic coast of Africa on the West to Iran to the East.[11] Geographically, this region can be divided into four main zones: firstly, the Nile River, Sahara Desert, and Atlas Mountains of North Africa; secondly, the Eastern Mediterranean region of the Levant comprising of a mild Mediterranean climate, mountains, valleys, and rivers; thirdly, the Mesopotamic region of Iraq, Iran, and Syria; and lastly, the Arabian Desert in the south comprising most of the Arabian Gulf States.[12] The economies and socio-political realities of the states of the MENA region are also diverse in nature – from the resource-rich countries of the Arabian Gulf, to the middle-income states comprising mostly North Africa and Levant, to fragile states such as Yemen, Syria, Iraq, and Lebanon.

Although the MENA region is diverse in many aspects, MENA states face similar environmental challenges. Extrapolation of historical data compounded with the current warming trends has grouped the MENA region as one of the most severely affected by global warming – with the effects dramatically increasing over the course of this century.[13] In fact, the coolest summers during the last three decades of this century (2070–2099) are expected to be warmer than the hottest summers recorded from 1960 to 1990.[14] This exponential rise in temperature has already led to serious environmental issues in the region.

The most threatening environmental issue is an acute scarcity of potable water resources.[15] The expedient growth witnessed due to sprawling urbanization and industrialization has come at the cost of poor agricultural practices (such as indiscriminate use of underground water and pesticides), improper water conservation,[16] and increased desertification,[17] resulting in a projection of severe water crisis in the near future.[18]

Similarly, and quite related to the water crisis, the decreased availability of arable land has also been a major concern, with just five per cent of the total land area (which is shrinking annually) available for agriculture.[19] Additionally, the health and existence of the biodiversity of the region is under acute pressure – both on land and sea.[20]

Many factors affect a country's response and adoption of legislative and regulatory measures to address climate change, as well as other environmental problems. Some of the major socio-economic and political factors are a country's economic performance, political stability, international obligations, international policy diffusion, as well as hosting an international climate summit.[21] States in the MENA region have also been active in adopting legislative and regulatory measures addressing climate change. This section analyzes climate change policies and environmental regulation instruments of one country from each of the main geographic areas of the broader region[22] based also on a socio-economic assessment: Morocco as a middle-income country from North Africa;[23] Lebanon as a country belonging to the Levant in addition to being a fragile state;[24] and the UAE as a country that is both resource-rich and located in the Arabian Gulf region.[25]

Addressing climate change through regulation

Regulatory design for climate change

Conventionally, environmental as well as climate change laws and regulations are comprised of CAC measures and MBAs.[26] Both of these measures are based on the assumption that individuals are rational. Rational choice theory assumes that individuals respond to opportunities, incentives, screens, and sanctions.[27] However, more recently, a novel form of regulatory instrument, called "nudge", is being considered by many states. Nudges are based on findings of cognitive psychology and behavioural economics that individuals are "boundedly rational", namely that they do not process information correctly all the time, they have bounded willpower, and are boundedly self-interested.[28]

Command-and-control (CAC)

CAC is the oldest and most widely utilized instrument for climate change regulation.[29] It is based on the assumption that individuals are rational and that sanctioning certain behaviours will deter actions that are not climate-friendly, and will therefore assist in the achievement of climate change targets. CAC, as applied in climate change policy, are regulatory instruments that enforce particular emissions standards or put a cap on the amount of hazardous pollutants that can be emitted, with penalties for violations of the set standards. CAC regulatory interventions usually impose uniform standards, mostly technology-based or performance-based, on all entities bound by the standards – irrespective of size or environmental footprint. While technology-based standards usually specify the type of technology, method, or equipment to be used, performance-based standards include a target a polluter must achieve – while often allowing for discretion on the manner in which this target is achieved.

CAC interventions are not flexible and may even sometimes be counterintuitive when it comes to achieving the goal of emission reduction itself. In many instances, CAC measures insist on the utilization of a specific technology to reduce pollution which could prove to be expensive for some small-sized firms,[30] resulting in a dearth of innovation in pollution-controlling technology as there was no incentive to do so. Furthermore, CAC may be expensive due to increased regulation and enforcement costs.[31] These challenges associated with CAC led to the development of market-oriented regulatory approaches.

Market-based approaches (MBAs)

Over the past decades there has been a gradual inclusion of MBAs to environmental regulation in almost all countries.[32] MBAs are "regulations that encourage behavior through market signals rather than through explicit directives regarding pollution control levels or methods".[33] The most common MBAs are incentives.

MBAs can be generally classified under four categories: pollution charges, tradable permits, market-barrier reductions, and government subsidy reductions.[34] While pollution charges are based upon the amount of pollution a source emits at regular intervals of time (e.g., yearly taxes on cars based upon horsepower of the engine), tradable permits (which can be traded with a few conditions) allow the holders to produce a certain amount of pollution. Market-barrier reductions aim to reduce the barriers to market activity that is environmentally friendly, while government subsidies for environmentally friendly innovation provide increased incentives for green technologies.[35] The international climate change regime includes three MBAs in the Kyoto Protocol to the United Nations Framework Convention on Climate Change (UNFCCC): the Clean Development Mechanism (CDM), Emission Trading (ET), and Joint Implementation (JI).[36] These three international market mechanisms give incentives to industrialized and developing countries to trade emission reduction credits or develop emission reduction projects.[37]

MBAs are designed to encourage innovation, and are more flexible, cost-effective, and business-friendly, as they do not require the government to set standards for each pollution source.[38] The government allocates the pollution control burden on all entities and allows the market to regulate. Especially after the implementation of the Emissions Trading Scheme in the EU in 2005, the market-based model in climate protection policies was replicated all over the world, now covering Europe, North America, and Asia-Pacific regions.[39]

However, MBAs also require high administrative costs, sometimes pre-existing mature markets, as well as a strong institutional capacity for effective implementation (although government subsidies have also resulted in economically inefficient practices).[40] Furthermore, governments sometimes set unrealistic expectations on new markets and may also face obstacles from the industry and the environmental community if the proposed policies are perceived to be contrary to their respective interests.[41]

Climate change nudges

Recently, a third regulatory tool has been increasingly adopted by many countries in order to complement CAC and MBA: green nudges. Green nudges, as regulatory tools, are relatively cheap and easy to administer. They are informed by the insights of cognitive psychology and behavioural economics that use salience,[42] default rules,[43] and social norms[44] to steer an individual's behaviour towards more environmentally friendly actions.

Green nudges can be divided into three categories, namely: pollution-reducing nudges (e.g., green footprints), energy efficiency nudges (e.g., energy efficiency labels, fuel economy labels), and climate change nudges.[45] Pollution reducing nudges are used for reducing pollution of an area. For instance, the Green Footprint initiative by the city of Copenhagen nudged people to follow "green footprints" to public trash bins and deposit their waste there – instead of littering.[46] This initiative proved to be quite successful – resulting in a decrease of littering, by almost half, in the designated areas of the city.[47]

Energy efficiency nudges aim at convincing people to use energy-efficient devices and consume less energy. For instance, green energy defaults, such as automatic enrollment of consumers for renewable energy providers was shown not only to result in an increase in uptake of energy from green providers but also consumers preferred to stay with green providers even when they were not very concerned about climate change issues.[48] Other instances of nudging consumers towards energy efficiency are the design of electricity bills with visually salient characteristics (e.g., emoticons) and/or comparison with neighbours' consumption, and the use of energy efficiency labels (e.g., with information on savings by use of energy-efficient devices).[49]

Lastly, climate change nudges refer to green nudges that aim to address the key drivers of climate change directly,[50] such as the greenhouse gas emissions (GHGs).[51] An example of a climate change nudge is the fuel economy label for vehicles adopted by many jurisdictions. In the past, such labelling had not been very successful in convincing new buyers to invest in fuel-efficient cars primarily due to their high upfront costs. Yet, upon changing the design of the labels from data with emissions to data with fuel economy, there was a change in the perception and receptivity of the consumers.[52]

It is important to note that nudges are non-binding, and remain a relatively new and largely untested instrument in many parts of the world. Moreover, what may work in one country might not necessarily work in the same way in others – largely due to cultural and semantic variance.[53]

Addressing climate change in the MENA region

Regulators in the MENA region have also adopted tools, such as CAC, MBAs, and climate change nudges to address climate change. This section discusses regulatory measures addressing climate change in three MENA countries: Morocco, Lebanon, and the UAE.[54]

Morocco

Morocco is located in Northeast Africa, and is quite vulnerable to climate change as its population is quite dependent on the agricultural sector and tourism[55] – both sectors heavily influenced by climate change. Morocco is one of the most responsive jurisdictions in the MENA region in the combat against climate change.[56] The fact that Morocco volunteered to host the Conference of the Parties 7 (COP7) as early as 2001 – and is the first of only two MENA countries having done so[57] – demonstrates an early involvement of the country in combating climate change. Upon ratifying the Kyoto Protocol, the Paris Agreement, and a number of international environment treaties, the country has published its Nationally Determined Contribution (NDC)[58] to the UNFCCC outlining its pledges for the global climate change reduction efforts.

Enacting Framework Law 99-12 on the National Charter for the Environment and Sustainable Development,[59] which called for the strengthening of capacities (institutional and otherwise) to tackle climate change, was a further development in the country's climate change policy. Although Law 47-09 on Energy Efficiency was enacted before the Framework Law, the Renewable Energy Law was modified (as Law 58/15) in 2015, later establishing the Moroccan Agency for Energy Efficiency in 2016 (Law 36/16). Subsequently, Law 57-09 created the Moroccan Agency for Solar Energy, which was renamed in 2016 as the Moroccan Agency for Sustainable Development. These agencies have mostly adopted CAC instruments to realize their objectives along with some MBAs, while they occasionally implement climate change nudges.

The Moroccan government mostly employs CAC measures to address climate change concerns. For instance, the 2010 Decree No. 2-09-631[60] deals with the regulation of limits on emissions into the air from stationary sources. This regulatory instrument defines the sources of pollution and sets hard limits (with exceptions) for industries to follow. Morocco has also instituted a ban on the import of vehicles older than five years, thus reducing emissions up to 60 $MtCO_2$ over a period of five years.[61] Non-compliance is dealt with by fines and sanctions in accordance with the law.[62] Similar regulations exist for water and waste. The List of Draft Standards for Chemical Regulations in potable water is the latest example of a CAC measure aimed at regulating the amount of chemicals present in drinking water.[63]

While CAC remains the most utilized instrument, Morocco has started applying the MBA to fulfil its domestic and international obligations and pledges in combating climate change. The primary market-based means to reduce the emission of pollutants in the environment is rolled out in the form of financial incentives for industries that apply less-polluting methods or technology, or subsidies in the form of grants, or eliminating subsidies for certain polluting sectors. Some examples of such grants and subsidies are the Fund for Industrial Depollution (FODEP),[64] the National Environment Fund (FNE),[65] subsidizing solar water pumps for agricultural use,[66] and taxing industry for the purification of polluted water.[67] An example of discontinuation of subsidies is the gradual phasing out of government subsidies on refined energy products – with the hope of achieving energy efficiency – which now follows international prices since the end of 2015.[68]

The country has also started implementing environment taxes, such as yearly road permits based on horsepower (thus impacting fuel consumption) of the vehicle, import of plastic products, and plastic packaging, among others, raising around US$25.4 billion from 2000 to 2017, with revenues rising each year.[69] Lastly, Morocco has also put in place a limited tradable-permit system, such as the pollution permits for the discharge of wastewater,[70] or for individual quotas for the commercial fishing of cephalopods.[71]

Similarly, although not as common as CAC or even MBA, Morocco has also employed climate change nudges largely for encouraging energy efficiency. Eco-Binayate is an energy-performance label for buildings, which was launched in the first quarter of 2020 targeting new construction.[72] However, there are plans underway for issuing the Eco-Binayate label to any property that satisfies the requirements of the Thermal Constructions regulation (RCTM) and the *Agence Marocaine pour l'Efficacite Energetique* (AMEE). AMEE has also developed the BINAYATE software for designing energy-efficient buildings.[73]

After implementing the Euro 4 Emissions Standard, which aims at reducing particulate matter and oxides of nitrogen (NOx) from cars, for all – imported or manufactured – light-duty vehicles in 2015, Morocco is also planning on introducing fuel efficiency labels for all new cars sold in the country.[74]

Lebanon

Lebanon is located in the Levant region bordering the Mediterranean Sea on the west. The country's economy is primarily dependent on tourism and agriculture;[75] both industries are quite vulnerable to climate change and Lebanon has been working towards implementing policies for mitigating climate change. Lebanon is a party to the Kyoto Protocol and the UNFCCC. It has signed, but not ratified, the Paris Agreement.[76] As part of its INDC submitted to UNFCCC, Lebanon has pledged to reduce the GHG emissions up to 15 per cent of the business-as-usual scenario, produce 15 per cent of its electricity through renewable means, and reduce its energy demand by three per cent by utilizing energy-efficient measures, by 2030.[77] The primary legislation addressing climate change is the Environmental Protection Law (Law No 444 of 2002), and the Law on the Protection of Air Quality (Law No 78 of 2018). Lebanon has planned on achieving its targets through using a mixture of CAC, MBA, and some climate change nudges.

Lebanon relies heavily on CAC methods to regulate the environment. Some CAC measures that directly impact Lebanon's climate change mitigation objectives are Law 78/2018 (Clean Air Act) that monitors, assesses, and surveils the level of pollutants in the atmosphere and then controls air pollution due to human activity. Law 243 of 2012 (Traffic Law) mandates the application of catalytic converters in all gasoline vehicles; Decree 8442 of 2002 establishes the standard for various chemicals (such as sulphur) present in fuel for vehicles; Ministry of Environment (MoE) Decision 8/1 of 2001 regulates the emission limit values (ELVs) from point sources (stack emissions and effluent discharges) from industries;[78] and Law 150/1992 bans the import of cars that are older than eight years from the date of manufacturing.

Lebanon has also started implementing MBAs to bolster its climate change efforts. A primary example is Decree 167 of 2017 under Law 444 of 2002,[79] which provides for tax incentives for renewable energy industries or on green products. Furthermore, Decree 8941/2012 incentivizes the use of public transport through exemptions and government-supported low-interest loans.[80]

The Lebanon Environmental Pollution Abatement Program (LEPAP) provides close to zero interest rate loans for industries that "implement industrial pollution abatement interventions", in addition to free technical assistance for industries that want support in environmental regulations.[81] Finally, Law 80/2018 provides for funding solid waste management through the "polluter pays principle",[82] which the government is planning on including as waste management taxes from households.[83]

Lebanon has an interesting history with the implementation of nudges, as a former economic advisor to the Lebanese Prime Minister's Office, Dr. Fadi Makki, was the first to implement nudge theory in the MENA region.[84] Nudge Lebanon, an organization established by Makki, is working closely with the Lebanese government to develop nudges in many areas, including environmental protection and climate change.[85] Nudge Lebanon has experimented with nudging individuals towards increasing composting, increasing use of communal glass recycling bins, increasing demand for leftover food at restaurants, using re-usable bags while shopping, and using less plastic cutlery, among others.[86] Based on results from these experiments, Nudge Lebanon will be partnering up with the government and other NGOs to implement nudges all over the country.[87]

Additionally, Lebanon is developing energy efficiency standards and labels for household appliances, in addition to educating the citizens on energy efficiency.[88]

United Arab Emirates (UAE)

The UAE is located in the lower eastern region of the Arabian Peninsula, and is comprised predominantly of barren coastal plain and rolling sand dunes, with rugged mountains along its border with Oman. By virtue of its geography, the UAE is one of the most vulnerable nations to climate change as the sea-level rises, increased temperatures, and water shortages will affect the country catastrophically.[89]

The UAE has played an important role in the combat against climate change in the region by adopting a host of measures and pledging stark interventions to tackle climate change, including (but not limited to) hosting numerous conferences and dialogues on climate change, voluntarily undertaking the task of reduction in GHG emissions, and the planning of two sustainable cities within the country.[90] The UAE is a party to the UNFCCC, Kyoto Protocol, and the Paris Agreement, and has pledged to produce 27 per cent of its energy from renewable sources by 2021.[91] The country has also committed to reducing its carbon footprint of power generation by 70 per cent and using renewable energy for the production of 50 per cent of its electricity by 2050.[92] Federal Law No 24 of 1999 (Environmental Law) is the primary environmental legislation of the country.[93] The UAE primarily resorts to CAC measures for regulating the environment and has started adopting or considering market-based incentives and climate change nudges.

Like the majority of countries globally and in the MENA region, the UAE employs a host of CAC measures to regulate the environment, with sanctions and fines for non-compliance of the measures. Ministerial Decree No. (98) of 2019 on the use of refuse-derived fuel includes a variety of CAC measures,[94] in addition to Federal Law No 24 of 1999. On the Emirate (State) level, the Emirate of Dubai has adopted a compulsory code pertaining to energy, water, waste, and other construction material standards, for all new construction starting from 2014.[95]

While there is still a preference for CAC measures, the UAE has started utilizing MBAs for environmental regulation as well. In addition to providing subsidies in the agricultural sector to use modern technology and for environmental protection,[96] the country has also set up the Mubdala Capital and the Dubai Green Fund to financially support and promote clean-tech, renewable energy, and green-economy projects.[97] Mubdala Capital has also set up Masdar, a "future energy" company established by the government of Abu Dhabi.[98] Its vision is to be the "world's reference for knowledge" in clean technology, renewable energy, and sustainable development.[99] Masdar City, a low-carbon city, was developed by Masdar. The buildings in Masdar City are made with low-carbon cement and 90 per cent of aluminium is sourced from recycled materials.[100] Masdar has been also very active in research, design, and the implementation of various clean energy projects in the UAE and throughout the MENA region.[101]

Additionally, as part of the Energy Efficiency Standardization and Labeling Program (EESL), the country has been providing financial incentives in terms of fee reductions (for issuance of energy efficiency labels) for manufacturers and sellers of higher performing appliances.[102] There has also been a gradual increase in tariff rates of utilities (which were highly subsidized) in order to decrease waste and over-consumption.[103] Despite the relative lack of taxes in other areas of activity, the UAE has also started introducing environmental taxes such as a tax on lorries carrying mineral resources.[104]

The UAE has been working closely with the Behavioural Insights Team (BIT) of the UK and other organizations to study and possibly develop its own behavioural insights unit.[105] Currently, the Emirate of Abu Dhabi has launched the "Tarsheed" campaign to influence a reduction in energy and water consumption, through prompts such as "Power is a Privilege" and "Water is a Privilege".[106] The EESL programme labels energy-efficient products.

The Estidama programme provides ratings for sustainability for the entire construction – right from design to planning and construction of new buildings – in three categories of developments: the Pearl Community Rating System, the Pearl Building Rating System, and the Pearl Villa Rating System.[107]

Assessing instrument choices in the MENA region

The choice of the right regulatory instruments, as well as their design, plays a very significant role in addressing climate change. This section of the chapter makes a holistic assessment of the endogenous and exogenous challenges to addressing climate change in the MENA region. The chapter closes with some suggestions on how to improve the effectiveness of regulatory interventions in MENA states.

Challenges to climate change regulation in the MENA region

All countries in the MENA region have been consistently amplifying their efforts to combat climate change by taking steps domestically as well as internationally. All of them have signed the Paris Agreement at COP 21, and almost all – except for Iran, Iraq, Libya, and Yemen – have ratified the treaty.[108] Domestically, they have all passed legislative or regulatory measures to combat climate change.[109]

The region is unique in the challenges it faces in combating climate change. MENA states face two types of challenges: endogenous and exogenous. Endogenous challenges relate to the climatic and geographical conditions of the MENA region that were presented above. The region faces also exogenous challenges, including unique socio-economic, political, and geopolitical uncertainties.[110] Apart from the uncertainties already discussed, some further challenges that could have a resounding impact on the region's, as well as individual states' ability to fight against climate change are the worsening political and economic crisis in Lebanon,[111] the looming water-sharing issue on the Nile River,[112] the Yemen war,[113] as well as the general health crisis due to the current COVID-19 pandemic, among many others. The special endogenous and exogenous conditions of countries in the MENA region have an impact on the effectiveness of the regulatory instruments used in different countries.

When it comes to instrument choice questions, CAC measures, while generally effective at least in the first years after their introduction, face issues of inflexibility, innovation stifling, bureaucratic costs, as well as enforcement. MBAs have contributed towards allaying the issues of inflexibility and dearth of innovation but are, in turn, sometimes expensive, have high administrative costs, and require an institutional capacity, as well as mature markets, that are not present in all countries. Climate change nudges fulfil the role of providing a flexible and cheap regulatory instrument, which is easy to implement and monitor, but are non-binding and have limited scope, as nudges could be very region and culture-specific due to regional and cultural variance. These regulatory challenges are present independent of countries and regions in which they are applied; they are thus expectedly encountered in MENA states as well. However, the specific endogenous and exogenous challenges of the MENA states exacerbate them.

As the examples of Morocco, Lebanon, and the UAE showcase, there is a heavy reliance in the MENA region on CAC instruments in the combat against climate change – even more than in other countries around the world. This may be due to the nature of the political systems of most MENA countries. In addition, the region's geopolitical uncertainties and the "survival-mode" under which many nations are operating makes regulating through familiar CAC measures more feasible to the governments of the region.

The MBAs in the MENA region usually take the form of grants and subsidies (for both the reduction of energy subsidies and the provision for subsidies for green projects). There is a limited use of tradable permits and pollution charges, and almost no utilization of market-barrier reductions. The limited degree of adoption as well as acceptance of climate change-related MBAs in the region could be traced to the fact that maturity and institutional capacity of markets in MENA is still in its infancy – especially when compared to other regions around the world such as Europe, North America, or Southeast Asia. Other adverse factors could be the particularly high costs associated with MBAs, design flaws in the development of the measures, unrealistic expectations on their effectiveness, and objections raised by industry as well as the environmental community.[114] Despite the non-linear progress in the adoption as well as the pushback against MBAs, there is a noticeable trend in the region to move towards increasing institutional capacity as well as developing a political will for designing various types of market-based instruments.[115]

While increasing institutional capacity and having a mature market can seem to be a longer-term strategy for increasing the use of MBAs to combat climate change in the MENA region, utilizing climate change nudges does not need the same amount of time or financial input, thereby offering a more viable regulatory option for MENA states. Having similar considerations in mind, governments around the world have established "nudge" or "behavioural insights units" to design and implement nudges – including climate change and green nudges. Globally, three types of models of such units seem to be prevalent: the centralized model (e.g., in Australia, France, and Germany), the decentralized model (e.g., in Denmark and the Netherlands), and the hybrid model (e.g., in the UK). While the centralized model favours a single unit, which applies behavioural insights across government sectors, the decentralized model promotes the idea of multiple units within various ministries that operate autonomously. The hybrid model consists of behavioural units that operate within individual ministries but coordinate with a central unit.[116] Almost half of the countries of the MENA region have either established government behavioural insights units, are in the process of instituting such units, or have non-governmental organizations working with nudges in collaboration with the government.[117] MENA states have also adopted these models. For example, while non-governmental behavioural insights units operate in Tunisia, Lebanon, Qatar, Saudi Arabia, and Kuwait have created centralized governmental behavioural insights units.

Overall, MENA states are not reticent about seeking assistance from behavioural science experts and governments that have already established behavioural insights units (while learning from their experiences and their past mistakes),[118] and have experience in administering climate change nudges. Apart from the European Union, no other multinational region has such a high concentration of nudge units;[119] and this statistic demonstrates the region's states' commitment to diversifying its cache of regulatory tools for combating climate change. However, it is important to note that considering various social and cultural differences between the West and MENA, the solutions that would work in other places might not work in MENA states.[120] There are cultural and semantic nuances that might not translate well if a nudge is directly picked up from the US or the EU and applied in MENA.

Improving the effectiveness of regulatory interventions in the MENA region

MENA states can take further steps towards improving the effectiveness of the adopted legislative and regulatory measures by intensifying the efforts towards combining all three regulatory approaches presented above, as well as increasing international and regional administrative cooperation in the field of climate change protection.

Countries around the world have been designing a variety of regulatory policies to combat climate change – usually a mixture of CAC with MBAs and sometimes climate change nudges. The reason for a worldwide preference of the use of all three instruments (in varied proportions) is that no instrument, on its own, is sufficient to address this complex issue – especially in the time frame that most of the nations have agreed upon during the COP 21 leading to the Paris Agreement. The three presented regulatory instruments thus complement one another and together form a triad for effective climate change regulation. Particularly in the context of the MENA region, simply relying on CAC measures is not enough, as doing so will not address the issues of inflexibility and increased regulatory costs. MBAs may be in the position to address some of the challenges of inflexibility, but MBAs are also expensive and require greater institutional capacity, particularly in the MENA context. It is here that climate change nudges may prove themselves as an effective tool against climate change, as they are both cheap and flexible. The challenge of their non-binding nature remains, but the untapped potential seems very significant. Hence, the specific endogenous and exogenous challenges prevalent in the MENA countries make the utilization of all three regulatory instruments more pertinent than in other regions in order to address the challenges of climate change, and improve the overall effectiveness of the regulatory interventions.[121] Given the level of maturity of the economic and green markets in MENA states, it may eventually be advisable to instead shift the immediate focus of their economic and other resources to the development of MENA-specific climate change nudges, alongside CAC measures, while slowly developing their institutional capacity for MBAs.

Besides the combination of all three regulatory instruments, the region seems to have another source of untapped potential to increase the effectiveness of the adopted interventions. There seems to be a lack of a broad-based collaboration among the governments of the MENA states. Intergovernmental efforts, particularly in the area of MBAs, may contribute towards improving regulatory interventions. The true hope lies though in the cooperation of administrative agencies. The collaboration of Masdar with the governments of six other MENA countries with a view to developing and implementing clean energy sources is one example.[122] Here again, developments in the area of climate change nudging may be the right way towards improving the effectiveness of regulatory interventions in the MENA region. Nudge units in the MENA region have been developed in collaboration with nudge units from outside the region, but also to a large extent via an intra-regional collaborative process. The cooperation between the nudge units of Qatar and Lebanon has produced some very useful insights in this regard. Such examples showcase the potential increased effectiveness of climate change mitigation measures if countries in the MENA region work together to find solutions to common issues.

Conclusion

Due to its geographical, political, and socio-economic conditions, the MENA region is acutely sensitive to climate change. The countries have recognized this vulnerability and are trying to develop regulatory solutions to mitigate climate change. While MENA states have been more active in designing and implementing CAC policies, they are currently working towards developing MBAs and climate change nudges. The endogenous and exogenous challenges faced by MENA states in dealing with climate change can only be addressed through an increased focus on designing regulatory interventions that are specific to the climatic, geographical, as well as political, socio-economic, and geopolitical conditions of the region, as well as developing further avenues of administrative cooperation among them.

Notes

1 See Chapter 1 of this book for a list of countries in the MENA region.
2 For an analysis of the correlation of industrialization and urbanization on carbon-emission in MENA region, see A. A. Abdallh and H. Abugamos, "A Semi-Parametric Panel Data Analysis on the Urbanisation-Carbon Emissions Nexus for the MENA Countries" (2017) 78 *Renewable and Sustainable Energy Reviews* 1350–1356. The problems of water and land scarcity, poor air quality, improper waste management, and loss of biodiversity that plague the Region can be directly attributed to

rapid industrialization and indiscriminate urbanization; see Iyad Abumoghli and Adele Goncalves, "Environmental Challenges in the MENA Region" (UNEP, 2019). https://wedocs.unep.org/bitstream/handle/20.500.11822/31645/EC_MENA. pdf?sequence=1&isAllowed=y (accessed 06 July 2020).

3   See United Nations Development Programme, "Climate Change Adaptation in Arab States – Best Practices and Lessons Learned" (Global Environment Facility 2018).

4   Climate Change Performance Index <www.climate-change-performance-index .org>. Morocco is ranked in the top three, while Iran and Saudi Arabia are at the bottom of the index.

5   B. Mirkin, "Population Levels, Trends and Policies in the Arab Region: Challenges and Opportunities" (2010) Arab Human Development Report Research Paper Series 2010, United National Development Programme Regional Bureau for Arab States.

6   The World Bank, "MENA Economic Update" (The World Bank 2019).

7   A. H. Cordesman, "Stability in the Middle East: The Range of Short and Long-Term Causes" (Center for Strategic and International Studies 2018).

8   B. Wahab, "What Will it Take to Repair Middle Eastern Economies?" in M. Rubin and B. Katulis (eds), *Seven Pillars: What Really Causes Instability in the Middle East?* (AEI Press 2019).

9   Ibid; Cordesman, n 7.

10   See generally on the use of CAC, MBA, and nudge interventions to promote climate change and more broadly environmental protection, P. Hacker and G. Dimitropoulos, "Behavioral Law & Economics and Sustainable Regulation: From Markets to Learning Nudges" in K. Mathias and B. R. Huber (eds), *Environmental Law and Economics* (New York: Springer 2017).

11   G. Emberling, "The Geography of the Middle East" (The Orient Institute-University of Chicago 2010).

12   Ibid (although we have divided the region into four zones, instead of five as envisioned by Emberling).

13   J. Lelieveld and others, "Climate Change and Impacts in the Eastern Mediterranean and the Middle East" (2011) 114 *Climate Change* 667–687.

14   Lelieveld and others, n 13, 682.

15   Abumoghli and Goncalves, n 2.

16   Eighty-six per cent of the groundwater is used in the agricultural sector. The total water withdrawal for agricultural purposes increased by three-fold in the past three decades; while total fresh-water withdrawal more than doubled (FAO Database, 2016 <www.fao.org/nr/water/aquastat/data/query/index.html?lang=en>).

17   A decrease in arable land is noticeable in all MENA countries (The World Bank, "Arable Land Data" 2017 <https://data.worldbank.org/indicator/AG.LND.ARBL.Z S?name_desc=false>).

18   A. Barton, "Water in Crisis-Middle East," (The Water Project) <https://thewate rproject.org/water-crisis/water-in-crisis-middle-east> accessed 6 January 2021; H. A. Amery, "Water Wars in the Middle East: A Looming Threat" (2002) 168 *The Geographical Journal* 313.

19   Abumoghli and Goncalves, n 2.

20   See International Union for Conservation of Nature, "Marine and Coastal Management Program" (IUCN) <www.iucn.org/regions/west-asia/our-work/marine-and -coastal-management-programme> accessed 6 January 2021.

21   S. Frankhauser, C. Gennaioli, and M. Collins, "Do International Factors Influence the Passage of Climate Change Legislation?" (2015) 16 *Climate Policy* 318.

22   The fourth geographical region of MENA – the Mesopotamian region comprising Iraq, Iran, and Syria – is not included in this analysis due to the ongoing political uncertainty.

23   African Development Bank, "Morocco: Country Results Brief 2019" (African Development Bank 2019).

24   Fund for Peace, "Fragile States Index" (Fragile States Index) <https://fragilestate sindex.org/country-data/> accessed 6 January 2021. Lebanon is in the 'High Warning' category.

25 Central Intelligence Agency, "UAE: The World Factbook" (CIA) <https://www.cia
    .gov/the-world-factbook/countries/united-arab-emirates/#introduction>   accessed
    6 January 2021.
26 Hacker and Dimitropoulos, n 10.
27 Screens usually refer to procedures that screen individuals based on some qualities
    that they have, either positively or negatively, such as appointments, constraints, eli-
    gibility for certain posts, etc. Sanctions and incentives work by the negative or posi-
    tive financial impact incurred on individuals. See P. Petit, "Institutional Design and
    Rational Choice" in R. E. Goodin (ed.), *Theory of Institutional Design* (Cambridge
    University Press 1996).
28 R. Thaler, "Doing Economics Without Homo Economicus" in S. G. Medema and
    W. J. Samuels (eds), *Foundations of Research in Economics: How Do Economists Do
    Economics?* (Edward Elgar 1996).
29 Hacker and Dimitropoulos, n 10.
30 G. Helfand, "Standards versus Standards: The Effects of Different Pollution Con-
    trols" (1991) 81 *American Economic Review* 634.
31 R. Stavins, "Market-Based Environmental Policies" in P. Portney and R. Stavins
    (eds), *Public Policies for Environmental Protection* (2nd edn) (Routledge 2000) 1.
32 This does not mean that the CAC interventions have been eclipsed since the intro-
    duction of MBAs. The use and even sometimes an amalgamation of both regulatory
    interventions may be seen in varying degrees and sectors in most countries.
33 Stavins, n 31; R. Stavins and B. Whitehead, "Market-Based Environmental Policies"
    in M. Chertow and D. Esty (eds), *Thinking Ecologically: The Next Generation of Envi-
    ronmental Policy* (Yale University Press 2018).
34 Stavins, n 31.
35 Ibid, 5.
36 D. Olawuyi, "From Kyoto to Copenhagen: Rethinking the Place of Flexible Mecha-
    nisms in the Kyoto Protocol's Post 2012 Commitment Period" (2010) 6 *Law, Envi-
    ronment and Development Journal* 21.
37 Ibid.
38 Stavins, n 31.
39 See, for a high-level account on the spread of market-based climate protection
    economies in the Asia-Pacific, T. Shinkuma and H. Sugeta, "Tax Versus Emissions
    Trading Scheme in the Long Run" (2016) 75 *Journal of Environmental Economics and
    Management* 12.
40 Stavins, n 31.
41 Stavins and Whitehead, n 33.
42 L. Reisch and J. Sandrini, *Nudging in der Verbraucherpolitik: Ansätze verhaltensbasierter
    Regulierung* (Nomos Verlagsgesellschaft 2015).
43 C. Sunstein and L. Reisch, "Green by Default" (2013) 66 *Kyklos* 398; C. Sunstein
    and L. Reisch, "Automatically Green: Behavioral Economics and Environmental
    Protection" (2014) 38 *Harvard Environmental Law Review* 127.
44 C. Sunstein, "Empirically Informed Regulation" (2011) 78 *University of Chicago Law
    Review* 1349.
45 Hacker and Dimitropoulos, n 10.
46 See iNudgeyou, "Green Nudge: Nudging Litter into the Bin" (iNudgeyou 2012)
    <https://inudgeyou.com/en/green-nudge-nudging-litter-into-the-bin/>.
47 Reisch and Sandrini, n 42, p. 113, as cited in Hacker and Dimitropoulos, n 10, p. 162.
48 M. Kaiser and others, "The Power of Green Defaults: The Impact of Regional Varia-
    tion of Opt-Out Tariffs on Green Energy Demand in Germany" (2020) 174 *Ecological
    Economics* 106685.
49 Hacker and Dimitropoulos, n 10, 163–166; C. Fischer, "Feedback on Household
    Electricity Consumption: A Tool for Saving Energy" (2008) 1 *Energy Efficiency* 79; I.
    Ayers, S. Raseman, and A. Shih, "Evidence from Two Large Field Experiments that

Peer Comparison Feedback can Reduce Residential Energy Usage" (2013) 29 *Journal of Law, Economics and Organization* 992.

50 Energy efficiency nudges could also be categorized as climate change nudges.

51 Hacker and Dimitropoulos, n 10, 167; D. Wuebbles and A. K. Jain, "Concerns about Climate Change and the Role of Fossil Fuel Use" (2001) 71 *Fuel Processing Technology* 99.

52 C. Codagnone, F. Bogliacino, and G. Veltri, "Testing CO_2/Car Labelling Options and Consumer Information: Final Report," European Commission, June 21, 2103, 9 as cited in Hacker and Dimitropoulos, n 10, 168.

53 For a discussion on cultural and semantic variance of Nudges, see E. Slinger and K. Whyte, "Competence and Trust in Choice Architecture" (2010) 23 *Knowledge, Technology and Policy* 461; T. Pinch, "Comment on 'Nudges and Cultural Variance'" (2010) 23 *Knowledge, Technology and Policy* 487.

54 See the section above discussing the reasons for the selection of the three countries as case studies.

55 Oxford Business Group, "Moroccan Authorities Target Diversification to Secure Long-Term Economic Growth" (Oxford Business Group 2018) <https://oxfordbusinessgroup.com/overview/transition-phase-authorities-target-diversification-secure-long-term-growth-0> accessed 6 January 2021.

56 Considering the primary source of energy is coal (64 per cent) and oil (22 per cent), it has still pledge to reduce its GHG emissions by 42 per cent by 2030. This target is to be achieved through the implementation of various National Programs, such as National Energy Strategy, National Logistics Strategy, National Household and similarly Water Program, National Liquid Sanitation and Wastewater Treatment Program, Morocco Green Plan, Preservation and Sustainable Forest Management Strategy, and Urban Public Transport Improvement Program.

57 Qatar hosted the COP 18 in 2012.

58 "Morocco Nationally Determined Contribution Under the UNFCCC" 18 September 2016.

59 Climate change legislation in Morocco is implemented at two levels. At the national level, the NCESD was enacted in order to set a national mandate in line with Morocco's obligations with UNFCCC.

60 Available in French at <http://dmp.uae.ma/textes_juridiques/air/decret_09_631_%20polluants_air.pdf>; in relation Law No. 13-03 on Air Pollution, available at <www.informea.org/en/law-no-13-03-prevention-air-pollution>.

61 International Energy Agency (IEA), "Energy Policies Beyond IEA Countries: Morocco 2019" (2019) 68.

62 It is to be noted that although fines and sanctions are stipulated in the law, they are not applied in the case of non-compliance. There is a deeper issue of implementation. EPRs are undertaken by United Nations Economic Commission or Europe (ECE), at the request of the state (UNECE, "Environmetnal Performance Reviews: Morocco," Environment Performance Review Series No. 38, 2014, UN Doc. ECE/CEP/170, 15).

63 Institut Marocain de Normalisation, "Liste des Projets de Normes en Enquete Publique" 2020. The public consultation has concluded.

64 A. Rattal, "Partnership for Sustainable Development in Morocco" (Ministry of Environment, 7 May 2008).

65 United Nations Economic Commission for Africa, "The Green Economy in Morocco" 18.

66 Ibid.

67 M. Tarradell, "Morocco Case Study: Analysis of National Strategies for Sustainable Development" (International Institute for Sustainable Development, June 2004).

68 IEA, n 61, 26. Butane is still heavily subsidized, and electricity tariffs are still lower than the generation costs. The government has planned to limit the subsidy to low-income household only but has not implemented it fully yet – mostly due to political repercussions.

69 OECD Data, "Environmental Tax" <https://data.oecd.org/envpolicy/environmental-tax.htm> accessed 6 January 2021.

70 Morocco Law No. 10-95 of 1995.

71 Ibid.

72 Econostrum, "Morocco Strengthens Its Housing Energy Efficiency System" (Econostrum, 24 January 2020) <www.econostrum.info/Le-Maroc-renforce-son-dispositif-d-efficacite-energetique-des-logements_a26385.html>.

73 Agence Marocaine pour l'Efficacite Energetique, "Energy Efficiency in Buildings" <www.amee.ma/en/node/118>; I. Merini and others, "Energy Efficiency Regulation and Requirements: Comparison between Morocco and Spain" in M. Ezziyyani (ed.), *Advanced Intelligent Systems for Sustainable Development* (Springer 2019) 204.

74 IEA, n 61, 68.

75 Invest in Lebanon, "Economic Performance" (Invest in Lebanon, 2019) <https://investinlebanon.gov.lb/en/lebanon_at_a_glance/lebanon_in_figures/economic_performance>.

76 Republic of Lebanon, Ministry of Environment, "Negotiations: Lebanon" <http://climatechange.moe.gov.lb/negotiations>.

77 Republic of Lebanon, "Lebanon's Intended Nationally Determined Contribution under the United Nations Framework Convention on Climate Change," September 2015.

78 Republic of Lebanon, Ministry of Environment, "Lebanon's National Strategy for Air Quality Management for 2030," December 2017.

79 OHCHR, "Questionnaire: Lebanon" 2018, available at <www.ohchr.org/Documents/Issues/Environment/SREnvironment/Pollution/Lebanon.pdf>; Republic of Lebanon, Decree No. 167 of 2017 to determine the minutes of application of Article 20 of the Law on Environmental Protection No.444 of 2002 (2017).

80 OHCHR, ibid, 4/7.

81 Ibid.

82 Republic of Lebanon, Law No. 80 of 2018: Integrated Waste Management, Article 8.

83 "Lebanese Environment Minister Fadi Jreissati: 'We Have to Implement the Polluter Pays Principle'" (Le Commerce, 13 July 2019) <https://www.lecommercedulevant.com/article/29191-lebanese-environment-minister-fadi-jreissati-we-have-to-implement-the-polluter-pays-principle>.

84 The MENA Region's first behavioural insights unit was established in Qatar as the Qatar Behavioral Insights Unit under the Supreme Committee for Delivery and Legacy. The QBIU was recently incorporated as a foundation under the Qatar Financial Center and renamed into B4Development <http://b4development.org/b4development/>.

85 Nudge Lebanon <https://nudgelebanon.org/>.

86 Nudge Lebanon, "Experiments" <https://nudgelebanon.org/category/experiments/>. It is to be noted that until now there have been no reported experiments that consider climate change nudges.

87 I. Stoughton, "Can Nudge Theory Change Citizen's Behavior, Government Policy in Lebanon?" *Al Monitor* (27 April 2018) <www.al-monitor.com/pulse/fa/originals/2018/04/lebanon-nudge-conference-economic-behaviour-shifting-policy.html>.

88 The Lebanese Center for Energy Conservation, "Projects" <http://lcec.org.lb/en/LCEC/Projects>.

89 The country has one of the largest per capita pollution, usage of energy, water, and waste-production. Most of the potable water is produced through desalination, and

energy (electricity and fuel) is available at a very low rate – despite the fact that almost 41 per cent of the country's energy emission originates from electricity and water generation (United Arab Emirates, Ministry of Energy and Industry, "United Arab Emirates: Fourth National Communication Report," 2018).

90   United Arab Emirates, Ministry of Climate Change and Environment, "National Climate Change Plan of the United Arab Emirates: 2017–2050".

91   In accordance with its National Determined Contribution. UAE, "Vision 2021: United Arab Emirates," www.vision2021.ae/en/national-agenda-2021/list/environm ent-circle; UAE, "Environment in Vision 2021" <https://u.ae/en/information-and-services/environment-and-energy/environment-and-government-agenda>.

92   UAE, "UAE Energy Strategy 2050" <https://u.ae/en/about-the-uae/strategies-init iatives-and-awards/federal-governments-strategies-and-plans/uae-energy-strategy-2050#:~:text=The%20strategy%20aims%20to%20increase,AED%20700%20bi llion%20by%202050>.

93   UAE is a federal state with several separate monarchies forming the Federation.

94   Ministerial Decree (98) of 2019 <www.moccae.gov.ae/assets/download/699b133e/M98-19.pdf.aspx?view=true>. See also "UAE Ministry of Climate Change and Environment: Legislations" <www.moccae.gov.ae/en/legislations.aspx>. This resource has a list of all the legislations and decrees regarding the regulation of various aspects of the environment, and it demonstrates the eagerness with which the country is pursuing environmental protection in recent years.

95   Dubai Green Building Regulations <www.dewa.gov.ae/~/media/Files/Consultants %20and%20Contractors/Green%20Building/Greenbuilding_Eng.ashx>.

96   B. Abdul Kader, "UAE Government Supports Farmers with Fund" *Gulf News* (10 May 2014) <https://gulfnews.com/uae/environment/uae-government-supports-farm ers-with-fund-1.1328143>.

97   UAE, n 92, 50–53.

98   "Masdar: A Mubadala Company" <https://masdar.ae/en>.

99   Masdar, "Vision, Mission and Values" <https://masdar.ae/en/about-us/management/vision-mission-and-values>.

100 Masdar, "The Source of Innovation and Sustainability – Investment and Leasing Opportunities at Masdar City" <https://masdar.ae/-/media/corporate/downloads/media/mas_mc_brochure_2020.pdf>.

101 Masdar, "Projects – Masdar" <https://masdar.ae/en/masdar-clean-energy/projects>.

102 UAE, n 90, 73.

103 Ibid, 75.

104 "Environmental Tax Enacted for Fujairah Lorries" *The National* (1 February 2016) <www.thenational.ae/uae/environment/environmental-tax-enacted-for-fujairah-l orries-1.195037>. As the country does not have an established tax-system, it is startling to see the adoption of environmental taxes.

105 For example, the Ministry of Cabinet Affairs and Prime Minister's Office launched one such study (Ministry of Cabinet Affairs and the Future, "UAE Government Launches Behavioral Interventions Studies to Improve Health and Education Outcomes" (MoCAF, 22 January 2017).

106 Tarsheed: Rethink <www.tarsheedad.com/en-us/Pages/Home.aspx>.

107 However, since late 2010, adhering to the rating system has become mandatory. For more information <www.upc.gov.ae/en/-/media/files/upc/media/prdm/prrs_v1.ashx>.

108 UN Treaty Collection, Paris Agreement: Depository: Status of Treaties <https://tr eaties.un.org/Pages/ViewDetails.aspx?src=TREATY&mtdsg_no=XXVII-7-d&chap ter=27&clang=_en>.

109 See Chapter 1 of this book.

110 M. Bishara, "Beware of the Looming Chaos in the Middle East" *Al Jazeera* (Doha, 3 August 2020) <www.aljazeera.com/indepth/opinion/beware-looming-chaos-middl e-east-200803042230463.html>; S. Tisdall, "Global Power Games Could Blow the Whole Middle East Apart" *The Guardian* (9 August 2020) <www.theguardian.com/ commentisfree/2020/aug/09/global-power-games-could-blow-the-whole-middle-east -apart>.

111 J. Macaron, "Lebanon is on Track to Become a Failed State" *Al Jazeera* (7 August 2020) <www.aljazeera.com/indepth/opinion/lebanon-track-failed-state-200806093 843287.html>.

112 R. Michaelson, "It Will Cause a Water-War: Divisions Run Deep as Filling of Nile Dam Nears" *The Guardian* (23 April 2020) <www.theguardian.com/global-developm ent/2020/apr/23/itll-cause-a-water-war-divisions-run-deep-as-filling-of-nile-dam -nears>.

113 R. Almutawakel, "Yemen: First Bombs, Soon a Coronavirus Epidemic" *Al Jazeera* (7 April 2020) <www.aljazeera.com/indepth/opinion/yemen-bombs-coronavirus-ep idemic-200406151530466.html>.

114 Stavins and Whitehead, n 33.

115 For example, countries such as Kuwait and Tunisia (among other MENA states) are working towards building institutional capacity for MBAs (UNDP, "Seeking Sustainability and Cost Efficiency: A UNDP Environment Programme for Kuwait," July 2011; United Nations Economic Commission for Africa, "The Green Economy in Tunisia" <www.uneca.org/sites/default/files/uploaded-documents/SROs/NA/AHEG M-ISDGE/egm_ge-_tunisa_eng.pdf>).

116 PWC World Government Summit, "Triggering Change in the GCC Through Behavioral Insights: An Innovative Approach to Effective Policymaking".

117 For instance: Morocco (MEN), Tunisia (Design Lab Tunisia), Nudge Lebanon (Lebanon non-governmental organization), Center for Strategic Development (Saudi Arabia), B4Development (Qatar, formerly: Qatar Behavioral Insights Unit), General Secretariat of the Supreme Council for Planning and Development (Kuwait), Behavioral Insights Unit (Oman), Egypt (National Council of Women used Behavioral Insights for gender-related concerns). See OECD, "Behavioral Insights and Public Policy", 2018 <www.oecd.org/gov/regulatory-policy/behavioural-insights.htm>.

118 Furthermore, states could also draw upon the experiences and best practices developed by international organizations such as the OECD; see, for example, OECD, "Behavioral Insights and Public Policy: Lessons from Around the World," 2017, 121–150, which provides some insights into the climate change and environmental nudges regulation in various jurisdictions. Furthermore, OECD, "Tools and Ethics for Applied Behavioral Insights: The Basic Toolkit," 2019 also provides some excellent insights.

119 Regions with a number of countries that host nudge units (inside or outside the government) – European Union: over 25; MENA: 7; Oceania: 2; Asia and Africa (excluding MENA): 6; North and Central America: 4; South America: 4. See OECD, ibid, for a detailed list of BIUs worldwide.

120 Slinger and Whyte, n 53.

121 It took decades for many countries to transition from CAC-only measures to a mixture of CAC and MBA, and eventually climate change nudges.

122 Masdar, n 101.

5 Climate change and the energy transition in the MENA region

S. Duygu Sever

Introduction

This chapter aims to examine the importance of decarbonization and energy transition for climate change mitigation in MENA. It evaluates the limitations of current energy transition strategies in the region, and then identifies policy frameworks for addressing existing barriers and challenges, in order to enhance climate change response across the region.

Identified as one of the most affected regions by global warming,[1] for the MENA region, climate change is a matter of survival. As the region gets significantly drier and warmer, freshwater scarcity[2] will increasingly deprive MENA of water and food by both quantity and quality.[3] Furthermore, potential militarization of key natural resources[4] in the face of scarcity risks contributing to political and military conflicts and deepening social vulnerability to climate change.[5]

In order to mitigate the risks of climate change, the Paris Agreement set the objective of limiting the temperature increase to 1.5°C above pre-industrial levels.[6] This requires a significant cut in emissions and a global decarbonization, for which the energy sector is critical.[7] The consumption of fossil fuels accounts for about two-thirds of global greenhouse gas emissions,[8] being the highly likely cause of climate change.[9]

The achievement of emission reduction targets entails a comprehensive transformation of the energy sector from fossil-based to zero-carbon systems by the second half of this century.[10] This transformation is called energy transition. The immediate call for this transformation on a global scale[11] includes strict measures such as the decarbonization of the electricity system[12] or leaving the fossil fuels in the ground.[13] The electrification of the economy, energy efficiency, and the integration of renewable energy resources into the existing energy systems are among major components of energy transition.[14] Renewable energy and energy efficiency measures alone have a potential to achieve 90 per cent of the energy-related carbon dioxide emission reductions required to accomplish climate goals.[15]

Building on this link between climate change and energy transition, this chapter discusses required policy and governance frameworks for advancing the decarbonization and energy transition agenda across the MENA region. The second section analyzes why MENA countries need to decarbonize in response to climate change and ongoing efforts in the region. The third section identifies some policy barriers or challenges hindering energy transition and decarbonization in MENA. The next section provides recommendations on how to address these barriers, and challenges for further progress towards a low-carbon, sustainable development in MENA. The final section offers concluding remarks.

Climate change and the need for energy transition in the MENA region

The spill-over effects of climate change that risk human security and political stability in MENA are the ultimate reasons why MENA countries need to address climate change mitigation as a priority. In this context, energy transition offers a dual purpose for MENA. First, in the long term, the success of the transition to zero-carbon economies will increase the MENA region's resilience by mitigating climate change-induced threats. Second, energy transition will help MENA to urgently address its energy security in a sustainable way. MENA's primary energy consumption is projected to grow 55 per cent between 2017 and 2040.[16] The region's CO_2 per capita also has been surpassing the world's average since the late 1990s (Figure 5.1), with Gulf Cooperation Council (GCC) states having significantly high scores. Moreover, fossil fuels maintain their dominance in total primary energy supply mix of MENA countries, as demonstrated in Appendix I.

So far, every MENA country has signed the Paris Agreement,[17] and agreed to be part of global climate mitigation efforts. Their intentionally determined contributions (INDCs)[18] have already demonstrated their political commitment for the transition to low-carbon, efficient, and sustainable systems.[19] Underlining the principles of "common but differentiated responsibilities" and "special national circumstances", MENA countries offered ambitious national policies so as to diversify their economies away from reliance on hydrocarbon extractive industries.[20]

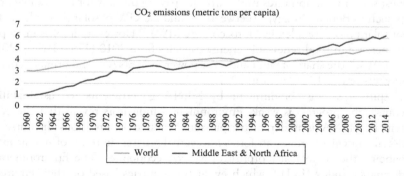

CO₂ emissions (metric tons per capita)

— World — Middle East & North Africa

Figure 5.1 CO_2 Emissions in the World and MENA. Compiled by the author. Data source: The World Bank, "CO2 Emissions (kt)".

Most of the MENA countries developed "National Visions" identifying their primary concerns and key targets.[21] In the majority of these national strategies, climate change has an explicit role. To illustrate, environmental development is one of the main pillars of the Qatar National Vision 2030 which aims for a proactive regional role in mitigating climate change.[22] Similarly, the Oman National Vision 2020 also underlines the necessity to increase resilience and capability for coping with climate change effects.[23]

A common feature of MENA's national visions is that they evolve around similar issue areas and objectives. For example, the United Arab Emirates (UAE) Vision 2021 targets a balanced, sustainable growth with an important role for renewable energy sources and energy-efficient technologies, and aims for the reduction of carbon emissions and ecological footprint.[24] In the same vein, the Saudi Arabia Vision 2030[25] also includes environmental sustainability with specific focus on efficiency, waste management, reduction of pollution, and desertification. The alignment of the strategies with the Sustainable Development Goals (SDGs)[26] emerges as another common theme in MENA government strategies, as clearly stated in the Kuwait Vision 2035: New Kuwait[27] or the Egypt Vision 2030.[28] Building on their national visions, some MENA countries also develop additional frameworks to exclusively address climate change, including Moroccan Climate Change Policy[29] and National Climate Change Plan of the UAE 2017–2050,[30] which depict climate change's potential impact for their country and list strategies to be followed.

As national strategies reveal, the political will to invest revenues from extractive industries into low carbon systems is present in some MENA countries.[31] In this context, the introduction of renewable energy technologies constitutes the central piece of energy transition, as countries set forth concrete renewable targets, summarized in Appendix I. Factors like cost reduction of solar and wind power generation, technological advancements in power storage, batteries, or electric vehicles, offer a window of opportunities to accelerate progress with regard to these low-carbon resources.[32] MENA has both the political willingness and physical potential for renewables. For example, the region has some of the highest solar irradiation rates in the world,[33] paving the way for remarkable projects such as Benban in Egypt (between 1.6 and 2.0 GW of solar power[34]) or the Noor solar power complex in Morocco (580 MW).[35] However, despite an 89 per cent increase in total renewable energy capacity from 2010 (17.6 GW) to 2019 (33.1 GW),[36] the share of wind and solar power in MENA's total primary energy supply was only 0.19 per cent in 2017.[37]

Despite a strong commitment by MENA governments to sustainability discourse, the progress is weak. Environmental indicators reveal that the ecosystem in MENA is under heavy pressure,[38] and that ambitious discourse in national strategies hardly transforms into actual progress in terms of significantly enhancing the ecological footprint of MENA economies. The Environmental Performance Index (EPI),[39] which evaluates countries based on their environmental performance,[40] offers relevant proof as MENA countries' scores range only between 28 and 70 out of 100 points. The ten-year change in MENA countries'

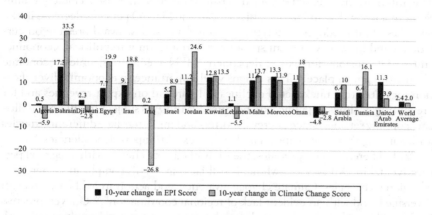

Figure 5.2 Ten year change in MENA's EPI scores. Compiled by the author. Data source: 2020 Environmental Performance Index.

environment and climate performances not only demonstrate a mixed progress but also reveal that the implementation of national strategies needs to be accelerated. Although the progress of some MENA countries' measures to combat climate change is higher than the world average (Figure 5.2), their scores only range between 12 and 67, out of 100.[41] Furthermore, data reveals that Algeria, Djibouti, Iraq, Lebanon, and Qatar perform worse than ten years before.

In the struggle against climate change, the gap between discourse and results remains, since implemented policies fall short of following the rapid pace of environmental degradation and ecological destruction.

Barriers for low carbon energy transition

The impediment of low carbon transition clearly affects MENA's success in climate change mitigation. This section identifies some policy barriers or challenges hindering energy transition and decarbonization in MENA.

Limited regional perspective

Every MENA country shares common climate change-induced threats.[42] Nevertheless, despite their equal vulnerabilities, MENA countries are at the same time very different too, due to their distinct domestic characteristics which enable or restrict their progress in energy transition. On the one hand, for some, military and political conflicts risk social harmony and human security, thus poverty prevails: as World Development Indicators (WDI) reveal, the percentage of the population with access to electricity in MENA is 97.7 per cent, meaning that 10.3 million people still lack access to electricity.[43] On the other hand, in MENA countries with hydrocarbon wealth,[44] the economy faces significant dilemmas.

The international quest for fuel diversification and increased efficiency aims result in a lower demand in oil. This can cause a 25 to 40 per cent decrease in the revenues of oil producers:[45] current account deficits, downward pressure on currencies, and lower government spending can transform into political, economic, and social tensions for export-dependent MENA countries, unless economic reforms are put in place.[46] This assigns vital importance to economic diversification for mitigating unexpected declines in hydrocarbon export revenues and for preventing declines in public expenditures.[47] MENA is subject to different political economy settings, since energy exporters and importers face different struggles for balancing their consumption and energy export/import dependencies.

With different country profiles, the study of climate change with a regional perspective, taking MENA as a whole, is still lacking.[48] This complicates emergence of regional strategies. Climate change awareness and action in MENA is "recent, diversified, and with an evident lack of regional coordination".[49] Uneven progress in the region's climate action and decarbonization is a natural result of political, economic, and social inequalities that persist across MENA. Differences in the social, political, and economic context in MENA determine not only how climate risks are distributed[50] but also the countries' capabilities to cope with them.

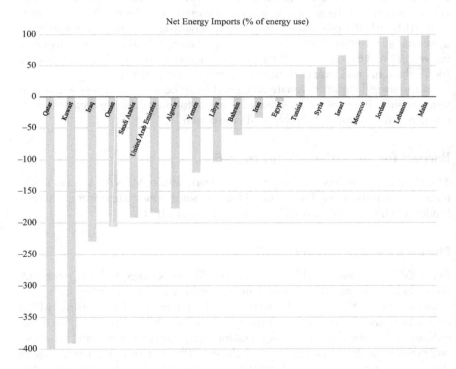

Figure 5.3 Net energy imports of MENA countries, 2014. A negative value indicates that the country is a net exporter. See: The World Bank, "Energy Imports, Net (% of Energy Use)".

Lack of holistic approach toward energy transition

As the effects of climate change at the water-energy-food nexus are multidimensional, the mitigation strategies, i.e. the energy transition, need to be acknowledged as integrated and interdependent. Energy transition is never only about energy supply portfolios. Integration of low-carbon resources into electricity generation is an important component of decarbonization, yet it is a limited strategy, unless economies, consumption habits, as well as mindsets, are broadly transformed into climate-friendly conduct. For a successful transition towards an environmentally sustainable future, interconnected objectives, integrated energy planning, and a holistic approach is required. The region lacks synchronized progress in different pillars of climate change mitigation.

Imbalanced policy focus between supply side and demand side management

The fundamental priority of MENA countries' energy strategies rests on the supply side, as additional power generation capacities are their major solution for meeting their increasing energy demand.[51] Between 2004 and 2017, the global funding provided for clean energy has increased from US$45 to 279.8 billion.[52] Although the amounts remained minor compared to China or Europe, investments in MENA for renewables had a considerable jump in this period, as Figure 5.4 indicates. On the other hand, as opposed to the investments in renewables, global efficiency investments have been stagnant with a less than one per cent increase from 2017 to 2018,[53] meaning that there is still a need for a substantial shift of investments from the supply side to the demand side.[54]

While diversifying their energy portfolios, MENA countries also need to target the demand side and consider the "hidden fuel", i.e. energy efficiency. At the discourse level, most MENA countries included energy efficiency targets into

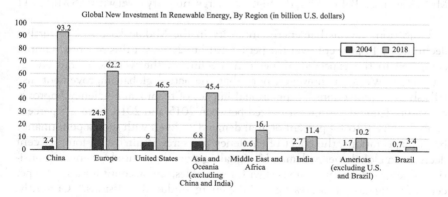

Figure 5.4 Global new investment in renewable energy by region 2004–2018. Compiled by the author (Bloomberg and others, n 52, p. 33).

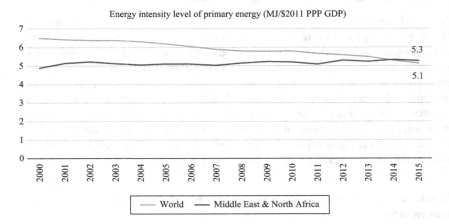

Figure 5.5 Energy intensity in the World and MENA. The data is compiled by the author from WDI 2020. World Bank defines the indicator as: "the ratio between energy supply and gross domestic product measured at purchasing power parity. Energy intensity is an indication of how much energy is used to produce one unit of economic output." (The World Bank, "Energy Intensity Level of Primary Energy (MJ/$2011 PPP GDP)" June 19, 2020.)

their national energy strategies,[55] yet energy efficiency improvements for MENA still need rapid and substantial action plans.[56] Figure 5.5 clearly demonstrates that energy intensity at the global level has a decreasing trend since 2000s, while the trend for MENA average is relatively constant even with a slight increase.

Certain countries in MENA, especially the GCC members and Iran, have significantly higher energy intensity compared to the rest of the region. Their economies are highly energy dependent since their energy intensity scores range between 5 to 10 MJ/$2011 PPP GDP, which is higher than both the world and MENA average. Bahrain has the highest energy intensity level with 9.8 MJ/$2011 PPP GDP.[57]

Projections reveal that energy intensity in the Middle East will be further declining by 11 per cent over the period 2017 to 2040, yet this progress is minor compared to the expected 37 per cent decline in the world's overall energy intensity.[58] We must, however, note that the actual global improvement also falls short of projections:[59] "the annual impact of technical efficiency improvements on demand has almost halved between 2015 and 2018, from 2.5 percent of final demand to 1.4 percent of final demand."[60] Even with such a performance below expectations, the effect of efficiency measures resulted in a four per cent decrease in global energy demand,[61] signaling a great potential for energy transition. Energy efficiency, combined with renewables, can account for up to 90 per cent of mitigation measures for curbing energy-induced emissions.[62] Currently, despite policy statements, projections, and high expectations, MENA's high energy intensity leaves this potential significantly untapped.

Rapid and unsustainable urbanization

Diverse environmental impacts of global warming, that will highly disrupt agriculture by land, quantity, and quality, will increasingly lead rural inhabitants to switch into off-farm employment and to relocate into cities for new income opportunities.[63] Overall, with desertification on one hand, and sea level rise on the other, climate change's destructive effects will lead 450 million people, globally, to migrate from rural to urban areas by 2050 due to scarce land resources.[64] Rapid and unplanned urbanization will render these populations further vulnerable to the impacts of climate change.[65] To be exact, globally, in 2018, 59 per cent of 1,146 cities with at least 500,000 inhabitants were at high risk of exposure to at least one of six types of natural disaster. This represents around 1.4 billion people.[66]

Urbanization is a process which changes the demographic and social structures by transforming lifestyles, major occupations, culture, and consumption behaviour.[67] Moreover, growing cities significantly expand their networks for transportation, trade, and information flow. Unless this expansion is planned and managed with adequate strategies, cities and their environment end up with pollution, environmental degradation, and unsustainable production and consumption patterns.[68] The fact that between 71 and 76 per cent of CO_2 emissions and between 67 and 76 per cent of global energy use come from cities[69] is proof that urbanization, indeed, offers considerable room for progress with regard to energy transition.

As of 2018, 65.37 per cent of MENA's total population lives in urban areas. On the country basis, with the exceptions of Tunisia (68.9 per cent), Morocco (62.4 per cent), Syria (54.1 per cent), Egypt (42.7 per cent), and Yemen (36.6 per cent), in all MENA countries, more than 70 per cent of the total population lives in urban areas.[70] Increasing urban populations require governments to meet increasing demands for fundamental goods such as electricity and water. To date, expensive desalination technologies have been offering energy-rich MENA countries a solution to supply water for their population.[71] To illustrate, 27 per cent of Oman and 87 per cent of Qatar's drinking water needs are reported to be provided via desalination.[72] The Middle East represents 90 per cent of the thermal energy used for desalination worldwide, led by the UAE and Saudi Arabia.[73] According to 2016 data, 67.2 per cent of the fuel input used for the region's water production from seawater desalination comes from fossil fuels and the fact that desalination is projected to account for 15 per cent of the region's total final energy consumption by 2040[74] demonstrates a critical example of the water-energy nexus in the context of urbanization.

Figure 5.6 illustrates that the highest pace with regard to the increase in urban population had already happened in the early 1990s for most MENA countries. Whether this increase was accompanied by sustainably growing cities is highly debatable given the current environmental records of these countries.

Urbanization is rapidly increasing in MENA countries. The forecasts estimate that Egypt and Iran will be among the top 20 countries worldwide with the

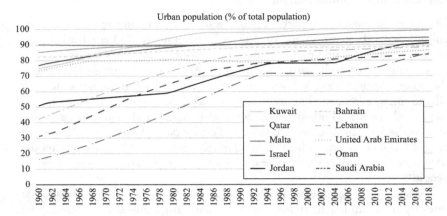

Figure 5.6 Top 10 MENA countries with highest urban population. Author's compilation (The World Bank, n 70).

largest projected urban population in 2050 with 85.3 and 80.4 million, respectively.[75] Therefore, how MENA's cities of today will be evolving into megacities of tomorrow with regard to sustainable development (especially climate resilience, resource consumption, and environmental protection), remains as a vital challenge.

Limited capacity and experience

Another important challenge in MENA is the limited availability of experience and technical know-how to apply new low-carbon technologies driving energy transition.[76] To illustrate, solar thermal cooling systems currently installed in Jordan are mainly planned, built, and monitored by companies or research institutions as merely scientific projects, decreasing the availability of this technology for household consumers.[77]

Moreover, under the transition scenario, jobs in the overall energy sector in MENA are projected to reach 7.3 million by 2050, 2.1 million of which will be coming from the renewables sector.[78] Although some of the MENA countries are preferred destinations for expats, indigenous expertise is still crucial for sustaining the transition strategies in the long term. Accordingly, educating local experts, both in technical and policymaking spheres is required.

Major disruptive crises

Major disruptive events, such as the global Coronavirus (COVID-19) pandemic, impose new challenges on modern systems. As of April 2021, the number of coronavirus cases reached more than 149 million people.[79] ILO revealed that due to the COVID-19 pandemic full or partial lockdown measures affected "almost 2.7 billion workers, representing around 81 percent of the world's workforce" and called the situation as "the most severe crisis since the Second World War".[80] The massive disruption caused by the coronavirus crisis has emphasized the role of electricity[81] for both human survival (i.e. for the operation of medical equipment in hospitals) and for the survival of our modern societies through internet, digital workspaces, communication, or online education under lockdown.

The long-term impacts of global crises, such as COVID-19 or the fall of oil prices, on the energy transition in MENA are yet to be seen. However, as such developments can disturb current trends and planned strategies, they will surely affect the resources committed to decarbonization as countries try to recover both financially and socially. The COVID-19 pandemic is likely to postpone the target years and project deadlines for renewables due to physical disruptions in supply chains and constructions under lockdown measures and due to economic pressure on public and private budgets which can hamper future investments.[82] The global energy investments are already projected to decline by 20 per cent in 2020.[83] Nevertheless, the structural changes in the energy systems and the shift to a low-carbon economy will still require long-term investments with high costs and technical specialties.[84] Since the COVID-19 pandemic is expected to redirect global investment patterns away from climate resilience projects,[85] it will severely challenge the states' financial and social capacity to cope with their ongoing climate vulnerabilities.

Advancing low carbon transitions in the MENA region: recommendations

The adoption of a holistic perspective toward energy transition as a climate change mitigation strategy allows for the emergence of several policy spheres which can complement the increase of renewable resources in energy supplies, and which can offer policy options to address the barriers and challenges identified in the previous section. This section offers a number of policy areas which can accelerate the comprehensive energy transition.

Optimizing the energy portfolios with low-carbon options

Diversification is one the fundamental strategies for energy transition in MENA countries since it addresses both their major economic activities and their energy portfolios. In the quest for replacing supplies from fossil fuels with renewable resources, several criteria should be evaluated while deciding which renewable resource is ideal for MENA countries. These mainly include: the evaluation of national geographical conditions, consideration of the intricate connection between energy systems and environmental (land, water, climate) conditions, and proper functioning of the economy. To illustrate, given the region's scarce water resources, investing in large-scale hydropower can create further pressure on water security.[86] Studies examining environmental and economic footprints of different resources reveal that wind, geothermal,[87] and solar photovoltaics (PV)[88] emerge as the most viable and most cost-competitive options for the region.

A recent study[89] finds that a 100 per cent renewable energy-based power system is 55 to 69 per cent cheaper than a business as usual scenario, in comparison to the cost of fossil-fuel dependence. While such a scenario is unrealistic given the role of oil and gas in MENA economies and in their socio-political structure, increasing the share of renewables offers great potential to curb MENA's carbon emissions, especially if concrete steps are taken such as reforming energy subsidies[90] or financing research and development in energy storage[91] and energy-saving technologies through hydrocarbon wealth. Moreover, since natural gas is identified as the most viable baseload resource to accompany intermittent renewable resources, natural gas production in the region can offer a comparative advantage to MENA[92] in transforming its image aligned with "dirty" fossil fuel into an accelerator of global decarbonization.

Building smart and sustainable cities

Cities emerge as one of the most important playgrounds for energy transition, since the pace of urbanization will further increase the demand for lighting, basic electrical appliances, and cooling systems. Air-conditioning should be a key sphere of innovation in MENA's energy transition agenda. Globally, the stock of air conditioners in buildings is already expected to increase from 1.6 billion in 2018 to 5.6 billion by 2050.[93] Global warming signals an increasing need for energy, to be used in air conditioning and cooling systems.[94] As an example, due to unbearable heat, Qatar has already begun to air-condition the outdoors[95] as part of its preparations for the 2022 World Cup. Similarly, Egypt's residential sector already uses 25 per cent of its energy consumption for space cooling and water heating.[96] Therefore, smart options for air conditioning that will offer energy-saving solutions are important. The incorporation of local renewable resources in the centralized cooling systems[97] emerges as a feasible option. In this regard, using solar energy in the cooling and heating of buildings is considered as a milestone in the quest for environment-friendly energy supply solutions.[98] Although renewables are growing very slowly in major energy-consuming sectors like buildings and industry,[99] MENA countries are eager to include more renewables in their buildings. To illustrate in its Voluntary National Review, Egypt indicates that 70 per cent of roof tops are to be covered by solar panels to increase the share of clean energy in consumption.[100] Similarly, Jordan is reported to be the first developing country to use solar thermal energy to cool buildings and to reduce cooling power consumption.[101]

It would be incomplete to consider the relationship between cities and energy transition only in reference to diversified energy consumption and supply. For MENA, the design of the cities, buildings, and the infrastructures should be as important as renewable targets. Studies reveal that the design of contemporary glass buildings and facades, which do not take into consideration climate conditions and the architectural identity[102] of these hot and dry countries, leads to an increased need for cooling and thus for more energy.[103]

Some progress is already on the way. Indeed, as the first country in GCC to implement energy conservation measures in air-conditioned buildings, Kuwait has been enforcing energy conservation measures since 1983.[104] In 2014, the Ministry of Municipal and Rural Development Affairs of Saudi Arabia also declared the installation of thermal insulation as mandatory for all new constructions, making thermal insulation a pre-requisite before electricity connection applications.[105] Back in the same year, the share of residential buildings with thermal insulation in Saudi Arabian cities was only 26.5 per cent on average and only 50 per cent of the homes in Riyadh had thermal insulation.[106] Since then, the market for thermal insulation has been considerably growing, and in 2021, the share of residential applications for thermal insulation is projected to account for 46.18 per cent of the market.[107]

With regard to environmentally responsible and resource-efficient designs of green and smart cities, LEED (Leadership in Energy and Environmental Design) certification offers an inspiring ground for future progress in the region. LEED is a globally recognized symbol of sustainability achievement through healthy, efficient, and cost-saving green buildings with reduced energy use and carbon emissions.[108] The National Museum of Qatar is a recent and remarkable example as the first national museum in the world which received a LEED Gold and a 4 Star GSAS (Global Sustainability Assessment System) sustainability rating.[109] Overall, the MENA region accounts for only 11.62 million of gross square meters (GSM) and 421 LEED certified projects. Compared to over 488.41 million GSM of certified LEED space and 36.886 certified projects in the US and Canada,[110] MENA countries' attempts to keep up with these standards remain minor. The criteria and philosophy under these "green" spaces hold many opportunities for MENA in reaching the goal of sustainable cities which can empower energy transition. Moreover, energy savings from such projects are also financially promising as between 2000 and 2016, worldwide energy savings derived from LEED-certified projects accumulated US$6 billion in energy savings from electricity.[111] While in Egypt alone, the replacement of 3.9 million streetlights with LED bulbs offered a saving of 872 million kWh of electricity,[112] the potential of a massive transformation of spaces and buildings in the region with green solutions is remarkable and not to be missed.

In 2017, UN-Habitat's New Urban Agenda declared that subnational and local efforts to apply affordable energy efficiency and renewables services in public buildings, infrastructure, and facilities must be promoted.[113] In this regard, the adoption of building performance codes and standards, and energy-efficiency labelling[114] should be encouraged. Moreover, "smart-grid, district energy systems and community energy plans to improve synergies between renewable energy and energy efficiency" must be prioritized in not only residential, commercial, and industrial buildings but also in other areas including waste management, sanitation, industry, and transport.[115]

To illustrate, by the end of 2020, car sales in MENA are projected to increase to 3.37 million from 2.35 million in 2013.[116] The fuel consumption and emissions from such an amount of cars is massive. It is necessary to direct the habits away from individual transportation towards low-carbon public transportation. Some substantial steps are present in MENA. In 2015, Morocco introduced Africa's first city bicycle hire scheme and extended its urban tram networks.[117] Similarly, Qatar's first-ever underground railway system opened to the public in 2019,[118] and by 2026, Qatar will be completing more than 60 stations.[119]

Additionally, the proper incorporation of renewable resources into the systems, the introduction of energy-efficient green and smart architecture into the cities, and the design and well-functioning of electricity grid rest on climate-resilient infrastructure. The electrification of the economy is doomed to fail unless relevant infrastructures with essential quality are present. The infrastructure quality in cities includes quality of roads, efficiency of transportation systems, electrification rate, power distribution, or reliability of water supply.[120] Ranking 12th with a score of 88.5 on a scale of 0–100, UAE is the only MENA country among top 20 countries according to the quality of infrastructure in 2019.[121] Without the establishment of quality infrastructure and/or upgrading of current systems, energy transition in MENA would remain slow and incomplete.

Overall, green buildings, low-carbon urban transportation, resilient infrastructures, and energy conservation and efficiency measures will be fundamental drivers of sustainable urbanization which constitute an important pillar of energy transition as well as healthy communities.

Development of effective digitalization strategies

Digitalization is another important pillar of energy transition. However, strategies that use digital technologies for energy efficiency[122] and consider digitalization as part of this transformation are rare. MENA region is no exception. With increasing electrification, digitalization should be a key tool for transformation in the power sector through optimization of complex power systems and management of demand-side energy efficiency.[123]

Digitalization and Internet of Things (IoT) can improve the flexibility and responsiveness of the grids. The internet penetration rate in the Middle East has increased from 28.8 per cent in 2009 to 67.2 per cent in 2019.[124] In Bahrain and in Qatar nearly the entire population (98 per cent and 94.29 per cent, respectively) has access to the internet.[125] In countries with high access to the internet, IoT can connect energy suppliers, consumers, and grid infrastructure, through bi-directional flow of data for demand-side management,[126] provided that data privacy and cybersecurity are ensured. Digitally connected buildings can communicate with the grid and offer consumption data to adjust the flexible load. Accordingly, the share of intermittent renewables can be further increased and the system can become more efficient and stable,[127] as the fluctuations in the residual load can be calculated more precisely. In this way, digitalization can reduce the global energy demand coming from buildings, by up to ten per cent between 2017 and 2040.[128] Similarly, in urban transport, digitally enabled innovative technologies, such as teleworking, shared mobility, or autonomous vehicles, can significantly lower CO_2 emissions in 2050 by more than 50 per cent.[129]

Additionally, smart technologies for buildings such as sensors or automated controls can further maximize energy efficiency for households. Table 5.1 offers examples from such technologies as identified by the International Energy Agency (IEA) and demonstrates their potential benefits.

Table 5.1 Selected technologies for buildings and their potential benefits on energy consumption[a]

Technology	Description	Potential Effect on Energy Consumption
Smart thermostat	Heating and cooling controlled remotely (or automatically) and temperature adjusted according to preferences or sensor inputs.	5–20% less energy use for heating or cooling.
Smart zoning	Individual rooms or zones heated or cooled to specific temperatures at specific times.	10% less energy use for heating or cooling.
Smart window control	Controls amount of light let through and can block heat or cold.	10–20% less energy use for heating or cooling.
Smart lighting (including occupancy control)	Remote control of lighting, automation, adjustment to occupancy.	1–10% less home energy use; 30–40% less lighting energy use.
Smart plugs	Turns unconnected products into connected devices.	1–5% less home energy use.
Home energy management system	Enhances control and automation of energy-using appliances and equipment.	8–20% less home energy use.

[a] The table is adapted from IEA Energy Efficiency Report, n 53, p. 86.

Improvement of investment and entrepreneurship environment

The promotion of public–private partnership in renewable energy generation and attracting foreign high-tech companies with special expertise in low-carbon technologies are fruitful strategies for MENA.[130] Consequently, a friendly investment environment, both in political and legal terms, is crucial. This is an area open for progress in MENA since, except UAE which ranks 16th among 190 countries, the ease of doing business scores in the region are not promising.[131] Nonetheless, this should not indicate relaxed regulations for, especially, environmental quality standards for foreign investments. Earlier trends where MENA has already attracted "dirty" foreign direct investment (FDI) flows in the context of fossil fuels, with negative effects on emissions, need to be replaced by green technology transfers.[132] Meanwhile, local talents and indigenous innovative projects should be equally encouraged by improving the business environment. Strikingly, statistics indicate that approximately 35.4 days are required to start a business in Kuwait, as opposed to 3.7 days in UAE,[133] which is also the top-rated country (24 per cent) in MENA by the Arab Youth to start their own business in the next five years.[134] Good practices for green entrepreneurship encouragement should be shared among MENA countries, to boost innovation and progress in energy transition.

Investing in human capital

Technical and economic dimensions of energy transition should be nourished with social, educational, and political planning. To be capable of fulfilling the expanding workforce in emerging sectors, with eligible professionals, MENA countries should invest heavily in their human capital so as to create their local occupational expertise, knowledge, and skills required by green technologies.

Major pre-requisites for the design, promotion, and implementation of climate-friendly energy policies include robust and coordinated institutions, clear and coherent legal framework governing alternative energy generation and supply,[135] as well as open-minded decision-makers ready to execute innovative climate solutions. Both the establishment and consolidation of this institutional framework and the development of relevant leadership skills for sustainability rest in education.

Cultural elements can influence the adoption of the technologies and practices,[136] such as electrical cars, digital efficiency measures, or recycling. Accordingly, behavioural and psychological barriers can affect successful implementation of climate change mitigation policies[137] at the demand-side of energy transition. Therefore, sustainable development education proves to be highly relevant in shaping what kind of new qualifications and attitudes are needed in today's and the future's citizens. Change in consumer behaviour, awareness, and readiness to transform consumption habits are parts of the human capital development which MENA needs.

Increasing regional coordination and cooperation

The lack of security and political stability in some MENA countries reveals that financial hurdles and technical limitations are not the only factors impeding energy transition. Consequently, benchmarking best practices and successful examples among regional partners that share similar climate vulnerabilities, and promoting solidarity to extend fruitful policies into disadvantaged populations of the region need to be prioritized. As a starting point, the SDGs can offer a common denominator to identify goals to work on in regional collaboration. Harmonization of transition policies at the national and regional level through increased communication and solidarity can also increase MENA's resilience against major disruptive crises. Cooperation rather than polarization should remain a major target that needs to be acknowledged by MENA leaders.

Conclusion

The business as usual needs massive transformation, as the effects of climate change are increasingly visible. Human security is under threat with disruptive crises, such as natural disasters or the Coronavirus pandemic, which is, for some, a "warning signal" from nature regarding man-made destruction of the environment, ecological balances, and biodiversity.[138] There is no magic solution to rapidly address the complex realities of MENA. However, energy transition can offer strategies to curb destructive ecological trends and increase the climate resilience of humans, nature, and economies. The best available solution for a sustainable future is a systemic transition to eco-friendly lifestyles and low-carbon economies.

This chapter demonstrated why an energy transition is required for MENA. By focusing on major playgrounds for energy transition, the study reviewed specific policy barriers and offered recommendations to meet MENA's potential in climate change mitigation from an energy perspective. The analysis revealed that although energy transition has taken its place in the national strategies and action plans of several MENA countries, for the targets to be achievable, concrete progress for transformation needs to be accelerated.

On the bright side, the MENA region has several policy areas to work on, which can potentially boost its climate change mitigation. Adopting a comprehensive approach to the notion of energy transition, MENA governments can further focus on sustainable cities with efficiency measures and digitalization, encourage indigenous talents and innovations through creating a green entrepreneurship environment, enhance regional cooperation, and invest in human capital with sustainable development education so as to change mind sets and consumption habits to transform the sustainability culture across generations.

Appendix 1: Country profiles[139]

Country Name	Renewable Share Target	Year	Total Renewable Energy (IRENA 2020) Capacity (MW)		Total Primary Energy Supply by Source, 2017 (%)						
			2010	2019	Natural Gas	Oil	Coal	Wind and Solar	Biofuels and waste	Hydro	Nuclear
Algeria	27% renewable electricity production	2030	253	686	64.6	34.9	0.4	0.1	0	0	0
Bahrain	5% renewable electricity generation	2020	1	7	86.0	14.0	0	0	0	0	0
Djibouti	87%–100% of renewable energy	2035	0	0	NA	NA	NA	NA	NA	NA	NA
Egypt	42% of renewable electricity mix	2035	3483	5972	51.1	45.1	0.4	0.3	2.0	1.2	0
Iran	10% of the country's energy demand	2021	8588	12933	66.6	31.4	0.5	0	0.2	0.5	0.7
Iraq	NA	NA	2274	2311	11.7	88.0	0	0	0.1	0.2	0
Israel	17% renewable electricity generation	2030	99	1500	36.7	39.7	21.1	2.3	0.2	0	0
Jordan	11% of renewable energy share in the total energy mix	2025	17	1401	37.9	56.8	1.8	3.1	0.4	0	0
Kuwait	15% electricity generation	2030	0	106	54.3	45.7	0	0	0	0	0
Lebanon	12% of energy demand	2020	282	321	0	95.7	1.9	0.6	1.5	0.3	0
Libya	7% of renewable electricity	2020	4	5	27.6	71.3	0	0	1.1	0	0
Malta	10% of final consumption	2020	1	158	0.0	91.7	0	5.0	3.3	0	0
Morocco	52% of installed electricity production capacity by renewables	2030	1560	3264	5.1	63.6	22.3	1.8	6.7	0.5	0
Oman	10% renewable electricity generation	2025	0	8	78.5	21.5	0	0	0	0	0

Country Name	Renewable Share Target	Year	Total Renewable Energy (IRENA 2020) Capacity (MW)		Total Primary Energy Supply by Source, 2017 (%)						
			2010	2019	Natural Gas	Oil	Coal	Wind and Solar	Biofuels and waste	Hydro	Nuclear
Palestine	10% renewable electricity generation	2020	0	43	NA	NA	NA	NA	NA	NA	NA
Qatar	20% (1.8 GW) of generation	2030	39	43	94.3	5.7	0	0	0	0	0
Saudi Arabia	54 GW of its installed capacity	2040	2	397	36.9	63.1	0	0	0	0	0
Syria	4.3% of primary energy demand	2030	857	1504	33.5	65.7	0	0	0.1	0.7	0
Tunisia	30% renewable electricity	2030	120	373	48.4	41.1	0	0.9	9.6	0	0
United Arab Emirates	44% of its power generation capacity from renewable resource	2050	11	1885	87.9	8.7	3.1	0.2	0.1	0	0
Yemen	15% of grid large- scale electricity generation mix (2600 GWh)	2025	1	250	15.9	76.0	2.4	1.9	3.7	0	0

Notes

1 V. Masson-Delmottee and others (eds), *Global Warming of 1.5°C: An IPCC Special Report on the Impacts of Global Warming of 1.5°C Above Pre-industrial Levels and Related Global Greenhouse Gas Emission Pathways, in the Context of Strengthening the Global Presence to the Threat of Climate Change Sustainable Development, and Efforts to Eradicate Poverty* (IPCC 2018).

2 M. A. Lange, "Impacts of Climate Change on the Eastern Mediterranean and the Middle East and North Africa Region and the Water–Energy Nexus" (2019) 10 *Atmosphere* 455; A. A. Kandeel, "In the Face of Climate Change: Challenges of Water Scarcity and Security in MENA" (Atlantic Council, 11 June 2019) <www.atlanticcouncil.org/blogs/menasource/in-the-   face-of-climate-change-challenges-of-water-scarcity-and-security-in-mena/>.

3 K. Tull, "The Projected Impacts of Climate Change on Food Security in the Middle East and North Africa" (UK Instiute of Development Studies, 21 February 2020).

4 J. D. Wilson, "A Securitisation Approach to International Energy Politics" (2019) 49 *Energy Research and Social Science* 114; B. San-Akca, S. D. Sever, and S. Yilmaz, "Does Natural Gas Fuel Civil War? Rethinking Energy Security, International Relations, and Fossil-Fuel Conflict" (2020) 70 *Energy Research and Social Science* 101690.

5 J. Sowers, "Understanding Climate Vulnerability in the Middle East and North Africa" (2019) 51 *International Journal of Middle East Studies* 621.

6 United Nations Framework Convention on Climate Change, *Adoption of the Paris Agreement*, 12 December 2015, UN Doc. FCCC/CP/2015/L.9.

7 International Energy Agency (IEA), "Climate Change" <www.iea.org/topics/climate-change 2020>.

8 European Environment Agency, "Energy and Climate Change", 10 December 2019 <www.eea.europa.eu/signals/signals2017/articles/energyandclimatechange/#:~:text=The%20EU%20ETS%20sets%20a,countries%20(%5B1%5D).&text=The%20European%20Environment%20Agency%20monitors,covered%20by%20the%20EU%20ETS>.

9 M. Stephenson, *Energy and Climate Change: An Introduction to Geological Controls, Interventions and Mitigations* (Elsevier 2018) 175–178.

10 International Renewable Energy Agency (IRENA), "Energy Transition" <www.irena.org/energytransition>.

11 V. Masson-Delmottee and others (eds), n 1.

12 J. Rogelj, G. Luderer, R. C. Pietzcker, E. Kriegler, M. Schaeffer, V. Krey, and K. Riahi, "Energy System Transformations for Limiting End-of-Century Warming to below 1.5°C" (2015) 5 *Nature Climate Change* 519 at 523.

13 F. Johnsson, J. Kjärstad, and J. Rootzén, "The Threat to Climate Change Mitigation Posed by the Abundance of Fossil Fuels" (2019) 19 *Climate Policy* 258.

14 RED Electrica De Espana, "Energy Transition and Climate Change" <www.ree.es/en/sustainability/decarbonisation-of-the-economy/energy-transition-and-climate-change>.

15 IRENA, "Climate Change and Renewable Energy," June 2019 <www.irena.org/publications/2019/Jun/Climate-change-and-renewable-energy>.

16 BP, "Energy Outlook – 2019: Insights from the Evolving Transition Scenario: Middle East," 2019, www.bp.com/content/dam/bp/business-sites/en/global/corporate/pdfs/energy-economics/energy-outlook/bp-energy-outlook-2019-region-insight-middle-east.pdf.

17 All MENA countries also ratified or accepted the agreement except Iran, Iraq, Libya and Yemen. See: United Nations Treaty Collection, "Paris Agreement Ratification Status as of 25-08-2020" <https://treaties.un.org/Pages/ViewDetails.aspx?src=TREATY&mtdsg_no=XXVII-7-d&chapter=27&clang=_en>.

18 UNFCCC, "INDCs as Communicated by Parties" <www4.unfccc.int/sites/submissions/INDC/Submission%20Pages/submissions.aspx>.

19 D. S. Olawuyi, "Can MENA Extractive Industries Support the Global Energy Transition? Current Opportunities and Future Directions" (2020) *Extractive Industries and Society*, 100685.

20 S. D. Sever, M. E. Tok, and C. D'Alessandro, "Global Environmental Governance and the GCC: Setting the Agenda for Climate Change and Energy Security" in L. Pal and M. E. Tok (eds), *Global Governance and Muslim Organizations* (London: Palgrave Macmillan 2019).

21 S. Tagliapietra, "The Impact of the Global Energy Transition on MENA Oil and Gas Producers" (2019) 26 *Energy Strategy Reviews* 100397.

22 General Secretariat For Development Planning, "Qatar National Vision 2030," July 2008 <www.psa.gov.qa/en/qnv1/Documents/QNV2030_English_v2.pdf>.

23 Oman Supreme Council for Planning, "Oman Vision 2040" <www.2040.om/>.

24 UAE, "UAE Vision 2021" <www.vision2021.ae/en/uae-vision>.

25 Kingdom of Saudi Arabia, "Saudi Arabia Vision 2030" <https://vision2030.gov.sa/en>.

26 United Nations, Department of Economic and Social Affairs, Sustainable Development, "The 17 Goals" <https://sdgs.un.org/goals>.

27 Ministry of Foreign Affairs, State of Kuwait, "Kuwait Vision 2015 'New Kuwait'" <www.mofa.gov.kw/en/kuwait-state/kuwait-vision-2035/>.

28 Arab Republic of Egypt, Ministry of Communications and Information Technology, "Egypt Vision 2030" <http://mcit.gov.eg/Publication/Publication_Summary/1020/>.

29 Kingdom of Morocco, Ministry Delegate of the Minister of Energy, Mines, Water and Environment, "Moroccan Climate Change Policy", March 2014 <www.4c.ma/medias/MCCP%20-%20Moroccan%20Climate%20Change%20Policy.pdf>.

30 UAE Ministry of Climate and Environment, "National Climate Change Plan of the UAE 2017-2050" <https://u.ae/en/about-the-uae/strategies-initiatives-and-awards/federal-governments-strategiesandplans/nationalclimatechangeplanoftheuae#:~:text=Objectives,a%20better%20quality%20of%20life>.

31 D. Olawuyi, "Advancing Innovations in Renewable Energy Technologies as Alternatives to Fossil Fuel Use in the Middle East: Trends, Limitations, and Ways Forward" in D. Zillman, L. Godden, L. Paddock, and M. Roggenkamp (eds), *Innovation in Energy Law and Technology: Dynamic Solutions for Energy Transitions* (Oxford: Oxford University Press 2018).

32 IRENA, "Global Renewables Outlook: Energy Transformation 2050", April 2020 <www.irena.org/publications/2020/Apr/Global-Renewables-Outlook-2020>; Tagliapietra, n 21.

33 M. Walton, "Desalinated Water Affects the Energy Equation in the Middle East", January 21, 2019 <www.iea.org/commentaries/desalinated-water-affects-the-energy-equation-in-the-middle-east>.

34 Arab Republic of Egypt Ministry of Planning, Monitoring and Administrative Reform, "Egypt's Voluntary National Review 2018" <www.arabdevelopmentportal.com/sites/default/files/publication/vnr-egypt-2018.pdf>.

35 C. Hicks, "Morocco Lights the Way for Africa on Renewable Energy," November 17, 2016 <www.theguardian.com/global-development/2016/nov/17/cop22-host-morocco-lights-way-africa-renewableenergy2020#:~:text=Morocco%20has%20no%20fossil%20fuel,around%2010%20gigawatts%20(GW)>.

36 IRENA, "Renewable Capacity Statistics 2020," March 2020 <www.irena.org/publications/2020/Mar/Renewable-Capacity-Statistics-2020>.

37 IEA, "Total Primary Energy Supply by Fuel, 1971 and 2018," 31 July 2020 <www.iea.org/data-and-statistics/charts/total-primary-energy-supply-by-fuel-1971-and-2018>.

38 Z. A. Wendling, J. W. Emerson, A. de Sherbinin, D. C. Etsy and others, "2020 Environmental Performance Index" <epi.yale.edu>.

39 Ibid.

40 The 2020 EPI provides a data-driven summary of the state of sustainability around the world by using 32 performance indicators across 11 issue categories. 180 countries

are ranked based on their scores out of 100, in order to evaluate leaders and laggards in environmental progress (ibid).

41　Ibid.
42　S. Borghesi and E. Ticci, "Climate Change in the MENA Region: Environmental Risks, Socioeconomic Effects and Policy Challenges for the Future" (2019) *IEMed. Mediterranean Yearbook 2019, 292.*
43　Djibouti (60.2 per cent), Libya (70 per cent), Syria (89.6 per cent) and Yemen (79.2 per cent) are still in need of improvement for access to electricity. See: The World Bank, "Access to Electricity (% of Population)" <https://data.worldbank.org/ind icator/EG.ELC.ACCS.ZS>.
44　W. Avis, "The Use of Fossil Fuels in the Middle East and North Africa," K4D Help-desk Report, University of Birmingham, February 21, 2020 <https://assets.publishing .service.gov.uk/media/5e9d713886650c031977ae65/763_Fossil_Fuel_Use_in_the _MENA_Region.pdf>.
45　IEA, "Outlook for Producer Economies," October 2018 <www.iea.org/reports/outlo ok-for-producer-economies>.
46　Tagliapietra, n 21.
47　M. Shehabi, "Diversification Effects of Energy Subsidy Reform in Oil Exporters: Il-lustrations from Kuwait" (2020) 138 *Energy Policy* 110966.
48　Tull, n 3.
49　Sever and others, n 20.
50　Sowers, n 5.
51　S. D. Sever, "Accelerating the Energy Transition in the Southern Mediterranean," Edito Energie, French Institute of International Relations (IFRI), September 16, 2019　<www.ifri.org/en/publications/editoriaux-de-lifri/edito-energie/accelerating -energy-transition-southern-mediterranean>.
52　Bloomberg, UNEP, and FS-UNEP Collaborating Centre, "Global Trends in Renew-able Energy Investment 2019" (UNEP, 2019) <https://wedocs.unep.org/bitstream/h andle/20.500.11822/29752/GTR2019.pdf> 32.
53　IEA, "Energy Efficiency 2019," November 2019 <www.iea.org/reports/energy-efficien cy-2019>.
54　Sever, n 51.
55　World Bank Group, "Delivering Energy Efficiency in the Middle East and North Af-rica: Achieving Energy Efficiency Potential in the Industry, Services and Residential Sectors," May 6, 2016 <https://documents.worldbank.org/en/publication/documents -reports/documentdetail/642001476342367832/delivering-energy-efficiency-in-the-middle-east-and-north-africa-achieving-energy-efficiency-potential-in-the-industry -services-and-residential-sectors>.
56　IRENA, n 32.
57　The World Bank, "Energy Intensity Level of Primary Energy (MJ/$2011 PPP GDP)" <https://data.worldbank.org/indicator/EG.EGY.PRIM.PP.KD>.
58　BP, n 16.
59　IEA, "World Energy Outlook 2019," November 2019 <www.iea.org/reports/world -energy-outlook-2019>.
60　IEA, n 53.
61　Ibid.
62　IRENA, n 32.
63　Tull, n 3.
64　M. Cherlet, C. Hutchinson, J. Reynolds, J. Hill, S. Sommer, and G. Von Maltitz (eds), *World Atlas of Desertification: Rethinking Land Degradation and Sustainable Land Management* (Luxembourg: Publications Office of the European Union, 2018).
65　United Nations Framework Convention on Climate Chance, *Initiatives in the Area of Human Settlements and Adaptation,* 25 April 2017, UN Doc. FCCC/SBSTA/2017/ INF.3.

66 United Nations, "The World's Cities in 2018" (2018) <www.un.org/en/events/citi esday/assets/pdf/the_worlds_cities_in_2018_data_booklet.pdf> 9.

67 M. R. Montgomery, R. Stern, B. Cohen, and H. E. Reed, *Cities Transformed: Demographic Change and Its Implications in the Developing World* (Abingdon, UK: Routledge 2013).

68 United Nations Department of Economic and Social Affairs (UNDESA), "2018 Revision of World Urbanization Prospects," May 16, 2019 <www.un.org/development/desa/publications/2018-revision-of-world-urbanization-prospects.html>.

69 Intergovernmental Panel on Climate Change, "Human Settlements, Infrastructure and Spatial Planning" in Ottmar Edenhofer and others (eds), *Climate Change 2014: Mitigation of Climate Change Working Group III Contribution to the Fifth Assessment Report of the Intergovernmental Panel on Climate Change* (Cambridge University Press 2014) 923–1000; United Nations Human Settlements Programme (UN-Habitat) "Hot Cities: Battle-Ground for Climate Change" (2011) <http://mirror.unhabitat.org/downloads/docs/E_Hot_Cities.pdf>.

70 The World Bank, "Urban Population (% of Total Population)" <https://data.worldbank.org/indicator/SP.URB.TOTL.IN.ZS>.

71 E. Borgomeo, and others, *The Water-Energy-Food Nexus in the Middle East and North Africa: Scenarios for a Sustainable Future* (Washington, DC: The World Bank Group 2018).

72 IRENA, "Renewable Energy Market Analysis: GCC 2019" <www.irena.org/publications/2019/Jan/Renewable-Energy-Market-Analysis-GCC-2019>.

73 Walton, n 33.

74 Ibid.

75 UNDESA, n 68.

76 M. Jaradat, M. Al-Addous, A. Albatayneh, Z. Dalala, and N. Barbana, "Potential Study of Solar Thermal Cooling in Sub-Mediterranean Climate" (2020) 10 *Applied Sciences* 2418; Olawuyi, n 31.

77 Ibid.

78 IRENA, n 32, 44, 47.

79 Worldometer, "Coronavirus," April 2021 <www.worldometers.info/coronavirus/>.

80 ILO, "ILO Monitor: COVID-19 and the World of Work," June 30, 2020 <www.ilo.org/global/topics/coronavirus/impacts-and-responses/WCMS_749399/lang--en/index.htm>.

81 F. Birol, "The Coronavirus Crisis Reminds Us that Electricity Is More Indispensable than Ever," March 22, 2020 <www.iea.org/commentaries/the-coronavirus-crisis-reminds-us-that-electricity-is-more-indispensable-than-ever>.

82 H. Bahar, "The Coronavirus Pandemic Could Derail Renewable Energy's Progress. Governments Can Help," April 4, 2020 <www.iea.org/commentaries/the-coronavirus-pandemic-could-derail-renewable-energy-s-progress-governments-can-help>.

83 IEA, "Sustainable Recovery: World Energy Outlook Special Report," June 2020 <www.iea.org/reports/sustainable-recovery>.

84 A. Alhamwi, S. Weitemeyer, T. Vogt, and D. Kleinhans, "Moroccan National Energy Strategy Reviewed from a Meteorological Perspective" (2015) 6 *Energy Strategy Reviews* 39.

85 C. Flavelle, "Here's How Coronavirus Could Raise Cities' Risk for Climate Disasters," April 24, 2020 <www.nytimes.com/2020/04/23/climate/coronavirus-cities-infrastructure-money.html>.

86 M. Mahlooji, L. Gaudard, B. Ristic, and K. Madani, "The Importance of Considering Resource Availability Restrictions in Energy Planning: What Is the Footprint of Electricity Generation in the Middle East and North Africa (MENA)?" (2020) 717 *Science of the Total Environment* 135035.

87 Ibid.

88 A. Aghahosseini, D. Bogdanov, and C. Breyer, "Towards Sustainable Development in the MENA Region: Analysing the Feasibility of a 100% Renewable Electricity System in 2030" (2020) 28 *Energy Strategy Reviews* 100466.

89 Ibid.

90 Sever, n 51.

91 Alhamwi et al., n 84.

92 Olawuyi, n 19.

93 IEA, *The Future of Cooling: Opportunities for Energy-Efficient Air Conditioning* (Paris, France: OECD Publishing 2018).

94 Jaradat and others, n 76.

95 S. Mufson, "Facing Unbearable Heat, Qatar Has Begun to Air-Condition the Outdoors," October 16, 2019 <www.washingtonpost.com/graphics/2019/world/climate-environment/climate-change-qatar-air-conditioning-outdoors/>.

96 P. Makumbe, "Arab Republic of Egypt: Egypt Energy Efficiency Implementation," June 15, 2017 <https://documents.worldbank.org/en/publication/documentsreports/documentdetail/578631498760292189/energy-efficiency-and-rooftop-solar-pv-opportunities-report-summary>.

97 M. Galindo Fernández, C. Roger-Lacan, U. Gahrs, and V. Aumaitre, "Efficient District Heating and Cooling Systems in the EU: Case Studies Analysis, Replicable Key Success Factors and Potential Policy Implications", European Union, 2016 <https://ec.europa.eu/jrc/en/publication/efficient-district-heating-and-cooling-markets-eu-case-studies-analysis-replicable-key-success>.

98 Jaradat and others, n 76.

99 IRENA, n 32.

100 Arab Republic of Egypt Ministry of Planning, n 34.

101 Jaradat and others, n 76.

102 F. Visser and A. Yeretzian, "Energy Efficient Building Guidelines for MENA Region: Energy Efficiency in the Construction Sector in the Mediterranean," MED-ENEC, Cairo, November 2013 <www.climamed.eu/wp-content/uploads/files/Energy-Efficient-Building_Guideline-for-MENA-Region-NOV2014.pdf>.

103 M. Ahmed Alaa El Din Ahmed and M. Anwar Fikry, "Impact of Glass Facades on Internal Environment of Buildings in Hot Arid Zone" (2019) 58 *Alexandria Engineering Journal* 1063.

104 H. Doukas, K. D. Patlitzianas, A. G. Kagiannas, and J. Psarras, "Renewable Energy Sources and Rationale Use of Energy Development in the Countries of GCC: Myth or Reality?" (2006) 31 *Renewable Energy* 755 at 762.

105 I. Mohammed, "All New Buildings Must Have Thermal Insulation," March 24, 2014 <www.arabnews.com/news/545031>.

106 HVACR Expo Saudi and TechSci Research, "Saudi Arabia HVAC-R Market Outlook, 2021" <www.caba.org/wp-content/uploads/2020/04/IS-2017-278.pdf> 15.

107 Ibid. Commercial applications correspond to 31.82 per cent, while industrial applications account for 22 per cent of the market.

108 Please see <www.usgbc.org/press/benefits-of-green-building>.

109 The Peninsula, "NMoQ Gets LEED Gold, GSAS 4 Star Sustainability Ratings," December 5, 2018 <www.thepeninsulaqatar.com/article/05/12/2018/NMoQ-gets-LEED-Gold,-GSAS-4-Star-Sustainability-ratings>.

110 S. Stanley, "The 2018 Top 10 Countries and Regions for LEED Outside US," February 13, 2019 <www.usgbc.org/articles/infographic-2018-top-10-countries-and-regions-leed>.

111 P. MacNaughton, J. J. Buonocore, X. Cao, and J. G. Cedeno Laurent, "Energy Savings, Emission Reductions, and Health Co-Benefits of the Green Building Movement" (2018) 28 *Journal of Exposure Science and Environmental Epidemiology* 307.

112 Arab Republic of Egypt Ministry of Planning, n 34.
113 United Nations Human Settlements Programme (UN-Habitat), "New Urban Agenda" 2017, Article 121.
114 Ibid.
115 Ibid.
116 A. Puri-Mirza, "Volume of Passenger Car Sales in the Middle East in 2018 and 2024 (in Million Units)," Augusy 26, 2020 <www.statista.com/statistics/1024614/middle eastpassengercarsales/#:~:text=In%202018%2C%20the%20volume%20of,2024%20in%20the%20Middle%20East>.
117 Hicks, n 35.
118 S. Aziz, "Qatar Rolls Out First-Ever 'landmark' Metro for Public," May 8, 2019 <www .aljazeera.com/news/2019/05/qatar-rolls-landmark-metro-public-190508104457971 .html>.
119 Ibid.
120 World Economic Forum, "The Global Competitiveness Report 2019," October 8, 2019 <www.weforum.org/reports/how-to-end-a-decade-of-lost-productivity-growth>.
121 Ibid. Other MENA countries with their worldwide rankings and scores are as the following: Israel (23rd, 83), Qatar (24th, 81.6), Oman (28th, 80.5), Bahrain (31st, 78.4), Saudi Arabia (34th , 78.1), Malta (47th, 75), Egypt (52nd, 73.1), Morocco (53rd, 72.6), Kuwait (66th, 68.4), Jordan (74th, 67.4), Iran (80th, 64.8), Algeria (82nd, 63.8), Tunisia (85th, 62.7), Lebanon (89th, 61).
122 IEA, n 53.
123 Ibid; IRENA, n 32.
124 Internet World Stats, "Internet Penetration Rate" <www.internetworldstats.com/ stats.htm>.
125 Statista, "Internet Penetration Rate of Selected Countries in the Middle East in 2016," August 26, 2020 <www.statista.com/statistics/265836/internet-penetration -in-middle-eastern-countries/>.
126 IRENA, n 32.
127 IEA, n 53.
128 IRENA, n 32.
129 IEA, n 53.
130 Olawuyi, n 31; Shehabi, n 47.
131 World Bank Group, "Doing Business 2020: Arab World" (2020) 4.
132 M. Shahbaz, D. Balsalobre-Lorente, and A. Sinha, "Foreign Direct Investment–CO_2 Emissions Nexus in Middle East and North African Countries: Importance of Biomass Energy Consumption" (2019) 217 *Journal of Cleaner Production* 603.
133 In Saudi Arabia 17.8; Bahrain 8.3; Qatar 7.7; Oman 6.3 days are required (The World Bank, "Time Required to Start a Business (Days)" <https://data.worldbank.or g/indicator/IC.REG.DURS>).
134 A. Burson-Marsteller, "Inside the Hearts and Minds of Arab Youth," 8th Annual Arab Youth Survey, 2016, 33.
135 Olawuyi, n 31.
136 B. K. Sovacool and S. Griffiths, "Culture and Low-Carbon Energy Transitions" (2020) 3 *Nature Sustainability*, 685–693
137 I. Alebaite, D. Streimikiene, and T. Balezentis, "Climate Change Mitigation in Households between Market Failures and Psychological Barriers" (2020) 13 *Energies* 1.
138 D. Carrington, "Coronavirus: 'Nature Is Sending Us a Message', Says UN Environment Chief" *The Guardian* (London, 25 March 2020).
139 Compiled by the author. Data Sources: (renewable targets) Mahlooji and others, n 86; IRENA, n 36; IEA, n 37.

Part II
Case studies

6 Enhancing the effectiveness of national and regional institutions in addressing climate change in the MENA region

Lessons From Morocco and the Maghreb

Riyad Fakhri, Laila Dalaa, and Saad Belkasmi

Introduction

This chapter examines the regulatory approaches employed to enhance the efficiency of national and regional institutions on climate change in the MENA region, especially in Morocco and Maghreb countries. The climate change issue is, undoubtedly, one of the key global impediments to development. In the MENA region, climate change is already affecting people's livelihoods and their well-being.[1] As a barrier to economic growth, climate change may also reverse many of the advancements that have been achieved during the last decades.[2] Climate fluctuations and changes threaten development by limiting human potential and robbing power from people and societies.[3] If immediate emergency actions are not undertaken to mitigate and adapt to climate change impacts, the situation will continue to deteriorate. For example, Maghreb countries may experience very high temperature levels, and the rainfall will decrease exponentially, leading to water scarcity throughout the whole region.[4] This situation is compounded by the population growth in the region; people will be unable to irrigate crops, support industry, or even provide drinking water.[5]

For instance, the cost of environmental degradation to Moroccan society has been estimated in 2014 at about DH 32.5 billion or 3.52 per cent of GDP. In addition, greenhouse gas (GHG) emissions cause damage to the global community estimated at 1.62 per cent of GDP in 2014. The impacts of environmental degradation at the national level (national costs) appear to be more than twice as high as those at the global level (global costs).[6] In response, Morocco has engaged in a wide consultative process with the involvement of all its institutions. The objective of this initiative is to review the policies and programmes that the country has in place to fight climate change and bring them into compliance with its commitments at the national, regional, and international levels.[7]

A key question arises relating to the efficiency of national and regional institutions to deal with climate change impacts and advance environmental protection. Strong and effective institutions are required to effectively implement and monitor climate change regulations and programmes. Drawing lessons from

Morocco, this chapter examines key challenges facing institutional efficiency in the supervision of climate change programmes in MENA countries.

This chapter is divided into five sections. After this introduction, the second section discusses the importance of institutional efficiency in addressing climate change in the MENA region. The third section surveys official or governmental institutions devoted to protecting the environment. It discusses key challenges facing institutions in the MENA region, especially the issue of diversity and overlapping competencies. The fourth section discusses key reforms and steps to enhance the effective supervision of climate change regulation and programmes in the MENA region.

Importance of institutional efficiency in addressing climate change in the MENA region

Climate change is one of the key challenges facing our world, this century. Populations, especially in developing countries such as those of North Africa, must adapt to climate change impacts.[8] Climate change affects disadvantaged social groups disproportionately more.[9] As the central influence on how different social groups gain access to, and are able to use, assets and resources,[10] local institutions must strengthen their processes and adaptability. "Institutions influence adaptation and climate vulnerability in three critical ways: a) they structure impacts and vulnerability; b) they mediate between individual and collective responses to climate impacts, and thereby shape outcomes of adaptation; and c) they act as the means of delivery of external resources to facilitate adaptation, and thus govern access to such resources."[11] In order to facilitate effective climate adaptation, local institutions require greater support for partnering with other private and civil society institutions, enhanced local capacities, improved institutional coordination, recognition of vulnerabilities and capacities when developing strategies, and the adoption of an adaptive perspective while developing their action plans.[12] To effectively deal with climate change impacts, MENA countries will need strong institutions that are capable of applying and implementing national policies on climate change.

Building effective climate change institutions can have significant impacts on how climate change policies, rules, standards, institutions, procedures, and financing mechanisms are applied and implemented. Furthermore, it could influence how decisions are made and implemented, the scientific information needed for decision-making, how the public and key stakeholders can participate in decision-making, the type of information that should be available, and how processes and systems are referenced.[13] The concept of accountability must be at the center of these actions, as is provided in the Moroccan constitution.[14] That said, any institution concerned with environmental affairs must act fairly in the coverage of national territory, guarantee the continuity in the performance of services, the good quality in services, transparency, accountability, and responsibility.[15] The implementation of public policies, national strategies, programmes, and plans concerned with the environment by the state or others must be subject to a set of principles agreed upon nationally and internationally, such as the principle of territorial integrity, the principle of solidarity, the precautionary principle, the principle of prevention, the principle of responsibility, and participation.[16]

Governmental institutions devoted to protecting the environment: the issue of diversity and overlapping competencies

Morocco has long been placing environmental issues at the heart of its national and governmental programmes.[17] For this purpose, many laws have been enacted, and several governmental institutions have been established.[18] His Majesty King Mohammed VI affirmed, "From here, we call on the government to determine the major directions of a broad dialogue concerning a comprehensive national charter for environmental protection and sustainable development, formulated in a plan of integrated action with goals that are achievable in all sectors".[19] Although efforts have been made by Moroccan officials, many practical problems still impede the institutions' effectiveness in dealing with climate change impacts. There are two main issues that emerge: the multiplicity of government institutions involved in environmental protection and the overlap of their competencies.

The multiplicity of government institutions involved in environmental protection

Morocco has established numerous ministries to deal efficiently with environmental issues, such as the Ministry of Energy, Mining and Environment,[20] the Ministry of Equipment, Transport, Logistics and Water,[21] as well as the Ministry of Agriculture, Fisheries, Rural Development, Water and Forests.[22] Alternatively, we find many local institutions, such as the High Commission for Water and Forests and Fight against Desertification, which aim to protect nature and preserve biodiversity. It plays the role of administrative police for the protection of the forest area, parks, and nature reserves.[23] Additionally, the National Environment Council is in charge of research relating to environmental issues. It can also issue guidelines and regulations on development projects. It contributes to spreading awareness on environmental issues through public sensitization programmes and educates the public on the importance of preserving the environment and all its components.[24] Similarly, the Supreme Council for Water and Climate is an advisory body that educates public authorities on the national policy where it relates to water resources management and weather and climate monitoring.[25] The Economic, Social and Environmental Council is a constitutional body that brings together all political, social, economic, and legal stakeholders, as well as representatives of civil society and the private sector. The purpose of this Council is to follow public policies, as well as to develop proposals that could contribute to sustainable development at the national and regional levels.[26]

In the same vein, the Kingdom of Morocco has launched the National Program for Sustainable Development, which is based on a set of strategies:

a. Strengthening the organization and effectiveness of the environment state department both at the national and territorial levels;
b. Encouraging the transition to a green economy;
c. Enhancing environmental protection and improving the living conditions of citizens;
d. Establishing environmental governance and sustainable development;
e. Strengthening monitoring, vigilance, prevention, and planning in the field of environment and sustainable development;
f. Mobilizing key actors and promoting sustainable development principles;
g. Ensuring the implementation of the national policy in the field of combating the climate change;
h. Ensuring the implementation of the national policy in the field of biodiversity conservation.[27]

The institutional pluralism is evidence of the great interest that the Kingdom of Morocco shows toward environment issues, and its awareness of the necessity to preserve all the components of the environment. However, the main challenge remains how to create strong synergies between all the stakeholders dealing with environmental policy.[28]

Similarly in Tunisia, there are numerous ministries dealing with the environment, such as the Ministry of Environment,[29] the Ministry of Agriculture, Fisheries and Water Resources,[30] the Ministry of Industry, Energy, and Small and Medium-Sized Enterprises,[31] in addition to a range of public institutions, such as the National Agency for the Protection of the Ocean,[32] the Agency for the Protection and Preparation of the Coastal Strip,[33] the Tunis International Center for Environmental Technology,[34] and the National Agency for Health and Environmental Control of Products.[35] Other local institutions concerned with the protection of environment, as well as some advisory and technical structures, include the National Commission for Sustainable Development,[36] the High Council for Energy,[37] as well as the National Agency for the Protection of the Environment.[38] In addition, there are several institutions which form a basic support of national strategies, programmes, and policies dealing with environmental components protection, such as the Ministry of Economy and Finance,[39] the Ministry of Interior,[40] the Ministry of Equipment, Housing and Spatial Planning,[41] and the Ministry of Higher Education and Scientific Research.[42] The same approach is followed by Mauritania,[43] Algeria,[44] and Libya, taking into account the political and economic conditions of each country.

The competencies of governmental institutions dealing with environmental protection: interference or complementarity

The multiplicity of institutions concerned with the environmental field intertwines the issue of environmental protection among many sets of programmes and policies adopted by several government agencies. A clear example of this multiplicity is the many institutions which include water resource management as a part of their portfolios. These include the National Water Program, the National Program for Liquid Purification,[45] and the National Program for Household Waste Management, which is important for the expansion of the national water network and monitoring the quality of water and reducing gas emissions.[46] This programme is monitored by the Ministry of Energy, Minerals and the Environment, in partnership with a group of actors, such as the Ministry of Economy and Finance, the Secretary of State in charge of Sustainable Development, the Ministry of Interior, the Ministry of National Education, Vocational Training, Higher Education and Scientific Research, as well as the Ministry of Equipment, Transportation, Logistics and Water.[47]

Financial mechanisms have also been established to deal with various projects, namely sustainable development projects. One of these mechanisms is the use of public funds supported by tax payments. For instance, the Energy Development Fund (FDE),[48] the Industrial Depollution Fund (FODEP),[49] and the National Fund for the Environment (FNE).[50] For Tunisia, its Ministry of Agriculture has been responsible, since its inception, for ensuring the sustainable development of the country and has made efforts to consolidate and support the environmental culture among Tunisian people.[51] After the revolution of 14 January 2011, it became necessary to follow a new approach in the social field in order to provide the appropriate conditions for sustainable and equitable development, on the basis of the values of solidarity between Tunisians. The main goal is to ensure the sustainability of economic growth, taking account of local, natural, and human resources, without putting additional pressure on the fragile natural resources. Accordingly, the work of these governmental institutions, despite their diversity and differences, raises many questions about the coherence, efficiency, and effectiveness of their programmes and actions.[52]

Improving the effectiveness of climate change institutions

This section discusses legal and policy frameworks for addressing concerns of regulatory overlap, lack of coordination, and inefficiency that have been identified in the implementation of climate change law and policies in North African countries.

Effective coordination between institutions

There is a need to create mechanisms for coordination between the various institutions involved in protecting the environment, nationally and regionally. One approach is to create an inter-ministerial council on climate change, involving representatives of all the various departments, ministries, and agencies that have roles to play in climate change mitigation and adaptation. This council could provide a platform for the relevant ministries and agencies to share experience, resources, and work towards improving their consistency and complementarity. It is also important to establish regional cooperation mechanisms across the region to deal with the challenges of climate change. Regional knowledge platforms could provide opportunities for cooperative efforts, as well as exchanging best practices to reduce the impact of climate change in MENA countries and enhance existing knowledge.

Improving legal and institutional frameworks

To promote coordination and efficiency, the underlying legal frameworks establishing the various departments, ministries, and agencies on climate change will need to be clarified and strengthened.[53] For example, there is a need for legislative provisions that clarify the roles of the respective agencies to avoid role duplication and overlap.[54] A legal framework is required that can ensure greater institutional coordination amongst multiple institutions with environmental supervision functions. Improving the quality of legal texts governing these institutions would enhance greater efficiency in their programmes and actions. Furthermore, legal texts need to be amended so that they can abide by international and regional agreements and obligations on climate change.

Resource and information exchange

In addition to clarifying the roles of climate institutions and agencies, there is a need to improve the exchange of information and communication amongst all relevant institutions. Communication exchange must be improved, through the integration of communication platforms, to make information on climate change affordable and accessible to all relevant institutions. Furthermore, many solutions must be developed, such as early warning and monitoring systems, as well as disaster risk reduction mechanisms.

Regional cooperation

As discussed earlier, climate change mitigation and adaptation efforts can be improved across the MENA region through regional cooperation. Given the interconnectedness of the economies and institutions of countries in the region, failure to properly combat climate change can have transboundary implications. MENA countries will, therefore, need to engage in mutually beneficial and transparent discussion about the future of the environment and all its components. There is also a need for the sharing of finance, knowledge, technologies, and resources, including the cancellation of financial debts by countries in the region to support poorer countries in the North African axis. Furthermore, regional knowledge exchange platforms and institutions should also be established to help provide training, resources, and best practices on high-leverage climate change mitigation and adaptation across the region.[55]

Conclusion

Despite the efforts made, issues relating to climate change and sustainable development still pose major challenges for the global community, especially for

Morocco and other MENA countries.[56] To deal efficiently with these problems, many actions have to be undertaken in order to enhance the efficiency of the local institutions with their crucial role in addressing climate change impacts. Hence, any government programme dealing with climate change impacts will be doomed to failure, in the absence of strong and effective institutions that are capable of its implementation in the field. All stakeholders, including governmental authorities, the private sector, as well as civil society, must be encouraged to contribute. The nature and severity of the challenges require more effort in terms of regional cooperation, knowledge exchange, and provision of financial resources and support to the poorer and more technologically challenged countries in the region. Morocco, like other MENA countries, remains engaged in a new dynamic to provide appropriate solutions to the multiple and complex challenges and promote South-South triangular cooperation to maximize the benefit from opportunities and exchange experiences between countries.[57]

In line with the above, many research avenues remain to be pursued, such as further inquiry into the efficiency and the roles of institutions, how stakeholders could develop more synergies between themselves and the institutions, and new tools to enhance institutional contributions to climate change adaptability. While institutions may not be able to stop the consequences of climate change themselves, their role in the adaptation process and support of populations remains an important pursuit.

Notes

1 J. Sowers, A. Vengosh, and E. Weinthal, "Climate Change, Water Resources, and the Politics of Adaptation in the Middle East and North Africa" (2011) 104 *Climatic Change* 599–627 at 600.

2 D. Olawuyi, 'Financing Low-Emission and Climate-Resilient Infrastructure in the Arab Region: Potentials and Limitations of Public-Private Partnership Contracts' in W. Filho and A. Meguid, *Climate Change Research at Universities: Addressing the Mitigation and Adaptation Challenges* (Springer 2017) 533–547.

3 Ibid.

4 The awareness, and recognition of the shared consequences, of environmental issues of the Arab Maghreb countries (Morocco, Mauritania, Libya, Tunisia, and Algeria) led to the *Maghreb Charter on Environmental Protection and Sustainable Development* <www.moqatel.com/openshare/Wthaek/Molhak/MalahekMag/AMalahekMagrab38_2-1.htm_cvt.htm>.

5 Olawuyi (n. 2). See also K. Born, A. H. Fink, and H. Paeth, "Moroccan Climate in the Present and Future: Combined View from Observational Data and Regional Climate Scenarios" in F. Zereini and H. Hötzl (eds), *Climatic Changes and Water Resources in the Middle East and North Africa* (Heidelber: Springer 2008) 29–45.

6 Croitoru Lelia and Maria Sarraf, "The Cost of Degradation of the Environment in Morocco" (World Bank group report number 105633-ma, 2017) <http://documents1.worldbank.org/curated/en/741961485508255907/pdf/105633-WP-P153448-FRENCH-PUBLIC-Maroc-Etude-CDE-Final-logo-Janv-2017.pdf> accessed 13 July 2020.

7 For more details on the Environmental and Sustainable Development Policy in Morocco, see <https://planbleu.org/sites/default/files/upload/files/5Politique_Env_DD_Maroc_SLV.pdf>; see also the official website of the Moroccan Ministry of Energy,

Mines and the Environment, Department of the Environment, available at <http://www.environnement.gov.ma/fr/>.

8 See Olawuyi, n. 2. See also G. Dumollard and A. Leseur, "L'Elaborationd'une Politique Nationale d'Adaptation au Changement Climatique: Retour sur Cinq casEuropeens, Etude Climat," March 2011 <www.i4ce.org/wp-core/wp-content/uploads/2015/10/11-03-Etude-Climat-n%C2%B027-Politiques-dadaptation_retour-sur-5-cas-europ%C3%A9ens_CDC-Climat-Recherche.pdf>.

9 A. Agrawal, "The Role of Local Institutions in Adaptation to Climate Change" (2008) Working Paper prepared for the Social Dimensions of Climate Change, Social Development Department, The World Bank, Washington, DC <http://hdl.handle.net/10986/28274> accessed 7 January 2021.

10 Ibid.

11 Ibid.

12 Ibid, pp. 3–4.

13 Ibid. See also Chapter 1 of this book.

14 The Constitution of the Kingdom of Morocco provides for a set of standards for the conduct of state institutions through Article 154 which stipulates: "The public services are organized on the basis of equal access of the citizens [feminine] and the citizens [masculine], of equitable covering of the national territory and of continuity of payments [prestations] rendered. They are submitted to the norms of quality, of transparency, of the rendering of accounts and of responsibility, and are governed by the democratic principles and values consecrated by the Constitution" (Morocco's Constitution of 2011 at <https://www.constituteproject.org/constitution/Morocco_2011.pdf>). Similarly, the Tunisian Constitution of 2014 states in Article 139: "Local authorities shall adopt the mechanisms of participatory democracy and the principles of open governance to ensure the broadest participation of citizens and of civil society in the preparation of development programmes and land use planning, and follow up on their implementation, in conformity with the law" (Tunisia's Constitution of 2014, at <https://www.constituteproject.org/constitution/Tunisia_2014.pdf>).

15 Article 154 of the Moroccan Constitution of 2011.

16 Agarwal, n 9.

17 For more details on the Environmental and Sustainable Development Policy in Morocco at <https://planbleu.org/sites/default/files/upload/files/5Politique_Env_DD_Maroc_SLV.pdf>.

18 See the official website of the Moroccan General Secretariat of the Government on <http://www.sgg.gov.ma>.

19 Speech of His Majesty the King Mohammed VI on the occasion of the Throne day, dated, July 30, 2010, available at <https://www.adrare.net/XYIZNWSK/discours41.htm>.

20 <https://www.mem.gov.ma/Pages/index.aspx>.

21 <http://www.equipement.gov.ma/AR/Pages/Accueil.aspx>.

22 <http://www.agriculture.gov.ma/>.

23 Decree No. 2-04-503 of 21 hija 1425 (1 February 2005) on the powers and organization of the High Commission for Water and Forests and the fight against desertification at <http://bdj.mmsp.gov.ma/Fr/Document/782-D%C3%A9cret-n-2-04-503-du-21-hija-1425-1er-f%C3%A9vrier-20.aspx?KeyPath=>.

24 Decree No. 2-93-1011 of 18 Chaabane 1415 (20 January 1995) on the reorganization of organizations responsible for the protection and improvement of the environment <http://adala.justice.gov.ma/production/html/Fr/liens/.%5C81887.htm>.

25 Law n ° 10-95 on water promulgated by the Dahir n ° 1-95-154 of 18 rabii I 1416 (16 August 1995) at <http://www.sgg.gov.ma/Portals/1/lois/Loi_36-15_Ar.pdf?ver=2018-11-28-152309-170>.

26 Royaume du Maroc, "Le Conseil Économique, Social et Environnemental" <http://www.cese.ma/Pages/Accueil.aspx>.

27 Royaume du Maroc, "Stratégie Nationale de Développement Durable (SNDD) 2030: Résumé Exécutif," October 2017 <www.environnement.gov.ma/PDFs/publication/Synthese-SNDD_FR.pdf>.
28 E. H. Miqdad, *Environmental Law* (1st edn, Imprimerie Najah ElJadida, Casablanca, 2012). For more details, see the official website of the Ministry of Energy, Mines and the Environment, Department of the Environment at <http://www.environnement.gov.ma/fr/>.
29 Decree No. 2933 of 2005, dated 1 November 2005 fixing the attributions of the Ministry of the Environment and Sustainable Development, available at <http://ext wprlegs1.fao.org/docs/pdf/tun60851.pdf>; Decree n °2006-898 dated 27 March 2006, relating to the organization of the Ministry of the Environment and Sustainable Development, available at <http://extwprlegs1.fao.org/docs/pdf/tun66808.pdf>; for more details see the official website of the Tunisian ministry of environment at <http://www.environnement.gov.tn/index.php/fr/>.
30 Decree No. 2001-419 of 13 February 2001, fixing the attributions of the Ministry of Agriculture, at <http://extwprlegs1.fao.org/docs/pdf/tun24288.pdf>; Governmental decree n ° 2018-503 dated 31 May 2018, amending and supplementing decree n ° 2001-420 dated 13 February 2001, relating to the organization of the Ministry of Agriculture, available at <http://extwprlegs1.fao.org/docs/pdf/Tun181267.pdf>; for more details see the official website of the Tunisian ministry of agriculture at <http://fr.tunisie.gov.tn/annuaire/2/9-minist%C3%A8re-de-l%E2%80%99agriculture-de-la-p%C3%AAche-et-des-ressources-hydrauliques.htm>.
31 <http://www.tunisieindustrie.gov.tn/>.
32 <http://www.anpe.nat.tn/Fr/>.
33 <http://www.apal.nat.tn/site_web/index.html>.
34 <http://www.citet.nat.tn/portail/accueilcitet.aspx?_lg=fr-FR>.
35 <http://www.ancsep.rns.tn/>.
36 Decree N° 93-2061 dated 11 October 1993 establishing the National Committee for Sustainable Development, in <file:///C:/Users/SAAD/Downloads/Prerogative_-_D%C3%A9cret_n%C2%B0_93-2061_du_11_octobre_1993.pdf>.
37 Decree No 89-945 dated July 11, 1989, establishing the High Council For Energy in <http://www.legislation.tn/sites/default/files/journal-officiel/1989/1989F/Jo05289.pdf>.
38 Law N°88-91 dated 2 August 1988 establishing the National Agency for the Protection of the Environment in <http://www.anpe.nat.tn/Fr/upload/1479290708.pdf>.
39 <http://www.finances.gov.tn/fr>.
40 <http://www.interieur.gov.tn/fr/>.
41 <http://www.mehat.gov.tn/index.php?id=2&L=2>.
42 <http://www.mes.tn/?langue=fr>.
43 <http://www.environnement.gov.mr/ar/>; see also the National Plan for Environmental Protection and Sustainable Development in Mauritania at <www.environnement.gov.mr/ar/images/pdf/sndd.pdf>.
44 <https://www.energy.gov.dz/?article=protection-de-lenvironnement-2>.
45 <http://www.environnement.gov.ma/fr/eau?id=207>.
46 <http://www.pncl.gov.ma/fr/grandchantiers/Pages/PNDM.aspx>.
47 <https://www.environnement.gov.ma/fr/strategies-et-programmes/prevention-risques/programmes-prevention/95-preventions-des-risques/programmes/192-programme-de-gestion-securisee-des-pcb-au-maroc?showall=1&limitstart=>.
48 The Moroccan state launched a $1 billion Energy Development Fund, whose capital share comes from donations from Saudi Arabia ($500 million), the United Arab Emirates ($300 million), and the Hassan II Fund for Economic and Social Development ($200 million). See R. Hallaoui, "Energie: 105 milliards de Dirhams d'Investissement" (Conjoncture.info, 20 November 2015) <https://www.cfcim.org/magazine/21934> accessed 7 January 2021.

49 In the framework of Moroccan–German cooperation, the State Secretariat for Water and the Environment has set up, in partnership with the German Agency for Financial Cooperation (KfW), the Industrial Depollution Fund (FODEP). This incentive instrument aims to encourage industrial and craft businesses to make investments in depollution or resource saving and introduce the environmental dimension in their activities to deal with the regulatory framework in preparation and the new developments in the globalization of trade. FODEP aims to:
    ensure compliance with environmental regulations; upgrade national industries in anticipation of the globalization of the international market; reduce liquid, solid, and gaseous industrial emissions; and save the use of natural resources.
    In <http://environnement.gov.ma/fr/service/fodep?id=390>.
50 <https://www.environnement.gov.ma/fr/secretariat-etat/2015-03-05-11-53-51/missions?id=174>.
51 <http://www.environnement.gov.tn/index.php/fr/>.
52 These programmes have made it possible to reduce pollution while creating added value and promising opportunities for green professions and the creation of 24 landfills and waste assessment centres in urban groups by 2022, as well as a high rate of waste valuation to reach 60 per cent of the produced waste in the year 2021. A national programme to purify the liquid water made it possible to reduce the percentage of waste water pollution in urban areas to 44 per cent in 2017 and 100 per cent by 2030. The jobs created by these two programmes will reach 25,000 by 2030. In Etude Stratégique: Système Hydraulique de la Tunisie à l'horizon 2030, Janvier 2014 at <http://www.onagri.nat.tn/uploads/Etudes/SYSTEME-HYDROLIQUE-DE-LA-TUNISIE-A-LHORIZON-2030.pdf>.
53 On the importance of enacting clear and comprehensive climate change laws, see D. Olawuyi, "Energy Poverty in the Middle East and North African (MENA) Region: Divergent Tales and Future Prospects" in I. Del Guayo and others (eds), *Energy Law and Energy Justice* (Oxford University Press 2020) 254–272.
54 Ibid. See also A. Averchenkova, *Legislating for a Low Carbon and Climate Resilient Transition: Learning from International Experiences* (Elcano Policy Papers, 2019) 78–81 <http://www.realinstitutoelcano.org/wps/wcm/connect/fe1aeb16-0561-46e4-a6a9-6e49033ae2a9/Policy-Paper-2019-Legislating-low-carbon-climate-resilient-transition.pdf?MOD=AJPERES&CACHEID=fe1aeb16-0561-46e4-a6a9-6e49033ae2a9> accessed 7 January 2021.
55 D. Olawuyi, "Advancing Innovations in Renewable Energy Technologies as Alternatives to Fossil Fuel Use in the Middle East: Trends, Limitations, and Ways Forward" in D. Zillman and others (eds), *Innovation in Energy Law and Technology: Dynamic Solutions for Energy Transitions* (Oxford University Press 2018) 354–370.
56 Olawuyi, n 53.
57 Y. Abourabi and J. Ferrié "La diplomatie Environnementale du Maroc en Afrique: un mix interieur-exterieur" (Telos, 7 June 2018) <https://www.telos-eu.com/fr/politique-francaise-et-internationale/la-diplomatie-environnementale-du-maroc-en-afrique.html> accessed 13 July 2020.

7 Economic sanctions and the effectiveness of the global climate change regime

Lessons from Iran

Mohsen Abdollahi[1]

Introduction

The aim of this chapter is to discuss the implications of economic sanctions on the global response to climate change. Drawing lessons from the Islamic Republic of Iran, it analyzes the need for environmental exemptions in the application of global economic sanctions if the world is to make any real progress in addressing global climate change.

Economic sanctions, whether international or unilateral, are among the prevailing realities of the current international life. While it is generally considered that sanctions are effective tools for enforcing the law by a wrongdoer state, there are many concerns about their negative consequences on the population of the targeted state.[2] The notions of "targeted sanctions" and the "humanitarian exemptions" were the UN Security Council (UNSC)'s responses to abovementioned concerns.[3]

Humanitarian exemptions are designed to mitigate the unintended harmful humanitarian consequences of the sanctions, while continuing to ensure their effectiveness.[4] These exemptions meet humanitarian needs; however, the negative implications of international economic sanctions on international environmental agreements remain challenging. Most of these environmental agreements set up international financial obligations and mechanisms, such as technical and financial assistance, transfer of technology, joint implementation, trade emission, and clean development mechanism.[5]

The international regime of climate change, including the Paris Agreement,[6] encompasses almost all these arrangements. The aim of the international climate change regime is to stabilize greenhouse gas concentrations in the atmosphere at a level that prevents dangerous human interference on the climate system.[7] However, the question is whether economic sanctions have negative ramifications on the international climate change regime and, if yes, how such unintended implications of sanctions should be addressed? As demonstrated in this chapter, economic sanctions undermine the global law of climate change by disrupting financial mechanisms of international environmental agreements and reducing the capacity of targeted countries to meet their commitments under international treaties. The recognition of "environmental exemptions" by any sanction regime may minimize the unintended implications of such sanctions to environmental obligations of targeted states including climate international instruments.

This chapter is divided into five sections. After this introduction, the next section discusses dual applications of the sanctions in the global regime of climate change. It raises the matter of the legality of the sanction in this regime as *lex specialis*. Drawing examples from Iran, third section demonstrates how sanctions undermine the regime of climate change. The fourth section offers proposals on how environmental exemptions can help reduce the negative ramifications of economic sanctions for global climate action. The final section is the concluding section.

Economic sanctions: one mean and two applications

As Rustler rightly noted, "economic sanctions are political instruments whose purpose is to isolate and hurt a target country's economy, forcing policy revisions or bringing about political change".[8] Economic sanctions are generally utilized to press authoritarian regimes to democratize and to respect human rights.[9] The term "sanctions" under international law generally refers to "coercive measures, taken by one State or in concert by several States, which are intended to convince or compel another State to desist from engaging in acts violating international law".[10]

Sanctions may have several dimensions such as military, diplomatic or political, cultural, and economic. The Charter of the United Nations provides a non-inclusive list of sanctions at the disposal of the UNSC.[11] The practice of economic sanctions is hardly new in international relations, but the 20th century is especially rich in sanction episodes.[12] These sanctions may include, *inter alia*, boycotts, embargoes, freezing assets, travel restrictions, financial restrictions on the flow of currencies, and the elimination of transportation, mail service and other means of communication to and from the target state.[13]

With respect to climate change, economic sanctions may be observed from two divergent points of view: a tool to foster cooperation in climate change global regime and a bar to meet the global climate goals. From the first point of view, trade sanctions, whether international or unilateral, in the form of import tariffs are one principal measure discussed as a means to foster cooperation. Economic sanction against countries without emission regulations is regarded as an effective means for incorporating these countries into the global regime.[14] To this end, the UNSC has increasingly been invited to address climate change as its mandate.[15] There is an increasing tendency to recognize climate change as a "threat multiplier" which could threaten international peace and security too.[16] Consequently, the UNSC has held open debates. The first debate took place in April 2007.[17] As a result of this meeting, three additional open debates were held in July 2011, July 2018, and in January 2019.[18] In all these debates, climate-related security risks were recognized and the UNSC was invited to integrate this into their outcomes.[19] Moreover, in March 2017, the UNSC adopted a resolution highlighting the need to address climate-related risks in order to tackle the conflict in the Lake Chad basin.[20] From the second point of view, economic sanctions can become key obstacles to meet global climate goals. Economic sanctions undermine the global climate change regime. The unintended consequences of accumulating sanctions can prevent the meeting of global climate goals, leading to devastating environmental impacts.

Economic sanction is one of the most controversial issues in contemporary international law. The general international law (*lex generalis*) does not explicitly prohibit states from taking such sanctions, and the UNSC, based on Article 41 of the UN Charter, may impose economic sanctions following a determination that there is a threat to or a breach of international peace and security.[21] On the contrary, unilateral sanctions are widely criticized as violating the principle of state sovereignty and the rule of law and holding the risk of breaking other principles of international law. Accordingly, unilateral sanctions could be considered as a challenge to the existing international legal order which is anchored in the UN Charter.[22] Putting the legality of economic sanctions under the general international law aside, for the purposes of this chapter the question is whether the legal regime of climate change as *lex specialis* prohibits economic sanctions or not.

As with general international law, there is not an appropriate regulation on economic sanctions in the climate regime. However, some of the provisions of the related international treaties touch on the matter strongly. According to one of the guiding principles of the United Nations Framework Convention on Climate Change (UNFCCC), "economic development is essential for adopting measures to address climate change" and "The Parties should cooperate to promote a supportive and open international economic system that would lead to sustainable economic growth and development in all Parties, particularly developing country Parties, thus enabling them better to address the problems of climate change".[23] Imposing economic sanctions is not in conformity with the duty to cooperate as written in this article.

Moreover, the developed country Parties to the UNFCCC "shall also provide such financial resources, including for the transfer of technology, needed by the developing country Parties" to meet their obligations under this convention. They "shall take all practicable steps to promote, facilitate and finance, as appropriate, the transfer of, or access to, environmentally sound technologies and knowhow to other Parties, particularly developing country Parties, enable them to implement the provisions of the Convention".[24] Indeed, economic sanctions are fundamentally contrary to these obligations.

Similar obligations are also repeated in the Kyoto Protocol.[25] This Protocol extends this obligation from the public domain to the private sector: all Parties are committed to "the creation of an enabling environment for the private sector, to promote and enhance the transfer of, and access to, environmentally sound technologies".[26] As will be further discussed later, economic sanctions are also a major obstacle against the Protocol Clean Development Mechanism (CDM).[27]

The Paris Agreement has paid full attention to all of the above requirements and emphasizes that "the efforts of all Parties will represent a progression over time, while recognizing the need to support developing country Parties for the effective implementation of this Agreement".[28] The Agreement underlines the importance of "making finance flowsconsistent with [...] climate-resilient development"[29] and "the importance of technology for the implementation of mitigation and adaptation actions under this Agreement and recognizing existing technology deployment and dissemination efforts, shall strengthen cooperative action on technology development and transfer".[30]

In sum, while the legal regime of climate change does not prohibit the imposition of economic sanction, economic sanctions are contrary to the obligations of the Parties to the agreements of the international climate change regime.

How sanctions undermine the regime of climate change

Economic sanctions weaken the implementation of salient international policies and measures that have been established for tackling climate change. The following are three ways that sanctions undermine the current international climate change regime.

Isolation and non-participation in global regime of climate change

International environmental law emphasizes the principle of cooperation. For example, Principle 24 of the Stockholm Declaration states that "international matters concerning the protection and improvement of the environment should be handled in a cooperative spirit by all countries, big and small, on an equal footing".[31] Similarly, Principle 7 of the Rio Declaration proclaims that "States shall cooperate in a spirit of global partnership to conserve, protect and restore the health and integrity of the Earth's ecosystem".[32]

The developed country Parties play a leading role in combating climate change,[33] along with the recognition of the "common but differentiated responsibilities and respective capabilities" of all Parties,[34] and the requirement of special provision to the developing countries, including additional financial resources and technology.[35] These principles and mechanisms, offered by the developed countries, build confidence in the climate change regime and help to encourage the participation of the developing countries in the treaties and conventions on environmental protection. The vast number of Parties to the Montreal Protocol,[36] UNFCCC,[37] and the Kyoto Protocol[38] proves the widespread global support for these mechanisms.

All sanctioned countries have become Party to these instruments, regardless of their economic sanctions.[39] This may be explained by the reality that the Montreal Protocol has afforded a considerable respite to the developing countries for meeting their emission reduction obligations, though there are very restrictive regulations on control of trade with non-Parties, and the Parties are banned from the import and export of the majority of their controlled substances.[40] A similar situation can be seen with regard to the UNFCCC and its Kyoto Protocol. These instruments have exempted the developing countries from any emission reduction obligations and adopted some commitments based on the principle of common but differentiated responsibility.[41] Hence, these countries had no concerns in acceding to the instruments. However, the Paris Agreement changed this pattern by ignoring the distinction between developed and developing Parties of the UNFCCC. The Paris Agreement builds upon the UNFCCC and for the first time brings all Parties into a common cause to undertake ambitious commitments to combat climate change and adapt to its effects, with enhanced support to assist developing countries to do so. In accordance with Article 4(2) of the Paris Agreement: "*Each Party* shall prepare, communicate and maintain successive nationally determined contributions that it intends to achieve. Parties shall pursue domestic mitigation measures, with the aim of achieving the objectives of such contributions."[42]

The participation of the sanction-targeted countries in the Paris Agreement is low in comparison with the previously mentioned instruments.[43] The study of the practices of the nations under the UN sanctions shows that these states were anxious about the Paris Agreement as among the 13 UN sanctioned states, 5 states have not ratified the Paris Agreement and in their Intended Nationally Determined Contributions (INDC) Iran,[44] Sudan,[45] and Somalia[46] mentioned their economic sanctions as a bar to attaining the financial sources and the required technologies for the successful implementation of their INDCs. Libya, as an exception to African states, did not disseminate its INDC.[47]

Among the above-mentioned states, the case of Iran is unique. Iran after six UNSC Resolutions,[48] which imposed a very strict regime of economic sanctions for restricting sensitive parts of Iran's nuclear activities, entered into hard negotiations with the five permanent members of the UNSC plus Germany.[49] The result was achieved on 14 July 2015. The 159-page document with five annexes, officially named the Joint Comprehensive Plan of Action (JCPOA), was endorsed by UNSC Resolution 2231 and adopted on 20 July 2015.[50]

In accordance with the JCPOA, Iran agreed to limit some sensitive parts of its nuclear activities in exchange for lifting UN, US, and EU economic sanctions. Simultaneously, Iran submitted its INDC on 23 November 2015 and signed the Paris Agreement on 22 April 2016. Iran's INDC declares its unconditional intention "to participate by mitigating its [greenhouse gas] (GHG) emission in 2030 by 4% compared to the Business As Usual (BAU) scenario".[51] In addition to its Unconditional Mitigation Action, Iran in the light of recently concluded nuclear deal, conditionally declared: "Subject to termination and non-existence of unjust sanctions, availability of international resources in the form of financial support and technology transfer, exchange of carbon credits, accessibility of bilateral or multilateral implementation mechanisms, transfer of clean technologies as well as capacity building, the Islamic Republic of Iran has the potential of mitigating additional GHGs emission up to 8% against the BAU scenario (i.e. 12% in total)."[52]

While the cabinet of ministers and the Majlis (parliament) in Iran had approved the Paris Agreement in November 2016, the Guardian Council, which constitutionally holds veto power over all legislations approved by the Majlis, did not ratify the Agreement.[53] There had been some concerns the deal may "act against [the] national interest" and as a result "[Iran] should proceed with extreme caution [...] [and] be wary of not being forced to make and legally binding commitments which may later result in *more sanctions and penalties* for [Iran]".[54]

There are two main reasons which may (unofficially) explain Iran's Guardian Council objection: the US withdrawal from the Paris Agreement[55] and the JCPOA. On 8 May 2018, President Trump announced that the US would withdraw from the JCPOA and reinstate US nuclear sanctions on the Iranian regime.[56] The re-imposition of the US suspended sanctions has put Iran's economy in the worst economic situation it has ever seen.[57] The widespread nature of the sanctions, which bans not only US citizens and entities from any deal with Iran, but also all non-US persons, including the companies of the remaining Parties of the JCPOA, have severely limited the availability of economic resources needed for climate action in Iran.[58] In the aftermath of the renewed sanctions several companies, including large EU companies, left Iran's market, thereby worsening the economic situation in Iran.[59]

Given the situation, there is no guarantee that Iran will comply with its commitments under the Paris Agreement. Economic sanctions against Iran have isolated the country, and its ability to meets its obligations under the international climate change regime, and in particular in the Paris Agreement, remain unclear.

Economic sanctions lead the targeted countries to unsustainable survivalist policy

Economic sanctions generally target daily, and even basic, needs of the population of the targeted state. In the short run, targeted countries may resort to survivalist mechanisms to avoid the sanctions' impacts. The states have primarily a responsibility to protect[60] their citizens' well-being against any external violence, including economic pressures. They may attempt to shield them from the economic sanctions' detrimental effects by ramping up domestic energy production or implementing unsustainable agricultural practices.[61]

While sanctions are not the main cause of sanctioned countries' environmental problems, they undoubtedly intensify the situation. Sanctions lead countries into unsustainable practices that will adversely affect the health and quality of life for generations. In poor economic times, priorities are shifted away from environmentalism and, in the long term, economic sanctions bar countries from accessing urgently needed paradigm changes enabling them to pursue sustainable and climate-responsible policies.[62]

The study of the environmental situation in targeted states shows more environmental degradation, including increased pollution, declining water resources, deforestation and desertification, as well as irreversible biodiversity loss.[63] Iran is an example in this regard. While Iran's daily consumption of gasoline was more than 62 million liters during the past UN sanctions time (2006–2013),[64] the US and EU sanctioned the gasoline supplied to Iran. As a result, to secure its energy supply, Iran has turned petrochemical factories into oil refineries and produced petrochemical gasoline.[65]

According to Paul Sampson, the sanctions forced Iran to be creative and to compromise quality: "They have been putting petrochemical components like benzenes and Methyl Tertiary-Butyl Ether into the gasoline, which has increased overall supplies but reduced the quality."[66] In other words, Iran's homegrown petrol contains ten times the level of contaminants found in imported fuel, and the sulphur content of its domestically produced diesel is 800 times higher than that of the international standard.[67]

Reduction of the capacity for compliance in targeted countries

Economic sanctions not only lead the targeted countries to unsustainable survivalist policies and decrease their participation in international environmental conventions but also would pose some risks of reducing cooperation by undermining capacity for compliance in targeted countries,[68] especially if applied to embedded carbon (carbon from energy used to produce traded goods)[69] or the technologies required for emission reduction.

In order to reduce emissions, environmental conventions commit their Parties to the use of the best technology or to the consumption of alternative, "green" materials, whose production requires the latest knowledge and technologies.[70] For the most part, the developing countries do not have access to such technology. Therefore, their compliance would depend on technology development and/ or transfer of technology from the developed Parties of the conventions.

This problem has been taken into account in many of the environmental conventions by creating obligations for the developed country Parties of capacity building of the developing country Parties to assist in their respective implementation of the agreements. For example, Article 11 of the Paris Agreement states: "All Parties enhancing the capacity of developing country Parties to implement this Agreement, including through regional, bilateral and multilateral approaches."[71]

The economic sanctions are flagrantly contrary to such an approach. The sanctions are a strong bar against any transfer of technology to developing Parties of environmental agreements and the sanctions decrease the affected developing countries from reaching their capacity for compliance. This can explain why the sanction targeted state Parties to the Kyoto Protocol have conditioned the vast part of their mitigation actions under the Paris Agreement, to the non-existence of sanctions, availability of international resources in the form of financial support, and effective technology transfer.[72]

Among these states, Iran, one of the most targeted states of multilateral and unilateral sanctions in the last decade, committed to reduce its carbon emissions by four per cent based on the business-as-usual scenario by 2030.[73] Given sufficient financial aid, Iran could reduce its emissions by a further eight per cent. Iran has estimated that $17.5 billion USD in investment is needed to ensure it meets its unconditional pledge, i.e. without international aid. To meet its conditional pledge of 12 per cent reduction, Iran would need $70 billion USD.[74]

After lifting the sanctions after the conclusion of the nuclear deal, the companies from EU states started to invest in Iran energy sectors. However, as soon as US sanctions were reimposed, the same firms' eagerness to invest quickly ended. The UK renewable energy firm Quercus abandoned a $570 million plan to construct a solar power plant, while companies such as Siemens stated they would not consider any new deals with Iran.[75]

Economic sanctions have already affected the developing targeted states interested in Kyoto Protocol's CDM projects. The current status of the CDM projects shows a trend of projects towards a few larger developing countries.[76] Contrary to the inclusion of the majority of the developing nations in the climate change process, present participation requirements of the CDM have unfortunately prevented 67 per cent of developing nations from engaging in CDM projects.[77] For example, the total number of the registered CDM projects in Iran is 22,[78] which is quite low when compared to countries like Brazil, Vietnam, and Malaysia.[79]

Figure 7.1 shows the current status of the distribution of CDM projects among the developing countries, whose economies are comparable with Iran and the rest of the sanctioned targeted states.[80]

As Figure 7.3 demonstrates, a number of the registered projects of targeted states have been aborted due to economic sanctions.

Economic sanctions, in addition to war, insecurity, and corruption, are one of the most important causes of this unequal distribution.[81] Iran's case makes clear that economic sanctions caused the termination of many CDM projects in one

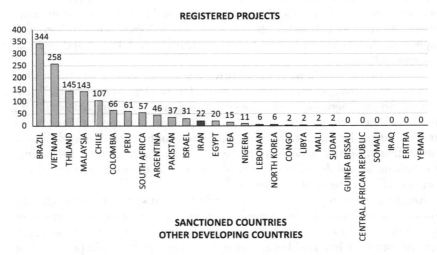

Figure 7.1 Distribution of CDM projects among developing countries including sanctioned countries. UNFCCC, n 78.

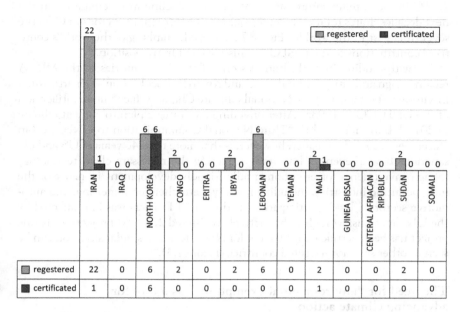

Figure 7.2 Aborted CDM projects in sanction-targeted countries. UNFCCC, n 78.

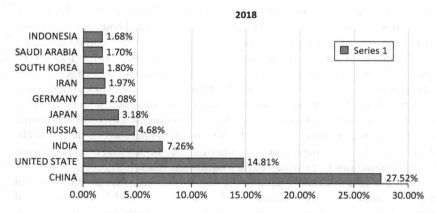

Figure 7.3 Ranking of the world's largest producers of territorial fossil fuel CO_2 emissions in 2018, based on their share of global CO_2. Source: T. Wang, https://www. statista.com/statistics/271748/the-largest-emitters-of-co2-in-the-world/ (Dec 16, 2019).

of the countries whose compliance is most significant, from the point of view of global climate change.

Iran's GHG emissions have increased by 7.3 per cent in the past decade and the average temperature has risen by 1.8°C since 1750, considerably higher than the global average of 1.1°C.[82] Consequently, Iran will experience an increase of

2.6 °C in mean temperatures and a 35 per cent decline in precipitation in the next decades. Iran's total GHG emissions equal 616,741 million tons of CO_2, the seventh largest in the world.[83] Figure 7.3 shows the ranking of the world's countries' contributions to the total CO_2 emissions and Iran's position.[84]

Notwithstanding its rank, Iran was one of the first countries in the MENA region to sign and ratify the UNFCCC and Kyoto Protocol.[85] Iran's Department of Environment established the National Climate Change Office[86] upon ratification of the UNFCCC in 1998.[87] After unveiling its strategic plan to combat climate in 2015,[88] Iran submitted its "Third National Communication to UNFCCC" in December 2017.[89] The reality, however, is that more than 40 years of US and EU sanctions have prevented Iran from attracting the necessary clean technology foreign investments for environmentally-friendly development in many of the most polluting industries, including refinery, petrochemical, smelting, and automotive sectors.[90] Iran's participation in the CDM Projects was also affected by the UN sanctions; while Iran has actively registered 22 CDM projects, only one project has been certificated.[91] Except for North Korea, a similar situation can be seen in other sanction-targeted countries (Figure 7.3).[92]

The need for "environmental exemptions" as a tool for advancing climate action

While the debate on sanctions is often centered on their political and economic implications, the environmental ramifications must be included in the discussion. A study of the environmental situations of the targeted countries has shown that the imposition of sanctions carries significant unintended consequences for the environmental policies of these countries and their mitigation actions on the climate changes.[93] Economic sanctions reduce the participation of targeted states in the global regime of climate change, lead the targeted countries to unsustainable survivalist policy, and undermine the capacity of these states for compliance with international commitments. Therefore, the economic sanctions undermine the global regime of climate change. Given the global and urgent nature of climate change, there is a need for environmental exemptions that can allow countries under sanctions to continue to play significant roles.

Regardless of the debates around the efficacy of sanctions in international law,[94] "environmental exemptions" could protect the world's environmental policies and measures from unintended implications of international sanctions. Based on the findings in this chapter, there follow two recommendations: (i) recognition of the detrimental effects of economic sanctions on the regime of climate change; and (ii) integration of environmental exemptions into the regimes of international economic sanctions.

Recognition of detrimental effects of economic sanctions on the regime of climate change: normative action

As mentioned previously, there is a great deal of research focusing on the effects of trade/economic sanctions and the resulting unwillingness of states to participate in, and comply with, the commitments of the climate change regime.[95] This research has revealed that economic sanctions have undermined the climate change global regime, *inter alia*, by reducing the capacity of the targeted countries to meet their commitments—a fact which has so far been overlooked. To avoid such unintended implications of economic sanctions, it is for the Conference of Parties (COP) of the UNFCCC to act in this regard in its upcoming meetings.[96] To this end, it is recommended that *the detrimental effects of economic sanctions on the capacity of the State Parties to meet their commitments under the climate change regime* are acknowledged and recognized properly in climate agreements and negotiations. Also, the UN bodies and all states are called to consider such implications in the former and any future regimes of sanctions.

This normative action is necessary for any further development. Intrestingly, international environmental law, in general, and the regime of climate change, in particular, are already rich in this regard. Many of the normative developments of environmental law have been achieved through similar mechanisms. Recognition of the "harmful impact of modification of the ozone layer on human health and the environment"[97] and acknowledging climate change as a common concern of humankind[98] are good examples. These achievements can be regarded as the cornerstones of subsequent lawmaking in the area of ozone layer protection and climate actions. The same course of action is recommended for protecting UN climate measures against unintended implications of economic sanctions.

Integration of environmental exemptions into the regimes of international economic sanctions

The negative effects of economic sanctions on the regime of climate change should be considered by the UNSC, regional organizations, such as the EU, and individual states when they decide to impose economic sanctions against a wrongdoer state. Therefore, regardless of the legality of economic sanctions under the global regime of climate change, as discussed above, environmental requirements, including all climate actions of the targeted states, should be considered as an exemption to the imposed sanctions.

This new exemption could be called an "environmental exemption". As argued earlier, the recognition of this *exemption* by international law, and especially by the UNSC, is a necessary action for the effective implementation of UNFCCC and its associated instruments. If the UNSC considers climate change as a new threat to international peace and security, as it has been called up, it should not undermine the UN legal campaign against it. [99]

The *exemptions* to the UNSC sanctions regimes are not unprecedented. As some commentators have argued, these exemptions are necessary for the effectiveness and legitimacy of UNSC resolutions.[100] The UNSC first introduced humanitarian exemptions to travel bans and commodity interdictions in 1968 within the Southern Rhodesian sanctions regime.[101] Since then, some form of exemptions for humanitarian purposes have been included in most of the sanction regimes.[102] Currently, two types of humanitarian exemptions exist: individual and sectoral exemptions.[103] The UNSC has also developed guidelines for requesting humanitarian exemptions in its sanctions regime.[104]

The concept of "environmental exemption" may be developed based on two main approaches. First, by accepting a new exception to sanction regimes similar to the humanitarian exemptions. This approach is preferred as it could give an independent legal status and more weight to environmental considerations. The UNSC would need to incorporate this exemption into its sanctions procedure and adopt a guideline for obtaining exemptions for compliance with environmental commitments. A second approach is to include the "environmental exemption" into the consolidated notion of *humanitarian exemptions*. The relationship between human rights and environmental law is a well-established notion, thus this inclusion will have sufficient theoretical foundations.[105]

The development of the notion of "environmental exemption", either as a new exemption or as a part of the humanitarian exemptions, may assist in the further development of international law. Extension of such exemptions into the unilateral regime of sanctions[106] is necessary for extensive protection of the earth's environment and its climate system.

Conclusion

Economic sanctions decelerate the effectiveness of the global climate change regime. They not only reduce the participation of targeted countries in the global climate change regime and undermine the capacity for compliance with emissions commitments, but also lead targeted countries to unsustainable survivalist policy. Therefore, environmental exemptions to sanctions are crucial for ensuring that the environmental protection actions of targeted states, including compliance with their climate commitments, are not negatively affected. To this end, it is recommended that the COP of the UNFCCC acknowledge *the detrimental effects of economic sanctions on the capacity of the State Parties to meet their commitments under the climate change regime* and that the UNSC take into account environmental considerations in its decisions to impose sanctions through a new exception of "environmental exemption" or as a part of the previous notion of humanitarian exemptions.

The focus of this chapter was on the Iranian experience of environmental actions under the regime of sanctions. Future research on the situations of other sanctioned states is necessary to illustrate a more profound understanding of this issue. This research is also critical in order to demonstrate how the climate efforts have been affected by unrealized and terminated CDM projects as a result of economic sanctions.

Notes

1   The author thanks Ms. Hanie Hejazi Moghadam for her assistance with research, and appreciates Ms. Farimah Kashkolli's support and valuable feedback.
2   S. S. Park, "Economic Sanctions as Human Rights Violations: Reconciling Political and Public Health Imperatives" (1999) 89 *American Journal of Public Health* 1510; M. Happold, "Targeted Sanctions and Human Rights" in M. Happold and P. Eden (eds), *Economic Sanctions and International Law* (1st edition, Hart Publishing 2016).

3 F. Giumelli, "Understanding United Nations Targeted Sanctions: An Empirical Analysis" (2015) 91 *International Affairs* 1351–1353; J. Gordon, "Smart Sanctions Revisited" (2011) 25 *Ethics and International Affairs* 315, 317–318.

4 High Level Review of United Nations Sactions, "UN Sanctions: Humanitarian Aspects and Emerging Challenges: *Chairperson's Report* (High-Level Review of United Nations Sanctions, 19 January 2015) <http://www.hlr-unsanctions.org/HLR_WG3 _report_final.19.1.15.pdf> accessed 7 January 2021.

5 For instance see: UN, *Montreal Protocol on Substances that Deplete the Ozone Layer* (as amended), 16 September 1987, 1592 UNTS 3, Art. 5(5) [Montreal Protocol]; UN, *Kyoto Protocol to the United Nations Framework Convention on Climate Change*, 11 December 1997, 2303 UNTS 162, Arts. 3(14) and 10(c) [Kyoto Protocol].

6 United Nations Framework Convention on Climate Change (UNFCCC), *Adoption of the Paris Agreement*, 12 December 2015, UN Doc. FCCC/CP/2015/l.9/Rev.1.

7 For example, see ibid, Arts. 2(1) and 4(1).

8 A. Rustler, "Are Sanctions Stopping Us from Achieving Global Climate Goals"? July 11, 2019 <https://jia.sipa.columbia.edu/online-articles/are-sanctions-stopping-u s-achieving-global-climate-goals>.

9 J. Ilieva, A. Dashtevski, and F. Kokotovic, "Economic Sanctions in International Law" (2018) 9 *UTMS Journal of Economics* 201.

10 C. Joyner, "Collective Sanctions as Peaceful Coercion: Lessons from the United Nations Experience" (1995) 16 *Australian Year Book of International Law* 242.

11 UN, *Charter of the United* Nations, 24 October 1945, 1 UNTS XVI, Art. 41 reads as follow: "The Security Council may decide what measures not involving the use of armed force are to be employed to give effect to its decisions, and it may call upon the Members of the United Nations to apply such measures. These may include complete or partial interruption of economic relations and of rail, sea, air, postal, telegraphic, radio, and other means of communication, and the severance of diplomatic relations."

12 M. Delevic, "Economic Sanctions as a Foreign Policy Tool: The Case of Yugoslavia" (1998) 3 *International Journal of Peace Studies* 183.

13 Joyner, n 10.

14 A. Hagen and J. Schneider, "Boon or Bane? Trade Sanctions and the Stability of International Environmental Agreements" (2017) ZenTra Working Paper in Transnational Studies 1.

15 United Nations, "Addressing Security Council, Pacific Island President Calls Climate Change Defining Issue of Next Century, Calls for Special Representative on Issue" (United Nations News, 11 July 2018) <https://www.un.org/press/en/2018/sc 13417.doc.htm> accessed 7 January 2021.

16 United Nations Peacebuilding, "Climate Change Recognized as 'Threate Multiplier'," UN Security Council debates its impact on peace" (United Nations) <www.u n.org/peacebuilding/news/climate-change-recognized-%E2%80%98threat-multipli er%E2%80%99-un-security-council-debates-its-impact-peace> accessed 7 January 2021.

17 F. Sindico, "Climate Change: A Security (Council) Issue?" (2007) 1 *Carbon and Climate Law Review* 29–34.

18 See n 15.

19 J. Sherman, "How Can the Security Council Engage on Climate Change, Peace, and Security?" (ReliefWeb, 20 June 2019) <https://reliefweb.int/report/world/how-can-se curity-council-engage-climate-peace-and-security> accessed 7 January 2021.

20 "Recognises the adverse effects of climate change and ecological changes among other factors on the stability of the Region, including through water scarcity, drought, desertification, land degradation, and food insecurity, and emphasises the need for adequate risk assessments and risk management strategies by governments and the United Nations relating to these factors." UNSC, *Resolution 2349 (2017) [on the situation in the Lake Chad Basin region]*, 31 March 2017, UN Doc. S/RES/2349 (2017).

21 UN, n 11. However these sanctions may violate international law, including human rights law. See I. Jazairy, "US Sanctions Violate Human Rights and International Code of Conduct. UN Expert Says," (UNHCR, 6 May 2019) <https://www.ohchr.or g/EN/NewsEvents/Pages/DisplayNews.aspx?NewsID=24566&LangID=E> accessed 7 January 2021.

22 Ilieva and others, n 9.

23 UN, *United Nations Framework Convention on Climate Change*, 9 May 1992, 1771 UNTS 107, Art. 3(4)–(5).

24 Ibid, Art. 4(3) and (5).

25 Kyoto Protocol, n 5, Art. 10(c) and (b).

26 Ibid, Art. 10(b).

27 See above, s 3(c).

28 UNFCCC, Paris Agreement, n 6, Art. 3.

29 Ibid, Art. 2(c).

30 Ibid, Art. 10(2).

31 UN Environmental Programme (UNEP), *Declaration of the United Nations Conference on the Human Environment*, 5–16 June 1972, UN Doc. A/CONF.48/14/Rev.1, Principle 24.

32 UN Economic and Social Council (UNESC), *Rio Declaration on Environment and Development*, 7–25 April 1997, UN Doc. E/CN.17/1997/8, Principle 7.

33 UN, n 23, Art. 3(1).

34 Ibid, Preamble, Art. 3(1), and Art. 4(1).

35 Montreal Protocol, n 5, Art. (5).

36 198 Parties <https://ozone.unep.org/all-ratifications>.

37 197 Parties <https://treaties.un.org/Pages/ViewDetailsIII.aspx?src=IND&mtdsg_no =XXVII-7&chapter=27&Temp=mtdsg3&clang=_en#1>.

38 192 Parties <https://treaties.un.org/Pages/ViewDetails.aspx?src=TREATY&mtdsg _no=XXVII-7-a&chapter=27&clang=_en>.

39 Ibid, n 37–39.

40 Montreal Protocol, n 5, Art. 4.

41 UNFCCC, Paris Agreement, n 6, Art. 4; Kyoto Protocol, n 3, Art. 10.

42 UNFCCC, Paris Agreement, n 6, Art. 4(2) [emphasis added].

43 The following UN sanctioned States have notare not Party to the Paris Agreeement: Iran, Iraq, North Korea, Yemen, Libya, Lebanon, Mali, Somalia, Guinea Bissau, Democratic Republic of Congo, Central African Republic, Sudan, Eritrea, Syrai, Burundi, Egypt, Guinea, Maldives, Myanmar, Russia, Bosnia and Herzegovina, Venezuala, Zimbabwe, and Cuba.

44 Islamic Republic of Iran, Department of Environment, "Intended Nationally Determined Contribution (INDC)", Department of Environment, November 19, 2015 <www4.unfccc.int/sites/submissions/INDC/Published%20Documents/Iran/1/INDC %20Iran%20Final%20Text.pdf>.

45 Republic of Sudan, "Intended Nationally Determined Contributions (INDCs)", October 28, 2015 <www4.unfccc.int/sites/submissions/INDC/Published%20Documents /Sudan/1/28Oct15-Sudan%20INDC.pdf>.

46 State Minister for Environment, Office of the Prime Minister and Line Ministries and Ministry of Planning, Federal Government of Somalia, "Somalia's Intended Nationally Determined Contributions (INDCs)", November 2015 <www4.unfccc.int/sites/sub missions/INDC/Published%20Documents/Somalia/1/Somalia's%20INDCs.pdf>.

47 Almost all African states (53 out of 54, Libya being the exception) have submitted an INDC, covering approximately 7.5 per cent of global emissions (African development Bank Group, Cliamte Investment Funds. "Transitioning From INDCS to NDCS In Africa", November 2015 <www.afdb.org/fileadmin/uploads/afdb/Docu ments/Publications/AfDB-CIF-Transitioning_fromINDCs_to_NDC-report-Nove mber2016.pdf> 4.

48 UNSC Resolutions 1696, 31 July 2006; 1373, 23 December 2006; 1747, 24 March 2007; 1803, 3 March 2008; 1835, 27 September 2008; and 1929, 9 June 2010 <www .armscontrol.org/factsheets/Security-Council-Resolutions-on-Iran>.
49 S. H. Mousavian and M. M. Mousavian, "Building on the Iran Nuclear Deal for International Peace and Security" (2018) 1 *Journal for Peace and Nuclear Disarmament* 179.
50 See Annex B of the UNSC. Res. 2231.
51 Islamic Republic of Iran, ibid, n 45.
52 Ibid.
53 M. Qarehgozlou, "Is Iran Pulling Out of Paris Agreement?" 20 May 2018 <www.t ehrantimes.com/news/423741/Is-Iran-pulling-out-of-Paris-Agreement>.
54 Ibid.
55 On 4 November 2019, the US Government notified the Secretary-General of its decision to withdraw from the Agreement which shall take effect on 4 November 2020 in accordance with Article 28(1) and (2) of the Paris Agreement <https://tr eaties.un.org/doc/Publication/CN/2019/CN.575.2019-Eng.pdf>.
56 The White House, Foreign Policy, "President Donald J. Trump is Ending United States Participation in an Unacceptable Iran Deal," May 8, 2018 <www.whitehouse. gov/briefings-statements/president-donald-j-trump-ending-united-states-participa tion-unacceptable-iran-deal/>.
57 BBC News, "Six Charts that Show How Hard US Sanctions Have Hit Iran," December 9, 2019 <www.bbc.com/news/world-middle-east-48119109>.
58 M. Mostatabi, "Sanctioning Iran's Climate," May 1, 2019 <www.atlanticcouncil.org /blogs/menasource/sanctioning-iran-s-climate/>.
59 Ibid.
60 International Commission on Intervention and State Sovereignty, "The Responsibility to Protect, December 2001, 17, 22, 32, and 52.
61 Rustler, n 8.
62 Ibid.
63 Ibid.
64 The US Energy Information Administration), "Iran: Gasoline Consumption" <www .theglobaleconomy.com/Iran/gasoline_consumption/>.
65 Mostatabi, n 60.
66 A. Merrat, "US Sanctions Have Made Iran's Air Quality Much Worse," (Vice, 15 January 2015) <www.vice.com/en_us/article/a38b38/us-sanctions-have-made-ira ns-air-quality-much-worse>; V. Balikhani, "Poor Quality Gasoline Deadly for Iranians," (Atlantic Council, 16 February 2017) <www.atlanticcouncil.org/blogs/irans ource/poor-quality-gasoline-deadly-for-iranians/>.
67 Rustler, n 8.
68 S. Murase, *International Law: An Integrative Perspective on Transboundary Issues* (Tokyo: Sophia University Press 2011).
69 R. Stavins and others, "International Cooperation: Agreements and Instruments" in O. Edenhofer and others (eds), *Climate Change 2014: Mitigation of Climate Change. Contribution of Working Group III to the Fifth Assessment Report of the Intergovernmental Panel on Climate Change* (Cambridge University Press 2014).
70 See, for instance: Montreal Protocol 1989 Arts. 4 (6), 9 (1)(a); UNFCCC, preamble 22; Kyoto Protocol Arts. 2 (a)(iv), 10 (b)(i) & (c).
71 UNFCCC, Paris Agreement, n 6, Art. 11(4).
72 See, for example, The State of Eritrea, "Eritrea's Intended Nationally Determined Contributions (INDCs) Report," September 2015 <www4.unfccc.int/sites/sub missions/INDC/Published%20Documents/Eritrea/1/ERITREA'S%20INDC%20RE-PORT%20SEP2015.pdf> 8 and 12; République du Congo, "Contribution Prevue Determinee au Niveau National dans le cadre de la CCNUCC Conférence des Parties 21", 21 September 2015 <www4.unfccc.int/sites/submissions/INDC/Published %20Documents/Congo/1/INDC_Congo_RAPPORT.pdf, 5>; Republic of Yemen

Intended Nationally Determined Contribution (INDC) under the UNFCCC (21 November 2015) 5-8.

73 Islamic Republic of Iran, "Third National Communication to United Nations Framework to Convention on Climate Change (UNFCCC)", December 2017 <https://unfccc.int/sites/default/files/resource/Third%20National%20communication%20IRAN.pdf> 92.

74 "Iran Unveils Strategic Plan to Combat Climate Change" *Financial Tribune* (Tehran, 17 May 2017) <https://financialtribune.com/articles/environment/64656/iran-unveils-strategic-plan-to-combat-climate-change> accessed 7 January 2021.

75 Mostatabi, n 60; Islamic Republic of Iran (n.73).

76 A. Silayan, "Equitable Distribution of CDM Projects among Developing Countries" (2005) Report No. 255 Hamburg Institute of International Economics <www.econstor.eu/bitstream/10419/32938/1/497849976.pdf> accessed 7 January 2021. See also D. Olawuyi, *The Human Rights Based Approach to Carbon Finance* (Cambridge University Press 2016).

77 Silayan, ibid.

78 See UNFCCC, *Project Search* <https://cdm.unfccc.int/Projects/projsearch.html>.

79 Ibid.

80 Ibid.

81 D. Olawuyi, "Achieving Sustainable Development in Africa through the Clean Development Mechanism: Legal and Institutional Issues Considered" (2009) 17 *African Journal of International and Comparative Law* 270–301.

82 Financial Tribune, n 74.

83 M. R. M. Daneshvar, M. Ebrahimi, and H. Nejadsoleymani, "An Overview of Climate Change in Iran: Facts and Statistics" (2019) 8 (7) *Environmental System Research*, 1–10.

84 T. Wang, "Largest Producers of Territorial Fossil Fuel CO_2 Emissions Worldwide in 2018, Based on Their Share of Global CO_2 Emissions", 16 December 2019 <www.statista.com/statistics/271748/the-largest-emitters-of-co2-in-the-world/> accessed 7 January 2021.

85 Iran ratified the UNFCCC and Kyoto Protocol on 18 July 1996 and 22 August 2005.

86 Islamic Republic of Iran, Department of Environment <www.doe.ir/portal/home/>.

87 Ibid.

88 Financial Tribune, n 74.

89 Islamic Republic of Iran, n 73.

90 Ibid.

91 Project 2422: *Soroosh & Nowrooz Early Gas Gathering and Utilization Project* (S&N project), Emission Reduction: 202,699 CO2e, (2009–2012) <https://cdm.unfccc.int/Projects/DB/SGS-UKL1236075151.6/view> accessed 28 February 2020.

92 Wang, n 84.

93 Rustler, n 8.

94 A. Mack and A. Khan, "The Efficacy of UN Sanctions" (2000) 31 *Security Dialogue* 279–292; The Graduate Institute Geneva, Programme for the Study of International Governance, "Effectiveness of UN Targeted Sanctions" (Targeted Sanctions Consortium, Watson Institute for International Studies, Brown University, November 2013) <https://repository.graduateinstitute.ch/record/287976/files/effectiveness_TCS_nov_2013.pdf> accessed 7 January 2021.

95 B. Gilley and D. Kinsella, "Coercing Climate Action" (2015) 57 *Survival* 7–28; H. Tian and J. Whalley, "Trade Sanctions, Financial Transfers and BRIC Participation in Global Climate Change Negotiations" (2010) 32 *Journal of Policy Modeling* 47.

96 The UN General Assembly can also play a constructive role. In its recent resolutions on "Combating sand and duststorms," the UN General Aassemblyrecognized the linkage between "sand and dust storms and [...] among other factors [...] climate

change" UNSC Resolutions 70/195 of 22 December 2015, 71/219 of 21 December 2016 and 72/225 of 20 December 2017; 73/237 of 15 January 2019.

97 UN, *Vienna Convention for the Protection of the Ozone Layer*, 22 March 1985, 1513 UNTS 293, Preamble.

98 UN, n 23, Preamble.

99 Addressing Security Council, Pacific Island President Calls Climate Change Defining Issue of Next Century, Calls for Special Representative on Issue, n 14.

100 J. Matsukuma, "The Legitimacy of Economic Sanctions: An Analysis of Humanitarian Exemptions of Sanctions Regimes and the Right to Minimum Sustenance" in H. Charlesworth and J. M. Coicaud (eds), *Fault Lines of International Legitimacy* (Cambridge University Press 2010) 375–376; A. Jones and others, "National Model United Nations" (New York Conference, NMUN-NY, 2017) <www.nmun.org/as sets/documents/conference-archives/new-york/2018/ny18-bgg-sc.pdf> 13–14.

101 UNSC, *Resolution 253 (1968) Question Concerning the Situation in Southern Rhodesia*, 29 May 1968, UN Doc. S/RES/253.

102 Jones, n 100.

103 K. King, N. K. Modirzadeh, and D. A. Lewis, "Understanding Humanitarian Exemptions: UN Security Council Sanctions and Principled Humanitarian Action" (Harvard Law School Program on International Law and Armed Conflict Counterterrorism and Humanitarian Engagement Project, April 2016) <https://dash.ha rvard.edu/bitstream/handle/1/29998395/Understanding_Humanitarian_Exemptions _April_2016.pdf?sequence=1&isAllowed=y> accessed 7 January 2021

104 UNSC, "Implementation Assistance Notice No. 7: Guidelines for Obtaining Exemptions to Deliver Humanitarian Assistance to the Democratic People's Republic of Korea," (UN, 6 August 2018) <www.un.org/securitycouncil/sites/www.un.org.securi tycouncil/files/1718_implementation_assistance_notice_7.pdf> accessed 7 January 2021.

105 UNEP, n 31; UNOHCHR, "Framework Principles on Human Rights and The Environment" (OCHR, 2018) <www.ohchr.org/Documents/Issues/Environment/SREn vironment/FrameworkPrinciplesUserFriendlyVersion.pdf> accessed 7 January 2021; A. Boyle, "Human Rights and the Environment: Where Next?" 23 *European Journal of International Law* 614.

106 Humanitarian exemptions have been extended to unilateral sanctions regimes, and states and regional organizations have considered these exemptions when they decide to impose sanctions. In this regard, the International Court of Justice (ICJ) recently ordered the United States to remove any impediments arising from the measures announced on 8 May 2018 on the free exportation to the territory of Iran of goods required for humanitarian needs, such as (i) medicines and medical devices; and (ii) foodstuffs and agricultural commodities; as well as goods and services required for the safety of civil aviation (*Alleged violations of the 1955 Treaty of Amity, Economic Relations, and Consular Rights (Islamic Republic of Iran v United States of America), Order of Provisional Measures, ICJ Reports* 2018, para 98).

8 Addressing climate change through the water–energy–food nexus

Lessons learned from Turkey

Ali O. Diriöz

Introduction

The aim of this chapter is to discuss how the water–energy–food (WEF) nexus approach can help address the challenges of climate change in the Middle East and North Africa (MENA) region through the integration of policies around low-carbon efficiency, renewable energy, water efficiency, and food security. Drawing lessons from Turkey, it discusses the scope and dimensions of the WEF nexus approach, emphasizing the preconditions for implementation, the barriers and difficulties for its application, and policy recommendations for addressing those challenges.

In the MENA region, there is an abundance of energy sources but insufficient food and water resources available. Climate change is having dramatic adverse effects on sectors such as water, food, and energy.[1] Rising temperatures accelerate freshwater evaporation, reducing the availability of water for households and for agricultural and industrial use, and posing a threat to critical investments such as dams. The declining water level in hydroelectric dams becomes a serious social and human rights issue, which is contested especially fiercely in transboundary water contexts.[2] Drought and heatwaves cause agricultural degradation, often affecting crops. At the same time, rising temperatures are melting polar ice caps, with consequences including rising sea levels and flash floods. The problems of multiple different sectors are interrelated in the areas of water, energy, and food, requiring the implementation of an integrated approach.

The WEF nexus approach is a method to ensure a broader and more holistic understanding of ecological balances.[3] Cross-sectoral integration of laws and policies, including renewable energy laws, can help address the problems of climate change. Sustainability in electricity generation, for example, promotes environmental conservation, low emissions, and energy efficiency.[4] Germany's policy of *Energiewende* is an example of a holistic approach to energy transition that also encompasses aspects of environmental security, water consumption, and agriculture.[5] The importance of a holistic approach has been highlighted by the global COVID-19 pandemic, which has shown how quickly disasters can affect multiple sectors.[6] Oil prices fell significantly between March and April 2020,[7] while renewables became more important. One common problem, as seen in the case of

Turkey, is the existing siloed approach which fragments and compartmentalizes WEF sectors. Even when water, energy, food, and environment security are considered collectively, one sector may be privileged over others.[8] This can be seen especially in discrepancies between intended regulations and actual practices. Renewable energy investments in Turkey have increased significantly and quite successfully over the last decade, but discrepancies remain in practice, partly because a WEF nexus approach was not effectively espoused and implemented. The lack of a holistic approach, as well as the lack of efficient mechanisms to coordinate the environmental portfolio between several ministries, is a weakness that needs to be addressed.

This chapter is divided into five sections. Following this introduction, the second section examines how the integration of energy, water, and environment-related policies and laws through the WEF nexus can address problems associated with climate change, and points to some successful examples which could be followed by MENA governments. The third section discusses the Turkish experience, in terms of both successes and challenges, and analyzes how actions such as local community engagement could better communicate a project to the local population and environmental groups. The fourth section examines how challenges to the implementation of the WEF nexus approach can be addressed, considering the lessons learned and making recommendations for Turkey and the MENA region. The final section offers some conclusions.

Drivers and importance of the WEF nexus for climate change mitigation and adaptation

Since the 1972 report, *Limits to Growth*,[9] neo-Malthusian approaches have drawn attention to issues around finite resources, the environment, and sustainability. While renewable energy technologies have since advanced, priority is still given to profitability rather than a long-term holistic approach to safeguarding the environment.[10] Profits-first approaches lack the regenerative feature of renewable investment that is necessary for an eco-centric approach that evaluates the environment as a whole over a long period of time.[11] The WEF nexus approach aims to achieve climate change adaptation objectives holistically, through integrated analyses,[12] in order to preserve and protect the well-being of the eco-system in the long run.

The WEF nexus includes innovations in the sectors of water, energy, and food, such as promoting renewable energy and non-conventional water resources, and increasing efficiency in water supply and agriculture,[13] with the aim of reducing carbon emissions and improving access to water and food.[14] Adapting WEF nexus policies could thus mitigate some of the consequences of climate change by promoting the efficient use of water, energy, and food resources, reducing greenhouse gases, and eliminating groundwater contamination through more resilient infrastructure. The WEF nexus model encourages policy makers to comprehend the water, food, and energy sectors as interlinked: this focus on interlinkages is vital to the success of sustainable development.[15] The WEF model provides a holistic perspective and thus strengthens the capacity of governments to provide climate-smart policies which concentrate on climate change adaptation.[16]

In most countries, water, energy, and food resources are governed by separate sets of laws, rules, and institutions,[17] leading to fragmentation. Within these siloed approaches, there may be discussions on the need for interdisciplinary and cross-sectoral cooperation—for example, a focus on energy might include an assessment of how water security connects to energy and food security—but this falls short of adopting a truly integrated approach.[18] Studies have illustrated and discussed certain governmental approaches to the water–energy nexus, particularly linked to electricity output from "anthropocentric" perspectives.[19] Others have focused on the nexus approach from a food security angle by taking account of interconnections between WEF resources,[20] recognizing synergies and trade-offs that arise from the way resources are managed, in order to achieve food security.

The WEF nexus approach is a framework for coherent, holistic, and integrated implementation of the United Nations Sustainable Development Goals (SDGs). Enhanced levels of legislation and rule linkage, development of common and coordinated approaches, together with information sharing on WEF-related decision making can contribute to a better-integrated governance of resources. According to Olawuyi:

> The nexus approach is characterized by UNECE to include five core features. These are the integration of: (1) Institutions; (2) Information sharing; (3) Instruments, laws and policies to address trade-offs and exploit synergies; (4) Infrastructure and technological solutions; and (5) International coordination and cooperation at regional levels. [21]

As Olawuyi notes, WEF resources are sensitive to the negative effects of climate change. Droughts, floods, and other climatic conditions are challenges for farmers in terms of agricultural food production, for municipalities in terms of water supplies and wastewater treatment, and for energy companies in terms of generating hydroelectricity. Failure to effectively mitigate these challenges thus increases vulnerabilities of all WEF resources. The Southern African Development Community (SADC), for instance, is a region where the effects of climate change are felt through extreme weather conditions and increased climatic volatility, particularly droughts and floods.[22] The WEF nexus approach offers opportunities to integrate climate change adaptation strategies and to assist in achieving the SDGs. Promoting the efficient use of water, renewable energy forms, and energy-efficient communities in urban areas are some of the SDG targets that are best addressed in a holistic and integrated manner. For example, one of the policies put in place in the SADC was to promote indigenous crops and seeds that are better adapted to local geographic and climatic conditions.[23]

The effects of COVID-19—declared a global pandemic in March 2020—have been far-reaching. In April 2020, prices of WTI/NYMEX type crude oil reached negative levels,[24] with an inevitable economic impact globally, including in MENA. Good practices such as the German model suggest that properly designed renewable energy policies can help address such economic volatility as well as climate-related problems. While MENA countries such as the United Arab Emirates had decided to continue with the development of nuclear energy following the Fukushima Daiichi disaster,[25] Germany adopted a policy to abandon nuclear energy and concentrate on developing renewable energy instead.[26] *Energiewende* is a bold policy, and the story of a significant and successful energy transition by one of the world's major economies.[27] *Energiewende* includes support schemes and mechanisms to promote investments, such as feed-in tariffs, or FIT, a type of fixed price paid to renewable energy producers. In addition to compliance with environmental rules, regulations, and laws, Germany promoted the development of renewables by the private sector. In the context of the MENA region, many significant investments in renewables have also been made, notably investments in solar energy in Morocco and the UAE.[28]

The next section assesses the integrated WEF nexus from the perspective of climate policy in Turkey. Turkey's status as candidate for the European Union (EU) and the incorporation of Turkish institutions into many European institutions set it apart from many MENA countries. Turkey's alignment effort with the EU *Acquis Communautaire* (transposition of EU legislation, rules, and regulations) was announced on 17 April 2007, aiming to make Turkish law and policy compliant with EU environmental legislation standards.[29] When putting Turkey in regional context, therefore, it is essential to highlight its resemblances to European countries in its economic, regulatory, legislative, and judiciary systems. Even before 2007, Turkey's public institutions were "westernized".[30] Turkey's position in the Middle East is also exceptional in terms of its energy-import dependence, and the size of its population.[31] The differences in Turkey's legal system and institutional framework, as well as domestic politics, can complicate transboundary regional cooperation.[32] In spite of these differences, however, there remain many cultural, geographical, and policy similarities to MENA countries, suggesting that lessons learned from the case of Turkey could still be valuable for the MENA region.

Integrating the WEF nexus in climate law and policy: lessons from Turkey

Turkey has a growing population, as well as an expanding industrial base, which drive up demand for energy year on year.[33] The population grew by around 11 million people in the last decade, from 72.6 million in 2009 to 83.2 million in 2019,[34] while energy consumption increased from 161,947 Gigawatt (GW) in 2008 to 258,232 GW in 2018, putting pressure on policy makers to secure new energy resources. Turkey is a net energy importer; it is currently diversifying its energy portfolio and is developing renewables to meet this growing demand and need. Viewed through the lens of WEF, Turkey has had success in some areas, but there are clear gaps and challenges in other areas. These are discussed in the subsections below.

Table 8.1 Power of installed power plants in Turkey, gross generation, and net electricity consumption

Year	Total power installed Megawatt (MW)	Gross generation Gigawatt/hour (GWh)	Net consumption
2008	41 817,2	198 418,0	161 947,6
2009	44 761,2	194 812,9	156 894,1
2010	49 524,1	211 207,7	172 050,6
2011	52 911,1	229 395,1	186 099,6
2012	57 059,4	239 496,8	194 923,4
2013	64 007,5	240 154,0	198 045,2
2014	69 519,8	251 962,8	207 375,1
2015	73 146,7	261 783,3	217 312,3
2016	78 497,4	274 407,7	231 203,7
2017	85 200,0	297 277,5	249 022,6
2018	88 550,8	304 801,9	258 232,2

Source: TETC Electricity Generation – Transmission.
Turkish Institute of Statistics – TUIK: Power installed of power plants, gross generation and net consumption of electricity in Turkey: <http://www.tuik.gov.tr/PreIstatistikMeta.do?istab_id=129> accessed 3 June 2020.

Success stories

Among MENA countries, Turkey has one of the most diversified economies, with a strong domestic industrial and agricultural production base. Turkey is able to use water to generate energy, while in many MENA countries, desalination consumes energy to generate water. The Eastern Mediterranean and Middle East regions are already amongst the most water-scarce regions, and they face the challenges of further global warming and climate change.[35] It is estimated that 12% of the world's population currently uses 85% of its fresh water, and it is likely that climate change will accelerate this inequality.[36] Given the chronic shortages, water-sector investments in MENA are often regional issues. Especially in the case of transboundary water, there are frequent disagreements as to how water is distributed to end-users (households, industry, or farmers).[37] Turkey's practices regarding water management, conservation, and raising public awareness on consumption, in addition to its geographical advantages, put Turkey in a more secure position compared to most MENA countries. However, although relatively better placed in terms of access to water resources than the rest of the Middle East,[38] Turkey is also facing the problem of having less water per capita available annually. According to FAO's Aquastat, Turkey's renewable water resources per capita fell from an average of 3,044m^3 in the period 2008–2012 to 2,811m^3 in 2013–2017.[39]

Some of Turkey's clearest successes are in the field of energy generation and hydroelectricity. Its diversified but energy import-dependent economy provided a primary motivation for strengthening national production capacity, including through renewables. From a business and energy security standpoint, Turkey's achievements include the passage of its comprehensive Renewable Energy Law (*Yenilenebilir Enerji Kanunu* – YEK), with a specific FIT rate; its significant and effective energy regulatory authority; and its stock market for energy exchange. However, the department for water—the State Hydraulic Works (DSI)—keeps shifting between the portfolios of different ministries. In terms of its markets, Turkey is mostly liberalized. Considerable rights are granted to companies, especially in hydroelectricity. Electricity Market Law No. 4628 is the basis for "water rights concession contracts" setting the principles for leasing the rights of water use from rivers for 49 years by the private sector to produce electricity.[40]

Turkey's model for the unbundling of energy markets is based on compliance with the EU's *Acquis*. The current market structure was created in 2001 with the Electricity Market Law No. 4628, to comply with the *Acquis* and to liberalize the market. Law No. 4628 included the establishment of an independent regulatory body, the Energy Market Regulatory Authority or EMRA (*Enerji Piyasası Düzenleme Kurumu* – EPDK).

The legislation in Turkey includes incentives for the use of renewable energy sources for generating electricity. Renewable Energy Law No. 5346 was passed in 2005 and was reinforced with generous FIT schemes in 2009–2011, which promoted the rapid development of renewables, including small-hydro and wind.[41] Additional support further encouraged local investors to use local equipment. The Renewable Energy Law was an important step in expanding investment into electricity generation from renewable energy sources, encouraging investors by providing them with a fixed price guarantee.

Low FIT levels in 2005 were the main reason for the failure to achieve the expected acceleration in renewables in the early stages.[42] Modifications were made in 2011 through Law No. 6094 and Law No. 5346, providing more generous incentives as shown in Table 8.2. In addition to the fixed price guarantee, provisions for unlicensed small-scale electricity generation and new investments incentives were introduced in 2011. This new initiative effectively accelerated

Table 8.2 Renewable energy, purchasing price guarantee rates in Turkey

Type of energy generation facility	Price guarantee (dollar cent / kWh)*	Additional maximum support if domestic equipment is used (dollar cent / kWh)
Hydroelectricity	7.3	2.3
Wind	7.3	3.7
Geothermal	10.5	2.7
Biogas/Biomass	13.3	5.6
Solar PV (Photovoltaic)	13.3	6.7
Solar (Concentrated Solar Plants – CSP)	13.3	9.2

* kWh = kilowatt per hour
Source: Ministry of Energy, EMRA, TSKB Economic Research, TOBB,

investments to more desirable levels. In 2019, Presidential Decree 1044 published in the Republic of Turkey Official Gazette No. 30772 allowed unlicensed electricity generation for renewable energy of up to five megawatts (MW). With this measure, the framework and incentives for rooftop and other types of small-scale production supplying electricity to the grid became possible. EMRA/EPDK promulgated its most recent regulation on unlicensed electricity in compliance with the presidential decree. This decree is expected to further encourage long-term development of renewables, building on the success of prior policies that resulted in a visible increase of all renewables, including hydro-power and non-hydro forms of renewables (see Table 8.3).

Turkey's incentives to private investors (Turkish or foreign) for developing renewable energy are pro-business and output-oriented. As shown in Table 8.3, the share of renewable energy and (burning) waste was below one per cent of the national energy mix in 2008. By 2018, the share of renewables and wastes had reached 12.7 per cent and hydroelectricity accounted for around 20 per cent. While the percentage of coal also increased steadily, the share of other hydro-carbons fell. The increase in the share of coal from 29 per cent in 2008 to 37 per cent in 2017 reflects the "national resources" policy of the Ministry of Energy and Natural Resources (MoE), as suggested by national energy discourses over the last decade.[43]

While Turkey's rapid development of renewable energy is a success story from the standpoint of electricity generation and emissions reduction, there are environmental pitfalls, as discussed later, which existing policies and legislation do not effectively address. Turkey's energy policy does not prioritize fulfilling international climate commitments,[44] but rather focuses on the use of all forms of resources, including renewables, to meet needs.

Researchers often highlight the potential of renewable energy in Turkey, such as hydropower.[45] The opening ceremony of the Ilisu Dam took place in May 2020, in the midst of the COVID-19 pandemic.[46] This builds on the image of Turkey remaining strong in the face of the pandemic, evoking a sense of national pride. The Ilisu Dam is located on the Tigris River; it is the largest hydropower project in Turkey, with a power station of 1,200 MW planned capacity expected to generate up to 3,800 GWh of power per year. The Ilisu Dam has caused some tensions between Turkey and Iraq, but Turkish authorities stress that the dam will not divert water for irrigation, and will only release water back into the Tigris.[47]

Besides major projects such as Ilisu Dam, small hydropower developments have been undertaken with the same ambition of meeting national energy needs.[48] Turkey's implementation of renewable laws and support schemes (especially for hydro) are comparable to EU and particularly German models, emphasizing investor friendliness. In the wake of Renewable Energy Law No. 5346, small hydropower plant projects (below 500 kW) increased by 15 per cent in 2007 compared to 2006.[49]

Turkey's greater emphasis on energy, compared to water, is evident from the different status of the two agencies. The energy regulator, EMRA/EPDK, is an autonomous agency with a robust *modus operandi*. Although it is closely linked

Table 8.3 Shares (percentages) of energy resources by type. Turkish Institute of Statistics –TUIK

Enerji kaynaklarına göre elektrik enerjisi üretimi ve payları

Electricity generation and shares by energy resources

Yıl	Toplam	Kömür	Sıvı yakıtlar	Doğal gaz	Hidrolik	Yenilenebilir Enerji ve Atıklar [1]
Year	Total	Coal	Liquid fuels	Natural Gas	Hydro	Renewable Energy and wastes[1]
	(GWh)				(%)	
1970	8,623	32.7	30.2	-	35.2	1.9
1971	9,781	30.4	41.2	-	26.7	1.7
1972	11,242	26.0	43.9	-	28.5	1.6
1973	12,425	26.1	51.3	-	21.0	1.6
1974	13,477	28.8	44.8	-	24.9	1.5
1975	15,623	26.3	34.5	-	37.8	1.4
1976	18,283	23.7	29.6	-	45.8	0.9
1977	20,565	23.8	33.4	-	41.7	1.1
1978	21,726	25.7	30.7	-	43.0	0.6
1979	22,522	28.6	25.1	-	45.7	0.6
1980	23,275	25.6	25.0	-	48.8	0.6
1981	24,673	24.9	23.6	-	51.1	0.4
1982	26,552	24.2	22.4	-	53.4	0.0
1983	27,347	31.4	27.1	-	41.5	0.0
1984	30,614	33.0	23.0	-	43.9	0.1
1985	34,219	43.9	20.7	0.2	35.2	0.0
1986	39,695	49.0	17.6	3.4	29.9	0.1
1987	44,353	39.8	12.4	5.7	42.0	0.1
1988	48,049	26.0	6.9	6.7	60.3	0.1
1989	52,043	38.9	8.2	18.3	34.5	0.1
1990	57,543	35.1	6.8	17.7	40.2	0.2
1991	60,246	35.8	5.6	20.8	37.6	0.2
1992	67,342	36.5	7.8	16.0	39.5	0.2
1993	73,808	32.1	7.0	14.6	46.1	0.2
1994	78,322	36.0	7.1	17.6	39.1	0.2
1995	86,247	32.5	6.7	19.2	41.2	0.4
1996	94,862	32.0	6.9	18.1	42.7	0.3
1997	1,03,296	32.8	6.9	21.4	38.5	0.4
1998	1,11,022	32.2	7.2	22.4	38.0	0.3
1999	1,16,440	31.8	6.9	31.2	29.8	0.3
2000	1,24,922	30.6	7.5	37.0	24.7	0.3
2001	1,22,725	31.3	8.4	40.4	19.6	0.3
2002	1,29,400	24.8	8.3	40.6	26.0	0.3
2003	1,40,581	22.9	6.6	45.2	25.1	0.2
2004	1,50,698	22.8	5.0	41.3	30.6	0.3
2005	1,61,956	26.6	3.4	45.3	24.4	0.3
2006	1,76,300	26.4	2.4	45.8	25.1	0.3
2007	1,91,558	27.9	3.4	49.6	18.7	0.4

(Continued)

Table 8.3 Continued

Enerji kaynaklarına göre elektrik enerjisi üretimi ve payları

Electricity generation and shares by energy resources

Yıl	Toplam	Kömür	Sıvı yakıtlar	Doğal gaz	Hidrolik	Yenilenebilir Enerji ve Atıklar [1]
Year	Total	Coal	Liquid fuels	Natural Gas	Hydro	Renewable Energy and wastes[1]
	(GWh)				(%)	
2008	1,98,418	29.1	3.8	49.7	16.8	0.6
2009	1,94,813	28.6	2.5	49.3	18.5	1.2
2010	2,11,208	26.1	1.0	46.5	24.5	1.9
2011	2,29,395	28.8	0.4	45.4	22.8	2.6
2012	2,39,497	28.4	0.7	43.6	24.2	3.1
2013	2,40,154	26.6	0.7	43.8	24.7	4.2
2014	2,51,963	30.2	0.9	47.9	16.1	4.9
2015	2,61,783	29.1	0.9	37.9	25.6	6.5
2016	2,74,408	33.7	0.7	32.5	24.5	8.6
2017	2,97,278	32.8	0.4	37.2	19.6	10.0
2018	3,04,802	37.2	0.1	30.3	19.7	12.7

Kaynak: TEİAŞ, Türkiye Elektrik Üretim - İletim İstatistikleri
Source: TETC, Electricity Generation - Transmission Statistics of Turkey
(1) Jeotermal, rüzgar, katı biyokütle, güneş, biogaz ve atık kaynaklarını içerir.
(1) Renewable energy and waste includes geothermal, solar,wind, solid biomass, biogas and waste.
Tablodaki rakamlar, yuvarlamadan dolayı toplamı vermeyebilir.

Figures in table may not add up to totals due to rounding.
Turkish Institute of Statistics – TUIK ; Electricity generation and shares by energy resources in Turkey: <http://www.tuik.gov.tr/PreIstatistikTablo.do?istab_id=1578> accessed 3 June 2020.

with the MoE, it is technically an independent regulatory authority. EMRA/ EPDK can, for instance, set price caps on petroleum. As indicated above, in order to attract investment to the sector, Turkey provides subsidies and other promotion schemes such as FITs for renewable energy. EMRA/EPDK regulates fixed price guarantee and support to renewable companies.

In theory, Turkey's Renewable Energy Law, with its comprehensive set of incentives through FITs, is a step towards the WEF nexus approach. In practice, however, companies continued to focus on profits rather than environmental impact and did not seek acceptance or collaboration from civil society and local communities. Such gaps and challenges, and the lessons to be learned, will be discussed next.

Gaps and challenges

Although Turkey has been successful in developing renewables, it demonstrates an "extractive" mentality, as in mining, rather than a "sustainable" outlook that encompasses the ecosystem as a whole. Energy output and profitability are the main driving forces. There is little effort to engage local populations during the environmental and social impact assessment stages, and there is often a lack of transparency.

In the development phase of renewable power plants, there is a tendency to treat environmental impacts as secondary compared to economic outcomes. Critical scholars point out that some key energy investments have been distributed to the same group of companies.[50] There are inconsistencies in the country's environmental regulations: although environmental impact assessment reports (*Çevre Etkileşim Raporu* – ÇED reports) are obligatory for all projects in Turkey, they are not implemented in a transparent manner. This often results in procedures being bypassed, as contractors try to accomplish "done deals" rather than effectively seeking environmental and social approval of the local population. There is insufficient emphasis on the environment.[51] Hydroelectric power stations (*Hidroelektrik Santraller* – HES), in particular, have been the site of many legal battles, especially in the Black Sea region, as local rivers dried up during the process, harming the ecosystem. In contrast to the expected beneficial outcomes, ineffective and output-oriented planning mean that renewable energy projects can sometimes be detrimental to the environment. This inefficiency and lack of transparency reflect the traditional siloed approach that has characterized the legislation and institutions over the last two decades.

In addition to government FIT schemes and international donors such as the World Bank, finance for energy projects has also come from the private sector. Since 2001, the Industrial Development Bank of Turkey (TSKB) has been providing loans for hydropower (including small hydropower) plants as well as other renewable sources for electricity generation in Turkey.[52] Many private companies involved in renewable energy projects fail to communicate with local communities and civil society, preferring to deal only with the government, in spite of the fact that their projects could, for example, affect land access for agricultural purposes and impact other areas such as food security, especially in rural communities.[53] Indeed, the 2009–2011 FIT schemes encouraged private companies to develop hydroelectricity without effectively communicating with local communities. What is needed now is a focus on measures related to preserving local ecosystems.[54] More transparent procedures, including in the environmental approval process, could also promote trust and increase international funding.

Turkey has recently undergone a major change in its government structure, shifting from a parliamentary to a presidential government system following the referendum of 16 April 2017, and Law No. 6771 of 21 January 2017. The president became head of state and head of government, as well as head of a political party. All executive power now rests with the president. While this transition is still ongoing, it may also be an opportunity for a new agency to adopt rules and regulations which align with the WEF nexus approach.

Despite its apparent success in developing renewable energy sources, this section has identified as areas of concern: weak legal structures, lack of civil society engagement, social exclusion, and insufficient transparency. Short-term thinking and the inconsistent application of mitigation standards are additional shortcomings in the current system.

The DSI is the primary agency in Turkey for planning, managing, and operation of national water resources; its main focus has been on building hydroelectric dams.[55] Unlike the energy regulator, EMRA/EPDK, the DSI is a general directorate, makes it a subordinate agency. For many years it came under the MoE, but it now falls under the Ministry of Agriculture and Forestry. Although its function is to deliver water across the country, its shifting status and lack of authority have often prevented environmental concerns from being prioritized. More generally, there is an overall lack of coordination between agencies involved in environmental issues in Turkey. There is no strong environmental agency or institutional body to consolidate environmental issues which are currently fragmented across the portfolios of three ministries, whose primary responsibilities are energy output, urban development, and agriculture, respectively (see Figure 8.1).

In some cases, this has led to the paradoxical situation of renewable energy investments being considered harmful for the environment, or being opposed by local populations and farmers due to contested water rights. There are examples of companies in Turkey that have adopted more transparent processes and CSR policies; their experiences suggest that there was less confrontation with local populations and less opposition to projects.

Thus one of the lessons to be learned from Turkey is to better engage local citizens when seeking environmental and social approval.[56] Compliance with tougher environmental regulations, as well as more transparent processes at the approval stages of environmental and social impact assessments, would improve outcomes for investors and governments alike. One successful example of foreign direct investment in hydroelectricity in Turkey is the construction of the Alparslan 2 hydropower dam on the Murat River by the Czech company Energo-Pro. The dam will also provide irrigation to the Muş area in Eastern Turkey. Energo-Pro adopted a more transparent process of approval for its project, with greater public communication.[57] The project is among the top ten hydro dams in Turkey, based on the size of the reservoir, and claims that it will become the largest reservoir project owned by the private sector. The Alparslan 2 dam is

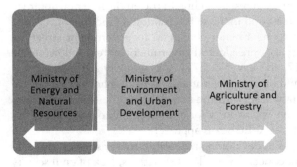

Figure 8.1 (original, created by author) Division of WEF and environmental portfolios

expected to start generating electricity by the end of 2020.[58] It is arguably one of the best examples of how transparent procedures and social engagement can reduce tensions and increase the success of a project.

Another example of good practice is the BTC company; while building the Baku Tbilisi Ceyhan crude oil pipeline, BTC agreed a rigorous "Human Rights Undertaking" in 2003. This has provided benchmarks for social and environmental impact assessments[59]—a precedent which should set the standards for other investing companies.

Advancing the implementation of a WEF nexus approach in the MENA region: recommendations

Based on the above analysis of the strengths and weaknesses of Turkey's approach, this section offers some recommendations moving forward.

Creation of a strong new institution for coordination of WEF nexus approach

The first recommendation is the creation of a strong institution such as an Environmental Regulation Agency (or a dedicated Ministry of Environment), which could implement the WEF nexus approach in a holistic manner. Establishing an independent and effective institution, with the power to impose rules and regulations, would help to balance the profitability of investors with long-term environmental safeguards as well as the needs of the local population. Furthermore, Turkey needs a robust water agency dedicated to coordinating and safeguarding the country's water supply, to replace the current piecemeal arrangements. In general, permanent bodies are most effective. Even when collaboration occurs in *ad hoc* project contexts, it is often limited to a couple of pilot projects achieving limited success.[60] What is needed is an institution with a strong mandate for coordination and the power to enforce rules and regulations.

Following the 2019 municipal elections, Turkey's three major metropolitan cities are now governed by the opposition party. Metropolitan municipalities and central government agencies must actively collaborate to address problems of climate change. The majority of the Turkish population live in cities, and face issues such as flash floods due to poor infrastructure.[61] An independent agency would also assume a coordinating role between the central and local governments. Considering the examples of Hamburg and Rotterdam, local governments' role would be to manage and govern many of the support schemes detailed above.[62]

More transparency for better engagement with citizens and civil society

The second recommendation is to improve transparency on environmental matters, which will establish public trust while reducing transaction costs in communication. There is no single successful model at the municipal or national government levels; however, there are many paths to governing that can engage the public with the aim of efficient climate change policy making and implementation. Again, the cases of Hamburg and Rotterdam illustrate different institutional frameworks, with Hamburg adopting more formal and Rotterdam more informal arrangements.[63] A good institutional framework could use a mix of formal regulations and incentives. To achieve effective climate change governance and adaptation in cities, well-designed institutional arrangements are needed at both local and national levels of government.

Mechanisms to engage civil society, local communities, and citizens in general are urgently needed. The role of CSR is important for engaging communities, but also for further strengthening the sense of "climate justice".[64] CSR norms and literature have much to say about social acceptance; if private companies are obliged to adopt these norms by the regulators, they would be better prepared to engage civil society and local communities more effectively. There are local NGOs in Turkey that could help with this.[65] Encouraging local citizens to participate in designing creative solutions to environmental challenges will increase public involvement and facilitate communication and public acceptance. The role of civil society would be to influence government decision making by awareness raising. Beyond raising awareness, local stakeholders can add benefits in addressing WEF nexus challenges at the research stage.[66] The water rights concession contracts established by the Electricity Market Law No. 4628 allow private companies to lease water use rights from rivers for 49 years to produce electricity.[67] When rights are being granted for such long periods of time, it is important to engage local citizens in the process.

Conclusion

Turkey can boast various successes, including a diversified energy mix and a series of legislative measures that legally comply with the EU's *Acquis Communautaire*. Many policies that align with the WEF nexus approach have been implemented, including the rapid development of renewable energy sources. Renewable energy can help to address challenges of climate change. Turkey's FIT scheme with price guarantees has been successful in promoting private sector investment in renewables. However, the policy was oriented towards meeting the growing energy demand and ensuring supply security[68] and profitability rather than encompassing a holistic environmental approach. The WEF nexus can offer a framework for integrating climate response into key sectors; it thus fits well with Turkey's diversified economy. However, in spite of some successes, output and profitability have been prioritized. Environmental assessment processes have often been hasty and insufficiently transparent.[69]

As demonstrated in this chapter, the existing piecemeal approach could be remedied by a strong independent environmental institution that could consolidate and coordinate environmental issues currently fragmented across the portfolios of several ministries. Such an institution could facilitate better engagement of civil society and enforce more transparent procedures. Positive engagement through policies of corporate, social, and environmental responsibility can boost the brand value and reputation of investors. The existence of a robust institution or independent environmental regulator should facilitate the enforcement of and compliance with environmental regulations.

Turkey's new presidential system is still in its early days. A period of institutional transition and re-shaping could be an opportunity for establishing new mechanisms to enforce rules and regulations on the environment, adopting a WEF nexus approach, including the creation of an independent institution. Further research is needed on how such an institution would operate and how it might coordinate the nexus approach.

Notes

1 For the Middle East, see G. Giordano and D. A. L. Quagliarotti, "The Water–Energy–Security Nexus in the Middle East" in Sh. Kronich and L. Maghen (eds), *Ensuring Water Security in the Middle East: Policy Implications* (Barcelona: European Institute of the Mediterranean 2020). For rising temperatures in other regions such as China, see J. Lewis, "Climate Change and Security: Examining China's Challenges in a Warming World" (2009) 85 *International Affairs* 1195.

2 D. Olawuyi, *The Human Rights Based Approach to Carbon Finance* (Cambridge University Press, 2016) 1-25.

3 See Giordano and Quagliarotti, n 1. Also see E. M. Biggs and others, "Sustainable Development and the Water–Energy–Food Nexus: A Perspective on Livelihoods" (2015) 54 *Environmental Science & Policy* 389.

4 M. Kurian, "The Water–Energy–Food Nexus: Trade-Offs, Thresholds and Transdisciplinary Approaches to Sustainable Development" (2017) 68 *Environmental Science & Policy* 97.

5 R. Beveridge and K. Kern, "The Energiewende in Germany: Background, Developments and Future Challenges" (2013) 4 *Renewable Energy Law & Policy Review* 3.

6 See IEA, "World Energy Investment 2020", <https://www.iea.org/reports/world-energy-investment-2020/key-findings> accessed 7 April 2021.

7 Energy Information Administration – EIA; NYMEX Future Prices <https://www.eia.gov/dnav/pet/hist/RCLC1D.htm> accessed 3 June 2020.

8 R. Staupe-Delgado, "The Water–Energy–Food–Environmental Security Nexus: Moving the Debate Forward" (2019) 22 *Environment, Development, and Sustainability* 6131–6147.

9 D. H. Meadows and others, *The Limits to Growth: A Report to the Club of Rome's Project on the Predicament of Mankind* (New York: Universe Books 1972).

10 Hultman, Bonnedahl, and O'Neill draw attention to the prioritization of business and profit. See, M. Hultman and others, "Unsustainable Societies–Sustainable Businesses? Introduction to Special Issue of Small Enterprise Research on Transitional Ecopreneurs" (2016) 23 *Small Enterprise Research* 1.

11 M. Vlasov, "In Transition toward the Ecocentric Entrepreneurship Nexus: How Nature Helps Entrepreneur Make Venture More Regenerative Over Time" (2019) *Organization & Environment* https://doi.org/10.1177/1086026619831448.

12 Staupe-Delgado, n 8.

13 Giordano and Quagliarotti, n 1.

14 D. S. Olawuyi, "Sustainable Development and the Water–Energy–Food Nexus: Legal Challenges and Emerging Solutions" (2020) 103 *Environmental Science & Policy* 1.

15 Biggs and others, n 3.

16 As in the example of Tanzania, analyzed by J. Pardoe and others, "Climate Change and the Water–Energy–Food Nexus: Insights from Policy and Practice in Tanzania"

(2018) 18 *Climate Policy* 863. See also D. Olawuyi, *Principles of Nigerian Environmental Law* (Ado Ekiti: Afe Babalola University Press 2015) 323–330.

17 Olawuyi, n 14.

18 Staupe-Delgado, n 8.

19 Erensu provides various examples from the Turkish context. See, S. Erensu, "Abundance and Scarcity Amidst the Crisis of 'Modern Water': The Changing Water–Energy Nexus in Turkey" in L. M. Harris, J. A. Goldin, and C. Sneddon (eds) *Contemporary Water Governance in the Global South* (New York: Routledge 2015).

20 See, V. De Laurentiis, D. V. L. Hunt, and C. D. F. Rogers, "Overcoming Food Security Challenges within an Energy/Water/Food Nexus (EWFN) Approach" (2016) 8 *Sustainability* 95.

21 Olawuyi, n 14, 3.

22 S. Mpandeli and others, "Climate Change Adaptation Through the Water–Energy–Food Nexus in Southern Africa" (2018) 15 (10) *International Journal of Environmental Research Public Health* 2306.

23 Ibid.

24 On 20 April 2020, WTI/NYMEX Crude Oil Prices reached negative levels at – 37$, EIA; NYMEX future prices <https://www.eia.gov/dnav/pet/hist/RCLC1D.htm> accessed 3 June 2020.

25 A. O. Dirioz and B. A. Reimold, "The Strategic Context of the UAE's Nuclear Project: A Model for the Region?" (2014) 21 *Middle East Policy* 71.

26 D. Jahn and S. Korolczuk, "German Exceptionalism: The End of Nuclear Energy in Germany!" (2012) 21 *Environmental Politics* 159.

27 For more on Energiewende, see Beveridge and Kern, n 5.

28 Masdar initiatives <https://masdar.ae/> attempted to create a zero emissions city, capitalizing on solar power. Morocco, likewise, was able to attract funding for solar power projects, including the 580 MW Noor-Ouarzazate complex. See, T. Makang, M. Nefer, and Z. Dobrotkova, *The Role of the Public Sector in Mobilizing Commercial Finance for Grid-Connected Solar Projects: Lessons from Seven Developing Countries* (Washington, DC: World Bank Group 2019), 16, 22, 23.

29 See, Ministry of Foreign Affairs – MFA Turkey, relations with EU in the field of environment <http://www.mfa.gov.tr/relations-with-the-european-union-in-the-field-of -environment.en.mfa> accessed 31 May 2020.

30 I. N. Grigoriadis, "Islam and Democratization in Turkey: Secularism and Trust in a Divided Society" (2009) 16 *Democratization* 1194.

31 According to World Bank data, in the Middle East, only Egypt has a larger population than Turkey, while Iran's population is similar to that of Turkey. <https://data.worldba nk.org/indicator/SP.POP.TOTL?locations=TR-IR>.

32 As discussed by Kurian, n 4, 102. See also D. Olawuyi, "Advancing Innovations in Renewable Energy Technologies as Alternatives to Fossil Fuel Use in the Middle East: Trends, Limitations, and Ways Forward" in D. Zillman and others (eds), *Innovation in Energy Law and Technology: Dynamic Solutions for Energy Transitions* (Oxford University Press 2018).

33 See, H. Ozturkler, "Türkiye'nin Enerji Sorunu ve Enerji Politikası-III" (2011) 5 *İktisat ve Toplum* 1–10.

34 Data from the Turkish Statistical Institute –TUIK's reported population census results reported on 4 February 2019 <http://www.tuik.gov.tr/HbGetirHTML.do?id=33705>.

35 Ministry of Foreign Affairs – MFA Turkey, n 29.

36 A. Badran, "Climate Change and Water Science Policy in Management" in A. Badran and others (eds), *Water, Energy & Food Sustainability in the Middle East* (Springer 2017).

37 V. Sumer, "A Chance for a Pax Aquarum in the Middle-East? Transcending the Six Obstacles for Transboundary Water Cooperation" (2014) 9 *Journal of Peacebuilding & Development* 83.

38 T. von Lossow and M. Shattat, "Less and Less: Water in the Middle East" in Kronich and Maghen, n 1.
39 These figures are (m³/per inhabitant/per year). See FAO AQUASTAT, Turkey, <http://www.fao.org/nr/water/aquastat/data/query/results.html> accessed 3 June 2020.
40 L. M. Harris and M. Islar, "Neoliberalism, Nature, and Changing Modalities of Environmental Governance in Contemporary Turkey" in Y. Atasoy (ed.), *Global Economic Crisis and the Politics of Diversity* (Basingstoke: Palgrave Macmillan 2014).
41 B. Dursun and C. Gokcol, "Impacts of the Renewable Energy Law on the Developments of Wind Energy in Turkey" (2014) 40 *Renewable and Sustainable Energy Reviews* 318.
42 See, O. Yılmaz and H. Hotunluoglu, "Yenilenebilir enerjiye yönelik teşvikler ve Türkiye" (2015) 2 *Adnan Menderes Üniversitesi Sosyal Bilimler Enstitüsü Dergisi* 74.
43 For national discourse descriptions, see Erensu, n 19.
44 D. Livingston, "Renewable Energy Investment in Turkey: Between Aspiration and Endurance" (2018) 17 *Turk Policy Q* 55.
45 M. Bilgili and others, "The Role of Hydropower Installations for Sustainable Energy Development in Turkey and the World" (2018) 126 *Renewable Energy* 755.
46 See, "Turkey Inaugurates Ilisu Dam" <https://www.hurriyetdailynews.com/turkey-inaugurates-ilisu-dam-154904> accessed 4 August 2020. The same report also points out that, in constructing the dam, "Turkey relocated many historic cultural assets in Hasankeyf, which sits on the banks of the Tigris River. People in the region were also resettled" (ibid).
47 See Ministry of Foreign Affairs – MFA, Turkey, Ilisu dam, <http://www.mfa.gov.tr/ilisu-dam.en.mfa> accessed 31 May 2020.
48 S. Kucukali and K. Baris, "Assessment of Small Hydropower (SHP) Development in Turkey: Laws, Regulations and EU Policy Perspective" (2009) 37 *Energy Policy* 3872.
49 Ibid.
50 See N. Pamir, *Enerjinin İktidarı* (Hayy Kitap, 2017) book available in Turkish; the tittle in English translates to *Energy in Power of the Government* or *Energy Ruling the Government*.
51 O. Urker and N. Cobanoglu, "Türkiye'de hidroelektrik santraller'in durumu (HES'ler) ve çevre politikaları bağlamında değerlendirilmesi" (2017) 3 *Ankara Üniversitesi Sosyal Bilimler Dergisi* 65.
52 Kucukali and Baris, n 48, 3877.
53 Olawuyi, n 14.
54 Kucukali and Baris, n 48.
55 The General Directorate of State Hydraulic Works, DSI, is currently under the Republic of Turkey Ministry of Agriculture and Forestry <http://en.dsi.gov.tr/> accessed 2 June 2020.
56 C. Hoolohan and others, "Engaging Stakeholders in Research to Address Water–Energy–Food (WEF) Nexus Challenges" (2018) 13 *Sustainability Science* 1415.
57 Environmental Impact Assessment Report is available in English on the company website <http://www.energo-pro.com/alpaslan2esia>.
58 See, Energo-Pro website (http://www.energo-pro.com/en#kde-jsme); and Anatolian News Agency <https://www.aa.com.tr/tr/ekonomi/su-tutmaya-baslayan-alparslan-2-baraji-enerji-uretimine-hazirlaniyor/1838623> accessed 12 May 2020.
59 H. Sahin, *Host Government Agreements and the Law in the Energy Sector: The Case of Azerbaijan and Turkey* (New York: Routledge 2018).
60 Pardoe and others, n 16.
61 Floods are a major challenge in Turkish cities, and account for the second highest number of natural disaster deaths, following earthquakes. See, A. Himat, M. Onucyildiz, and S. Dogan, "Urban Drainage Design According to Turkish Rainwater Harvesting and Disposal Guideline" (2019) 2 *International Journal of Environmental Pollution and Environmental Modelling* 219; *Daily Sabah*, 22 June 2020, "Flash Floods

Kill 5 in Turkey's Bursa Province, 1 Missing" <https://www.dailysabah.com/turkey/f lash-floods-kill-5-in-turkeys-bursa-province-1-missing/news>.

62 J.-T. Huang-Lachmann and J. C. Lovett, "How Cities Prepare for Climate Change: Comparing Hamburg and Rotterdam" (2016) 54 *Cities* 36.

63 Huang-Lachmann and Lovett, n 66, demonstrate how local companies in Hamburg had to comply with rules, while Rotterdam promoted informal incentives and points in bidding proposal award processes. While a combination of both measures is necessary, starting with formal processes may be more efficient.

64 D. S. Olawuyi, "Climate Justice and Corporate Responsibility: Taking Human Rights Seriously in Climate Actions and Projects" (2016) 34 *Journal of Energy & Natural Resources Law* 27.

65 The Corporate Social Responsibility Association of Turkey <http://csrturkey.org/> is one such NGO that could potentially guide companies to adopt policies more compatible with CSR principles.

66 Hoolohan and others, n 56.

67 Harris and Islar, n 40.

68 For Turkish Energy Security, see, E. Ersen and M. Celikpala, "Turkey and the Changing Energy Geopolitics of Eurasia" (2019) 128 *Energy Policy* 584.

69 See, Urker and Cobanoglu, n 51.

9 Urban law and resilience challenges of climate change for the MENA region

Robert Home

Introduction

The chapter examines law and policy frameworks through which MENA countries can better integrate climate vulnerability assessment and resilience planning into urban law and management. It explores the MENA region's "urban crisis", emerging holistic views about urban resilience, and issues of urban land governance, using case studies from across the region.

Urban law, as a subset of environmental law, has been around for half a century; and it provides law and governance frameworks for the effective management and planning of cities.[1] Ongoing international and regional responses to the existential challenges of climate change, especially its impacts on cities, are adding new impetus and dimension to urban planning and management. The United Nations Framework Convention on Climate Change (UNFCCC), as well as the 2030 Agenda on Sustainable Development with its 17 Sustainable Development Goals (SDGs), all accentuate the need for urgent and drastic response if global cities are to be able to manage and cope with the risks of climate change.[2]

SDG 13 specifically urges countries to "take urgent action to combat climate change and its impacts".[3] Furthermore, since "the sustainable development fight will be won or lost in our cities",[4] "SDG 11: Inclusive, safe, resilient and sustainable cities" is also engaged, and indeed has been claimed as "the heart of the SDGs",[5] and additionally supported by UN-Habitat's New Urban Agenda (NUA).[6] All the SDGs are ultimately relevant, directly or indirectly, to climate change, given the complex interdependencies involved, and this chapter will also refer specifically to two others. To promote urban resilience to climate change, lawmakers will have particular regard to "SDG 16: Promote peaceful and inclusive societies, access to justice, effective, accountable and inclusive institutions".[7] Similarly, "SDG 17: Partnerships to achieve the goal"[8] includes, among its many targets, enhancing the global partnership for sustainable development, and sharing knowledge, expertise, technology, and financial resources.[9]

This chapter examines the progress made and the challenges that remain in the MENA region with integrating climate vulnerability assessment and resilience planning into urban law and management. The approach can be labelled inter-disciplinary or even post-disciplinary, and so refers to a wide range of relevant literature, not only in law, but also in other allied disciplines such as history, geography, governance, urban planning, and environmental sciences.

The chapter is divided into five sections. After this introduction, the second section situates climate change and vulnerability within the long history of the region, and reviews development (or lack of it) of climate change law and practice in MENA. The third section explores the MENA Region's "urban crisis" and the emerging holistic view of urban resilience, and issues of urban land governance, using specific city cases drawn from recent research. The fourth section explores directions for future legal reforms and changes in social behaviour. The final section draws conclusions.

Climate change vulnerability and the MENA region

The MENA region has been profoundly important in human history for many reasons, including the fact that the first humans to leave Africa probably crossed into it via the Horn of Africa hundreds of thousands of years ago. Within the past five millennia great civilisations and religions have risen and fallen in the region, long-distance trade routes and pandemics have passed through it, and it contains the so-called "fertile crescent", where settled farming and the first cities emerged.[10] MENA societies' centuries-old traditional adaptations to deal with water scarcity and hot climate offer a valuable repository of human knowledge related to climate change, exemplified by the Al-Ain oasis in Abu Dhabi, designated a World Heritage site for its ancient underground irrigation system (*falaj*).[11]

The 20th century was arguably the most turbulent one that the region has experienced, and the convulsions continue into the present century, with serious impact upon its geo-politics and institutions which restrict the possibility of serious climate change action. Some half of MENA countries have been profoundly damaged in recent years by civil war and political instability, and even the Gulf Cooperation Council (GCC) may not be immune.[12] The region hosts an estimated 12 million refugees and internally displaced persons, and some 80 million people live in slums or informal settlements.[13] Conflict-affected Syria, Libya, and Yemen have suffered a sudden fall in development ranking, while Lebanon is affected by waves of refugees from Syria's civil war, and Jordan has an estimated 630,000 Syrian refugees living in informal settlements and camps, as well as half a million Palestinian refugees originally displaced from Israel in 1948. Such disruptive forces contribute to high migration rates to, from, and between MENA countries (averaging 1.5 migrants per 1,000 population), and Lebanon has the highest net emigration (40.3 emigrants per 1,000 population). Massive anti-government pro-democracy protests in the so-called Arab Spring 2010–2012 were renewed in 2018–2020 ("Arab Spring 2.0"), demanding basic infrastructure, such as adequate water and electricity, as well as deep political reforms.[14]

As threats from climate change become more extreme, competition over diminishing resources can be expected to cause increased future conflict, food shortages, and economic pain. The region's present drought, probably intensified by global warming, is the worst in over a century, and has been better termed as a "drying up" (i.e. a new normal and permanent climatic pattern) rather than "drought" (a normal and reversible pattern).[15] Desertification is a perennial and growing problem, and dust- and sand-storms increase in frequency and intensity, causing serious health problems and degrading infrastructure. Prolonged heat waves (summer temperatures are expected to grow at more than twice the global average) and droughts may make parts of the region increasingly less inhabitable. The Tigris and Euphrates rivers will disappear this century, and more frequent extreme weather events (Qatar in 2018 experienced almost a year's worth of rain in one day) will create new emergency relief needs.[16] The North African countries, for example, have average annual fresh-water availability of less than 1,000 cubic meters per capita, which is the common benchmark for water scarcity, and far below the global average of about 7,000. Jordan, for example, is one of the most water-scarce countries on the planet, with per capita freshwater availability falling from 3,600 cubic metres per year in 1946 to 135 cubic metres in 2017, largely because of abstraction from the River Jordan by neighbouring countries. Rising sea levels in the Mediterranean can be expected to make urban coastal zones more flood-prone; a temperature increase of 1–3 degrees could expose millions of people to coastal flooding, with Tunisia, Qatar, Libya, UAE, Kuwait, and Egypt at particular risk.[17]

International and regional responses to climate change generate close scrutiny from legal academics. Progress with climate change law operates within a complex and changing international legal context, and involves actions and measures by national and local governments, regional and global organisations, that embrace legislative and regulatory reforms, policies, and resourcing.[18] Several MENA governments, however, have often been more preoccupied with managing unrest and conflict within their borders, and this reduces their capacity to devise strategies and act on climate change mitigation and adaptation.

The London School of Economics in 2015 gathered evidence on national laws and policies relating to climate change mitigation and adaptation in 99 countries, nine of which were in the MENA region.[19] It found that, between 1997 (when the Kyoto Protocol was adopted) and 2014, the number of laws and policies grew from 54 to over 800, and most countries developed some form of climate change risk assessment and strategy, in accordance with UNFCCC reporting requirements. Framework legislation was encouraged to provide a comprehensive and unifying basis for climate change policy and action, addressing issues of climate change mitigation or adaptation (or both) in a holistic manner. Of the top 20 GHG-emitting countries, however, Saudi Arabia was one of only two (the other being Canada) that still had no framework policies for either mitigation or adaptation in 2014, and resists international climate change action, while attending climate talks.[20]

The UNFCCC and SDGs seek to mobilize a Green Climate Fund of $100 billion annually by 2020 to reduce emissions and adapt and mitigate climate change effects. The fund was to be resourced by states through a "national designated authority" and "nationally determined contributions", with projects implemented by "accredited entities".[21] Most MENA countries have designated a national authority, which is usually a Ministry of Environment, but sometimes also includes Natural Resources or Sustainable Development. Saudi Arabia designated its Ministry of Petroleum and Mineral Resources (although without committing itself to reducing GHG emissions),[22] while tiny Comoros designated its already multi-tasked Ministry of Production, Environment, Energy, Industry and Handicrafts.[23] The UAE, whose reliance upon oil and gas made them poorly placed to commit to the UNFCCC's targets for GHG emission reduction, did not participate in the Green Climate Fund, and only a minority of MENA countries are involved in its projects, which are classified as either mitigation or adaptation, and are mostly "cross-cutting" with multiple partners.[24]

Progress with supporting national climate change legislation has been slow globally, and particularly so in the MENA region.[25] A 2017 public opinion survey across the Arab countries found that most respondents (80 per cent) felt that the environmental situation was deteriorating, and 95 per cent that their country was not doing enough to tackle environmental challenges.[26] Qatar did, however, show support for the UNFCCC by hosting the 2012 UN Climate Change Conference (COP 18) in Doha, and in 2015 the International Islamic Climate Change Symposium in Istanbul adopted a declaration to support COP 21 in Paris.[27]

Complex issues of global and local climate change call for international and regional co-operation and supporting institutions (as SDG 17 urges[28]). The Climate Action Network brings together over 1,300 non-governmental organizations (NGOs) in more than 120 countries, and has an Arab World network since 2015, with a coordinator for each of the two MENA Regions.[29] The Arab Forum for Environment and Development (AFED) was inaugurated in 2006, and now has over a hundred institutional members, comprising government agencies as observers, corporate bodies, NGOs, universities and research centres, and media organizations. Its first baseline report (2008) on environmental issues identified six regional priorities: water, energy, air, food, green economy, and scientific research.[30]

Law reform has to come through existing legal systems, which in MENA countries draw from many sources, through statutory codes, constitutions, and bills of rights. Some of these countries only came into existence in the 20th century, with boundaries and legal institutions set by European colonial interventions: the Sykes-Picot treaty of 1916 has had long-term consequences for international boundaries fixed under the legal principle of *uti possedetis*.[31] The French civil or Napoleonic code, itself inspired by the sixth century Byzantine Code of Justinian, was introduced into Egypt as a system of mixed courts during the late 19th century, and was transferred across much of the MENA region. In the same period the Ottoman civil code (the Mecelle) was developed, incorporating the Hanafi school of Islamic law, and this survived the Ottoman empire's collapse, in former provinces that became separate nation states (notably Jordan, Lebanon, Syria, and Kuwait). Such mixed influences upon MENA legal systems make a comparative law approach especially important in the present age of internationalism, democratization, and SDGs.[32] To harmonise such legal codes with new concerns over environment and climate change is, however, unlikely in the present uncertain political climate, as MENA countries' minimal democratic reforms limit the scope for legislative and policy change. The Islamic *sharia* system of laws, under different schools of Islamic jurisprudence, may, however, be moving towards new hybrid models of Islamic constitutionalism.[33]

The realization of a climate change agenda would mean legislative and perhaps even constitutional changes within the countries signed up to it.[34] One of the targets for SDG 17 was to "respect each country's policy space and leadership to establish and implement policies",[35] and few states have achieved the SDG target of formulating a framework climate change law. The UK's Climate Change Act 2008 did impose a legal duty upon its present and future governments to reduce GHG emissions and make the UK a low-carbon economy, but has proved difficult to implement in practice;[36] Germany produces more than half of its energy from solar power, but the cost is high.[37] The UN Environment Programme in 2019 produced an online Law and Climate Change Toolkit for assessing a country's climate change legislation, and formulating legal instruments for planning, mitigation, and adaptation planning.[38]

Nation states may organise themselves into supra-national regional groupings. The European Union is considered to have the most extensive environmental laws of any regional body, with sustainable development a legal objective under its European Treaties, and a system of regulations and directives binding its members, of which framework directives (notably in the water and waste sectors) have been important environmental instruments.[39] Yet even the EU has not matched its ambitions with ability to deliver in practice, and is experiencing its own structural and political problems.

Regional integration in the MENA region is less advanced compared with other regions, and some half of its countries struggle with existential challenges to security and political legitimacy. There are, however, two notable regional bodies. One is the GCC six-state grouping, which are among the richest countries in the world, and the most polluting in GHG emissions. Their electricity consumption increased in the period 1980–2000 by an annual average of ten per cent (against a world average of three per cent), and subsidized electricity prices encourage extravagant air-conditioning in buildings and massive energy usage for water desalinization.[40] They have relatively small populations, and their monarchical system of government limits more inclusive governance, making a lead from them on climate change law unlikely.[41] The other regional body, to which the countries of North Africa belong, is the geographically vast 55-state African Union, but its institutional structure is weak, and, while the African Charter and Court of Human and Peoples' Rights have resulted in some potentially significant environmental and land resource cases, enforcement against non-compliant states remains ineffective.[42]

If the scope for realizing SDGs and climate change through legislative and constitutional reform at national or regional level is limited, climate change action and basic equality rights may also be progressed through changes in urban governance and local institutions, as explored in the following sections.

Responding to the MENA Region's "urban crisis"

"Cities in the Middle East are in crisis."[43] That statement was made in a book published nearly a quarter-century ago, and since then the situation has only gotten worse, as all MENA countries are becoming more urbanized.[44] The region's population grows at an average annual rate of 1.56 per cent, well above the global average of 1.1 per cent, and has been estimated will double in the next 30 years.[45] The urban share of that population grew from 48 per cent in the 1980s to 60 per cent in 2000, and may by now exceed 70 per cent. The region's highest urbanization rates (76 per cent) are found in the countries of the MENA Southern sub-region, aggravated by conflict, environmental degradation, drought, and rural poverty.[46] The MENA region faces severe droughts, exacerbated by human actions. Extreme drought between 2007 and 2012 in the fertile lands around the Euphrates, Tigris, and Nile rivers caused disastrous crop failures, with groundwater resources reduced by unsustainable, water-intensive agriculture; adverse conditions in north-eastern Syria, which previously accounted for over 60 per cent of the effective cultivated land, forced an estimated 300,000 people to move into cities such as Damascus and Aleppo, contributing to political unrest and civil war.

MENA cities will feel climate change impacts directly upon their physical environments, and indirectly from population displacement.[47] The urban heat island effect means temperatures are higher in urban than surrounding rural areas, with poor air quality and adverse public health consequences. Dubai, with one of the hottest climates on the planet, has built many high-rise structures of concrete and glass, and black-top roads and car-parks to accommodate its vehicles, darkening what was near-white sand, thus absorbing and releasing more heat. Another effect of large population numbers is experienced in the holy city of Mecca, Saudi Arabia, whose population of two million increases three-fold during the *hajj*; a crowd collapse in 2015 killed over 2,000 pilgrims, who were moving slowly on foot at densities of about six/seven persons per square meter, at a time of peak combined temperature and humidity levels.[48]

The plans and actions of individual cities, often struggling with political instability and conflict, show little capacity or appetite for climate change action, apart from some research and development in alternative sources to carbon-based energy. When the World Economic Forum produced a research report on city competitiveness in 2014 it identified six wider "global megatrends": urbanization and demographics; rising inequality; sustainability; technological change; industrial clusters and global value chains; and governance. Climate change issues were subsumed within sustainability, and the report included short case studies of four cities in the MENA region, discussed below. These show that investment has concentrated in high-income real estate and global business competitiveness, rather than addressing longer-term issues such as climate-change action.[49]

The wealthy GCC sub-region comprises city-states and countries with strong central direction of urban affairs, reliance upon motor vehicles, dispersed land use, high energy consumption, and environmental pollution. Among the United

Arab Emirates (UAE), Abu Dhabi has since 2007 undertaken an ambitious urban project in Masdar City, a self-proclaimed living ecological laboratory funded by the sheikh with US$20 billion in capital and developed as a public-private partnership. International design consultants Foster and Partners prepared its master plan for up to 50,000 people, inspired by traditional Islamic architecture, and seeking to achieve a compact urban design that could maximise passive shading and air circulation in the extreme dry, hot, and windy desert conditions. After a promising start, the enterprise now appears to be faltering, and urbanists and political scientists have criticized its limited environmental achievements and undemocratic nature, with individual projects not assessed against the development plan or for their environmental impact upon regional ecosystems. For instance, visitors to the city travel overwhelmingly by private transport, park around the periphery, and transfer to a free personal rapid transit system of electric-powered cars; and most of the housing is air-conditioned and reserved for high-income workers.[50]

The World Economic Forum report included a case study of another UAE city, Dubai, which has planned for life after oil by becoming a regional, and indeed global, business and travel centre. It grew its population from a few hundred thousand in 1985 to over 4 million by 2014, attracted major global firms, and created a competitive infrastructure and free trade environment. Governance is by a non-elected executive council reporting to the ruler and largely comprised of ruling family members, while the Dubai Group (also controlled by the ruling family) engages in real estate projects such as business parks, conference facilities, and hospitality investments, with climate action little mentioned.[51] Neighbouring Qatar (2.6 million population in 2017) has its Qatar National Vision 2030 and master plan documents. It is developing Energy City Qatar, which is being marketed as an international showcase for technology innovations, especially in renewable energy (such as concentrated solar/wind power, and new power generation for seawater desalination).[52] While the UAE cities can draw upon vast financial resources and import skilled workers who remit money to their home countries, the pace of growth is causing significant environmental damage. Their average carbon dioxide emissions are among the highest in the world, directly attributable to oil and gas extraction, but also to provision subsidised electricity for air-conditioning and water desalinization. The UAE's two dozen golf courses demand thousands of cubic meters of water daily, using treatment chemicals, and both depleting and polluting precious groundwater resources.[53]

Three other case studies in the World Economic Forum report relate to MENA cities. Firstly, Casablanca, the largest city in the Maghreb region (with 4 million people), represents about half of Morocco's GDP, investment, and labor force. It has invested heavily in transport infrastructure: highways, a high-speed rail link to Tangiers (and Europe), an international airport and aviation hub, and the second largest port in Africa, with short shipping lanes to European ports. Critics have questioned the size of external investment (nearly $1 billion from France and $500 million from oil-rich MENA countries), the close supervision by central government, and damage to cultural and architectural heritage.[54] Secondly, the holy city of Medina (Saudi Arabia) (1 million population) received granted land and billions of dollars from the late King Abdullah to be remade into a centre for spiritual and intellectual enlightenment, with new higher education institutions, conference tourism to supplement religious tourism, knowledge-based industries, 20,000 jobs, housing for 150,000 people, and a high-speed rail line to Mecca. The project involved much destruction of physical heritage (some 300 historic sites lost) and has been hampered by poor co-ordination among real estate developers, city leaders, and Saudi visionaries.[55] Finally, Manisa (Turkey, population 1.7 million in 2019) a centre of industry and services near the port city of Izmir, with a fertile agricultural hinterland, attributed its competitive success to political leadership and broad-based collaboration between public and private sectors and business associations.[56]

Some tentative general findings can be made from these city case studies. Big urban projects were often driven by a narrowly-based group of decision-makers (perhaps the ruler or ruling family), were typically planned top-down and centrally directed, and may have subsequently suffered from a lack of co-ordination or poor implementation between sectors or levels of government. The desire to appear modern may have meant an emphasis upon expensive prestige projects (such as rapid transit systems, or high-income business and housing schemes) rather than broad-based improvements to improve living conditions of the wider population. Longer-term and difficult issues received less priority, such as climate change, urban resilience, or reducing social inequalities. The NUA principle of leaving no-one behind[57] is not much followed, for instance the need to provide basic services, and to engage women and young people in development projects that create jobs. This is unfortunate, since MENA has a young population (median age 26.8), and that demographic creates an age bulge and high youth unemployment. In North Africa, for example, people under 25 years comprise 48 per cent of the total population, and most unemployment is among men aged 15–25 with only a primary education.[58] Governance in the MENA region is thus weak at involving its citizens, as the next section explores further.

It is against this unpromising background that the MENA Region's governance resources and response to global climate change will be discussed in the next section, presenting a challenge for human ingenuity in the future.

Advancing urban law, governance, and resilience in the MENA region

The concept of governance concerns the processes of interaction and decision-making between the institutions and traditions by which authority is exercised. There is no shortage of aspirations and exhortations in SDGs about climate change and associated issues, and progress in the urban areas where population is concentrated depends upon local as well as national leadership. SDG 11 ("make cities and human settlements inclusive, safe, resilient and sustainable"[59]) is particularly relevant, and Table 9.1 summarizes its targets. The NUA offers directions that urban governance could pursue and emphasizes sustainable development

Table 9.1 Targets for SDG 11 (by 2030)

SDG target numbers	Targets (abridged from original)
SDG 11	**"make cities and human settlements inclusive, safe, resilient and sustainable"**
11.1	Access for all to affordable housing and basic services
11.2	Access to safe and sustainable transport systems
11.3	Inclusive and sustainable urbanization and participatory planning
11.4	Protect and safeguard cultural and natural heritage
11.5	Reduce losses from disasters, protecting the poor and vulnerable
11.6	Reduce the adverse environmental impact of cities, including air quality and waste management
11.7	Access to green and public spaces

Source: UN Global 2030 Agenda (2015)

principles of "leaving no one behind", ensuring sustainable and inclusive urban economies, and ensuring environmental sustainability.[60] UN-Habitat has drawn particular attention to the potential role of "urban law", whether national laws or local regulations.[61] United Cities and Local Governments, the largest organization of sub-national governments in the world, is lobbying to increase the role and influence of local government, perhaps through legislative change and decentralization of functions and responsibilities through applying subsidiarity principles.[62]

SDG 11, the NUA, and SDG 13 presume supportive institutional structures and social action, which are topics of other SDGs, especially 16 and 17. Governance in the land sector, managing social relationships in land, is now receiving more attention in the NUA and the international development community.[63] UN-Habitat recommends higher urban densities (15,000 people per square kilometre), rather than the low-density car-dependent urban forms found particularly in the wealthy GCC countries. Its Global Land Tool Network has developed a number of innovative instruments,[64] which often would require specific legislation. Another UN-Habitat priority is better access to and control over land by women, especially in Islamic and patriarchal societies.[65] Bringing more land into use for affordable housing, basic infrastructure, and provision of services remains challenging for deeply unequal MENA countries, and local government work in informal settlements can identify appropriate land tenure and housing opportunities.[66] Protecting and adapting MENA's cultural heritage for future generations (target 11.4) is also important, given its long urban history, as is provision of green and public spaces in urban areas, which the recent COVID-19 pandemic has exposed in the context of social distancing measures. Cities should also be empowered to create new finance opportunities for debt management and public-private partnerships in urban development projects.

Calls for disaster risk management in national and local strategies make the climate change connections with urban law and governance, which appear in SDG 11 (target 11.5), SDG 13, and other SDGs, as well as in the Sendai Framework for Disaster Risk Reduction 2015–2030.[67] Approaches to disaster risk reduction formerly focused on such threats as natural disasters, terrorism, and climate change, and on physical adaptation measures, early warning systems, and emergency responses to reduce impacts. These approaches are now expanding into longer-term urban resilience strategies, and generating growing academic interest, with a genealogical method tracing how urban resilience practices develop over time, and the values and power relations underpinning them.[68] Urban resilience has been broadly defined as the capacity of individuals, communities, institutions, and systems within a city to survive, adapt, and grow, no matter what kinds of chronic stresses and acute shocks they experience.[69] Such stresses come not only from climate change and natural hazards, but from wider issues of basic infrastructure and services, poverty reduction, and social inclusion/exclusion.

As the policy discourse moves toward a more holistic and pro-active approach, cities are becoming seen as complex adaptive systems or networks, interacting with political and institutional processes. Urban risk management thus involves an interplay between complex development processes (social, economic, environmental), vulnerable settlements in high risk zones, achieving disaster-resilient building construction, and a governing strategy of learning and adaptation. The recent shock of the COVID-19 pandemic is increasing attention to these issues of urban governance, resilience, and disaster preparedness. Unfortunately, city planning and management regimes are still often disconnected from disaster risk and resilience building, while countries lack legislation to integrate city resilience into broader development planning.[70]

New international thinking on urban resilience is emerging. The Rockefeller Foundation launched a resilience index in 2013, while UN-Habitat has formed an urban resilience "hub" based in Barcelona. Its "city resilience profiling tool" identifies four dimensions—leadership and strategy, health and well-being, economy and society, and infrastructure and environment—while its work projects focus on vulnerability and hazards assessment, risk root cause analysis, and governance and planning.[71]

Shifting towards an integrated urban resilience approach necessarily involves participatory strategies for civil society. The NUA aspires toward "peaceful, inclusive and participatory societies" (item 37),[72] SDG 16's target 16.7 concerns inclusive, participatory, and representative decision-making,[73] and SDG 17 promotes partnerships, both global and local, between stakeholders.[74] Urban resilience depends upon local actions and behavioural change by citizens, neighbourhoods, and communities, private sector participation in urban development, and engaging place communities on the ground, and mobilizing new-style social movements.

SDG 16's target 16.6 is to "develop effective, accountable and transparent institutions at all levels".[75] Governance stakeholders can be a broad spectrum of state and non-state, customary and religious authorities, and private and professional interests. Urban leaders, planners, and developers have to make local decisions on plans, regulatory change, infrastructure projects, economic diversification, and revenue collection. Arab countries need to develop financing strategies and action plans, with private finance as a pool of resources for implementing the SDGs.[76] Meanwhile the deep-rooted political crisis may yet force meaningful change; it affects many MENA countries, is particularly evident in cities, and concerns basic needs as well as human rights and democratic representation.

One aspect of growing regional co-operation is the emergence of expert conferences and fora on themes of common interest, with the UN a partner or equal participant.[77] Such gatherings have tended to be for major geographic regions (Asia or Africa particularly) but are growing in the MENA region, often hosted by its wealthier GCC countries, as three examples can testify. Firstly, recent world and regional urban forums have included WUF 10 in Abu Dhabi in 2020,[78] and the Marmara Urban Forum in Istanbul in 2019.[79] Secondly, Dubai hosted the first Arab Land Conference in 2018. Its objective was to exchange knowledge, promote regional cooperation, and develop capacities and innovation in land governance and real estate, and key topics were access to land for sustainable business and investment, land and conflict, and access to land for women.[80] Thirdly, Qatar in 2018 hosted the first conference of the Association for Environmental Law Lecturers in Middle East and North African Universities (ASSELLMU), followed by a second in Casablanca.[81] While the impact of the COVID-19 pandemic may restrict such physical face-to-face meetings in future, the format can adapt with improved online connectivity.

A brief case study of the city of Beirut shows some of the extreme governance challenges in building urban resilience.[82] Beirut is Lebanon's capital, a regional business and financial centre with a population of about 2 million in a country of 7 million. Earthquakes occurred in 2006 and 2014, with little response planning in place, and other significant hazards have been civil war and occupation by neighbouring countries.[83] A quarter of Lebanon's population are urban refugees, most recently from the Syrian civil war, living in informal settlements and contributing little or nothing in taxes. The Interior Ministry is concerned more with security than urban development, and there is no planning or housing ministry. The municipal budget comes from central ministries rather than locally raised revenues, and is allocated mainly for maintenance, with most services operated by private companies under central ministry control. The urban plan follows a master plan dating from the 1930s, ill-equipped for the recent challenges of rapid urbanization and civil war damage, and does not provide for much-needed urban green spaces. A Beirut Comprehensive Urban Resilience Master Plan was launched in 2016, prepared by international engineering consultants with World Bank support, but at only 24 pages long was hardly comprehensive, and was not enough to prevent public demonstrations in 2019 that exposed a collapse of trust in the whole political system.[84] Years of waste mismanagement had detrimental effects on public health, and a new waste incinerator as a solution was blocked in 2019 by civil society activists because of its air pollution implications.[85] Much of Lebanese society operates through clientelism, nepotism and corruption, and civil protests, initially triggered by planned taxes, quickly expanded into a country-wide condemnation of sectarian rule, lack of basic services, stagnant economy and unemployment, and legislation seen as shielding the ruling class from accountability. With the legislature and executive grid-locked and unable to govern, these troubles combined with the COVID-19 pandemic to provoke an economic crisis and request for external financial intervention.

Conclusion

Many MENA countries are experiencing deep-rooted political crises, a measure of which is how many of them appear high on the Fund for Peace's State Fragility Index.[86] Such stresses are particularly evident in MENA cities, and hamper realistic attempts to deal with climate change challenges while basic needs and human rights are still not addressed. As the SDGs (especially SDGs 11, 13 and 16) and the NUA highlight, meaningful legal reform could start with constitutional or legislative change to redistribute statutory responsibilities and powers between levels of government under the principle of subsidiarity.[87] This can form a basis for local institutions and leadership in such matters as urban resilience, disaster risk management, and service delivery.

Climate change action can come not only through legislation, but also through regional co-operation, policies, local governance, and social mobilization. Local action can embed the needed transitions in governance and social behaviour, including the involvement of young people, civil society organizations, the private sector, academia, and other stakeholders.

UN-Habitat's recent report on responding to the COVID-19 pandemic which erupted in early 2020 further emphasizes the need for local community-driven solutions, and mitigating local economic impacts.[88] Several of the recommendations in the report, with respect to how cities can cope with COVID-19 related disruptions, are equally important in the context of climate change. The report identifies as immediate responses better vulnerability assessments, online data-gathering and risk-mapping to inform municipalities and communities, improving water, sanitation, and hygiene, and ensuring safe mobility and transport. As medium-term responses it recommends improving water, sanitation, and infrastructure, supporting communities that have lost livelihoods, and building capacity for integrated health actions.[89] MENA countries can reinvigorate urban resilience to climate change impacts by integrating these disaster risk response measures into urban development, planning, and management.

Notes

1 N. M. Davidson, "What is Urban Law Today? An Introductory Essay in Honor of the Fortieth Anniversary of the Fordham Urban Law Journal" 40 *Fordham Urban Law Journal* 1579, 1589.
2 United Nations Economic and Social Council, "Universality and the 2030 Agenda for Sustainable Development from a UNDG Lens," UNDG Sustainable Development Working Group, 2016 <www.un.org/ecosoc/sites/www.un.org.ecosoc/files/files/en/qcp r/undg-discussion-n-on-universality-and-2030-agenda.pdf>; United Nations General

Assembly (UNGA), *Transforming Our World: The 2030 Agenda for Sustainable Development*, 21 October 2015, UN Doc. A/RES/70/1.

3 United Nations Department of Economic and Social Affairs, Sustainable Development, "Goal 13" <https://sdgs.un.org/goals/goal13>.

4 J. de Boer, "The Sustainable Development Fight Will Be Won or Lost in Our Cities", (World Economic Forum, 24 September 2015) <www.weforum.org/agenda/2015/09/the-fight-for-sustainable-development-will-be-won-or-lost-in-our-cities/> accessed 7 January 2021.

5 C. Doll, "Cities Should Be at the Heart of the SDGs", (United Naitons University, 19 September 2015) <https://unu.edu/publications/articles/cities-heart-of-sdgs.html>.

6 United Nations General Assembly (UNGA), *New Urban Agenda*, 25 January 2017, UN Doc. A/RES/71/256.

7 Ibid, Goal 16.

8 Ibid, Goal 17.

9 Ibid.

10 The term fertile crescent refers to the area from the southeast corner of the Mediterranean, its centre north of Arabia, and the north end of the Persian Gulf. It originated with the American archaeologist J. H. Breasted in the early 20th century, who characterized its history as a long struggle between the mountain peoples of the north and the desert wanderers of the south to control the resource-rich lands of the Persian Gulf.

11 E. Yildirim and S. El-Masri, "Master Planning for Conservation in Al Ain Oasis" (46th ISOCARP Congress, 2010) <www.isocarp.net/Data/case_studies/1705.pdf>.

12 K. Cerny, "Great Middle Eastern Instability: Structural Roots and Uneven Modernization 1950–2012" (2018) 31 Journal of Historical Sociology 53–71.

13 UN-Habitat, "UN-Habitat's COVID-19 Response Plan," 2020 <https://unhabitat.org/un-habitat-covid-19-response-plan>.

14 A survey by Egypt's National Center for Social and Criminological Research shortly before the 2011 uprising found that people's primary goals were basic: clean drinking water, improved sewage systems, and solutions to unemployment. "Arab Forum for Environment and Development" <www.afedonline.org/en>.

15 H. M. S. Al-Maamary, H. A. Kazem, and M. T. Chaichan, "Climate Change: The Game Changer in the GCC Region" (2017) 76 *Renewable and Sustainable Energy Reviews* 555–576.

16 M. Lange, "Impacts of Climate Change on the Eastern Mediterranean and the Middle East and North Africa Region and the Water–Energy Nexus" (2019) 10 *Atmosphere* 455.

17 UN Economic Commission for Africa, "Land Policy in Africa: North Africa Regional Assessment," African Union, African Development Bank, December 2010, <www.uneca.org/sites/default/files/PublicationFiles/regionalassesment_northafrica.pdf>; D. Olawuyi, "Financing Low-Emission and Climate-Resilient Infrastructure in the Arab Region: Potentials and Limitations of Public-Private Partnership Contracts" in W. L. Filho and A. A. Meguid (eds), *Climate Change Research at Universities: Addressing the Mitigation and Adaptation Challenges* (New York: Springer 2017).

18 B. Mayer, *The International Law on Climate Change* (Cambridge, MA: Cambridge University Press 2018); G. P. Overhauser, *Constructing Climate Change Legislation: Background and Issues* (Hauppauge, NY: Nova Science Publishers, 2009).

19 M. Nachmany, S. Fankhauser, J. Davidová, N. Kingsmill, T. Landesman, H. Roppongi, P. Schleifer, J. Setzer, A. Sharman, C. S. Singleton, J. Sundaresan, and T. Townshend, "The 2015 Global Climate Legislation Study: A Review of Climate Change Legislation in 99 Countries: Summary for Policy-makers," Grantham Research Institute on Climate Change and the Environment, London School of Economics. GLOBE – The Global Legislators Organisation, the Inter-Parliamentary Union, 2015 <uneca.org/sites/default/files/PublicationFiles/regionalassesment_northafrica.pdf>.

The nine MENA countries were Egypt, Iran, Israel, Jordan, Kuwait, Morocco, Saudi Arabia, Turkey, and the UAE.

20 Ibid, 28.
21 Green Climate Fund, "About GCF" <www.greenclimate.fund/about>.
22 Green Climate Fund, "Saudi Arabia" <www.greenclimate.fund/countries/saudi-arabia>.
23 Green Climate Fund, "Comoros" <www.greenclimate.fund/countries/comoros>.
24 Green Climate Fund, "Areas of Work: Themes" <www.greenclimate.fund/themes/>.
25 I. A. Gelil, "History of Climate Change Negotiations and the Arab Countries: The Case for Egypt," Issam Fares Institute for Public Policy and International Affairs, American University of Beirut, July 2014, <www.aub.edu.lb/ifi/Documents/publications/research _reports/2014-2015/20140723_Abdel_Gelil.pdf>; D. Olawuyi, "Advancing Climate Justice in International Law: An Evaluation of the United Nations Human Rights Based Approach" (2015) 8 *Florida A and M Law Review* 103.
26 N. Saab and A. K. Sadik (ed), *Financing Sustainable Development in Arab Countries: Annual Report of Arab Forum for Environment and Development* (Beirut: Technical Publications 2018).
27 J. J. Kaminski, "The OIC and the Paris 2015 Climate Change Agreement: Islam and the Environment" in L. Pal and M. Tok (eds), *Global Governance and Muslim Organizations* (London: Palgrave Macmillan 2019).
28 See n 8.
29 Climate Action Network International, "Arab World" <www.climatenetwork.org/ policyinformation/publication/regions/354>.
30 "Arab Forum for Environment and Development" <www.afedonline.org/en>.
31 S. Lalonde, *Determining Boundaries in a Conflicted World: The Role of Uti Possidetis* (Montreal, QC: McGill-Queen's University Press 2002).
32 H. P. Glenn, *Legal Traditions of the World* (Oxford: Oxford University Press 2010); J. Griffiths, "What is Legal Pluralism?" (1986) 24 *Journal of Legal Pluralism and Unofficial Law* 6.
33 A. E. Mayer, "Conundrums in Constitutionalism: Islamic Monarchies in an Era of Transition" (2002) 1 *UCLA Journal of Islamic and Near Eastern Law* 183.
34 M. Belov (ed), *Global Constitutionalism and Its Challenges to Westphalian Constitutional Law* (London: Hart Publishing 2020).
35 *SDG* 17, n 6.
36 *Climate Change Act 2008*, UK Public General Acts, 2008 c 27.
37 B. Wehrmann, "Solar Power in Germany – Output, Business and Perspectives," (Clean Energy Wire, 16 April 2020).
38 United Nations and the Commonwealth "Law and Climate Change Toolkit" <https:// lcc.eaudeweb.ro/>. For the concept of "hegemonic international law", see S. Wiessner, "The New Haven School of Jurisprudence: A Universal Toolkit for Understanding and Shaping the Law" 18 (2010) *Asia Pacific Law Review* 45–61.
39 European Commission, Environment, "Sustainable Development" <https://ec.europa .eu/environment/sustainable-development/index_en.htm>.
40 A. N. Khondaker, M. A. Hasan, S. M. Rahman, K. Malik, M. Shafiullah, and M. A. Muhyedeen,, "Greenhouse Gas Emissions from Energy Sector in the United Arab Emirates – An Overview" (2016) 59 *Renewable and Sustainable Energy Reviews* 317–1325.
41 Ibid.
42 O. Amao, *African Union Law: The Emergence of a Sui Generis Legal Order* (Abingdon, Oxfordshire: Routledge 2019).
43 M. E. Bonine (ed), *Population, Poverty, and Politics in Middle East Cities* (Florida University Press 1997).
44 Y. Elsheshtawy (ed), *Planning Middle Eastern Cities: An Urban Kaleidoscope* (Washington, DC: Earthscan 2004); Y. Elsheshtawy (ed), *The Evolving Arab City: Tradition, Modernity and Urban Development* (Abingdon, Oxfordshire: Routledge 2008.

45 MENA has eight cities of over five million population (2016). In descending size order they are Cairo (with some 20 million in its metropolitan area), Istanbul, Tehran, Baghdad, Riyadh, Dubai, Ankara, and Alexandria.

46 "MENA Countries: Urbanization in 2018" <www.statista.com/statistics/804824/urbanization-in-the-mena-countries/>.

47 Olawuyi, n 17.

48 The Saudi government subsequently built a permanent air-conditioned tent city (grouped by nationality) for use by *Hajj* pilgrims ("Mint Tent City – Saudi Arabia" <www.seeleyinternational.com/me/project/mina-tent-city-saudi-arabia/>).

49 World Economic Forum, "The Competitiveness of Cities: A Report of the Global Agenda Council on Competitiveness" (WEF, August 2014).

50 F. Cugurullo, "Exposing Smart Cities and Eco-Cities: Frankenstein Urbanism and the Sustainability Challenges of the Experimental City" (2018) 50 *Environment and Planning A: Economy and Space* 73.

51 World Economic Forum, n 49, 28.

52 State of Qatar, Government Communications Office, "Qatar National Vision 2030" <www.gco.gov.qa/en/about-qatar/national-vision2030/>; "Energy City Qatar" <www.energycity.com/>.

53 F. Wiedmann, A. Salama, and A. Thoerstein, "Urban Evolution of the City of Doha" (2012) 29 *Journal of Faculty of architecture (JFA) Middle East Technical University (METU)* 35.
   D. Olawuyi, "Advancing Innovations in Renewable Energy Technologies as Alternatives to Fossil Fuel Use in the Middle East: Trends, Limitations, and Ways Forward" in D. Zillman, M. Roggenkamp, L. Paddock, and L. Godden (eds), *Innovation in Energy Law and Technology: Dynamic Solutions for Energy Transitions* (Oxford University Press 2018).

54 H. Diab, *Nationally Determined Contributions of the Kingdom of Morocco* (Berlin: Urban Pathways 2018); I. Berry-Chikhaoui, "Major Urban Projects and the People Affected: The Case of Casablanca's Avenue Royale" (2010) 36 *Built Environment* 216.

55 World Economic Forum, n 49, 21.

56 Ibid.

57 UNGA, n 6, para 14.

58 See n 49.

59 United Nations Department of Economic and Social Affairs, Sustainable Development, "Goal 11" ><https://sdgs.un.org/goals/goal11>.

60 UNGA, n 6.

61 M. Glasser and S. Berrisford, "Urban Law: A Key to Accountable Urban Government and Effective Urban Service Delivery" (2015) 6 *World Bank Legal Review* 209.

62 Subsidiarity is the principle of allowing the individual members of a large organization to make decisions on issues that affect them, rather than leaving those decisions to be made by the whole group. UNDP, "Decentralization: A Sampling of Definitions", working paper prepared in connection with the Joint UNDP–Government of Germany evaluation of the UNDP role in decentralization and local governance, October 1999.

63 The NUA promotes the "ecological and social function of land" (UNGA, n 6, para 69), and "planning and managing urban spatial development" (UNGA, n 6, paras 93–125).

64 "GLTN at UN-Habitat, a world in which everyone enjoys secure land rights" <https://unhabitat.org/gltn>.

65 S. Sait and H. Lim, *Land, Law and Islam: Property and Human Rights in the Muslim World* (London: Zed Publishers 2006), UN-Habitat, "Women and Land in the Muslim World," 2018 <https://unhabitat.org/women-and-land-in-the-muslim-world>.

66 M. Ababsa, B. Dupret, and E. Denis (eds), *Popular Housing and Urban Land Tenure in the Middle East: Case Studies from Egypt, Syria, Jordan, Lebanon, and Turkey* (American University in Cairo Press 2012).

67 L. Pearson and M. Pelling, "The UN Sendai Framework for Disaster Risk Reduction 2015–2030: Negotiation Process and Prospects for Science and Practice" (2015) 2 *Journal of Extreme Events* 1571001.

68 M. Borie, M. Pelling, G. Ziervogel, and K. Hyams, "Mapping Narratives of Urban Resilience in the Global South" (2019) 54 *Global Environmental Change* 203–213; Prasad and others, n 67.

69 W. N. Adger, "Vulnerability" (2009) 16 *Global Environmental Change* 268.

70 J. Saghir, "Urban Resilience: The Case of the Middle East and North Africa," Payne Institute for Public Policy, Colorado, April 2019 <https://payneinstitute.mines.edu/wp-content/uploads/sites/149/2019/04/Saghir-Resilience_Comment.pdf>.

71 UN-Habitat, "Urban Resilience Hub" <https://urbanresiliencehub.org/>.

72 UNGA, n 6.

73 SDG 16, n 7.

74 SDG 17, n 8.

75 SDG 16, n 7.

76 N. Saab and A. K. Sadik (eds), "Financing Sustainable Development in Arab Countries: 2018 Report of the Arab Forum for Environment and Development," 2018 <www.greengrowthknowledge.org/resource/financing-sustainable-development-arab-countries>.

77 *International Urban Forums and Conferences*, (2019) special edition of Sehir and Toplum Journal, Marmara Municipalities Union.

78 <www.worldbank.org/en/events/2020/02/08/world-urban-forum-wuf10>.

79 <www.marmaraurbanforum.org>.

80 <https://arabstates.gltn.net/conferences/>.

81 <www.hbku.edu.qa/en/middle-east-environmental-conference>.

82 Beirut was the MENA city included as one of ten partner cities in UN-Habitat's urban resilience initiative (see fn 71). This text was written before the explosions that destroyed much of inner Beirut on 4 August 2020 and came to symbolize the failures of Lebanese governance.

83 K. V. Monroe, *The Insecure City: Space, Power, and Mobility in Beirut* (Chicago, IL: Rutgers University Press 2016).

84 BuroHappold Engineering, "Comprehensive Urban Resilience Masterplan for the City of Beirut" 2016; J. Liu, "Financing and Implementing Resilience with a Systems Approach in Beirut" in F. Gatzweiler (ed), *Urban Health and Wellbeing* (New York, NY: Springer 2020).

85 "Beirut Municipality Postpones Waste Incinerator Decision amid Protest", 4 July 2019 <www.naharnet.com/stories/en/262239>.

86 Several MENA states appear on the 2020 index in the alert, high alert, or very high alert categories: Yemen, Syria, Sudan, Southern Sudan, and Libya. Common indicators include a state whose central government is so weak or ineffective that it has little practical control over much of its territory; non-provision of public services; widespread corruption and criminality; refugees and involuntary movement of populations; and sharp economic decline ("Fragile States Index" <https://fragilestatesindex.org/>).

87 For a South African example: R. Malherbe, "The Constitutional Distribution of Powers" in B. De Villiers (ed), *Review of Provinces and Local Governments in South Africa: Constitutional Foundations and Practice* (Berlin: Konrad Adenauer Stiftung 2008).

88 UN-Habitat, n 13.

89 Ibid.

10 Integrating public health into climate change law and policies in the MENA region

Current issues and future directions

Rasha Abu-El-Ruz and Karam Turk-Adawi

Introduction

This chapter discusses the major implications of climate change for public health across the Middle East and North Africa (MENA) region, and how the countries can effectively integrate public health response into climate change policies and planning.

Without good health, humankind may not live life to the fullest.[1] Public health therefore focuses on well-being. One of the greatest and most complex public health concerns facing the current generation is climate change. Since the early decades of the 21st century, the world has witnessed a dramatic industrial shift, which directly influenced several global environmental and climate changes. The list of diseases that are linked to climate change is long; however, respiratory, cardiovascular, genetic, and infectious diseases stand out.[2] Several risk factors are well-documented in literature to be associated with disease development, due to climate change, such as age, socioeconomic resources, and location of residence.[3]

Climate change can create potentially suitable environments for several infectious diseases and deaths.[4] The World Health Organization (WHO) estimates an increase of 250,000 deaths due to climate change by 2050.[5] Food and water borne diseases have been a major public health concern. Since the beginning of the 21st century, more than 91 million individuals have developed food and water-borne illness in Africa.[6] Floods play an important role in altering the water systems in vulnerable countries.[7] Malaria is prevalent in Africa; the low socioeconomic conditions, heat, and floods create the ultimate environment for the Anopheles mosquito (the parasite vector) to breed and transmit more *Plasmodium* parasite. Zoonotic Cutaneous Leishmaniasis (ZCL) is also a vector-borne disease, caused by a parasite called *Leishmania* that is transmitted from infected rodents to humans through the sand flies. It is well-documented that ZCL is endemic in the Middle East, and Sub-Saharan and North Africa. The ecological models indicate that desertification is significantly correlated to ZCL.[8]

An increased prevalence of pollen allergies has also been linked to the increasing temperatures due to climate change.[9] Several molecular models reveal interactions between climate change, air pollution, and allergic diseases.[10] These interactions can adversely affect the immune system, either directly or indirectly, leading to the development of intolerance or allergies.[11] Climate change and other environmental factors play a role in inducing chemical modifications of allergens, increasing oxidative stress in the human body, and stimulating the immune system.[12] Pollutants or allergens often act as adjuvants, which are immunogenic molecules.[13] Other adverse health outcomes that may result from climate change include: inflammatory disorders, asthma, eczema, skin irritation, dermatitis, rhinitis, sinusitis, conjunctivitis, and anaphylaxis.[14] Some public health strategies employ critical adaptation efforts to reduce the effect of climate change on human health, such as humidity control, air filtration, and proper ventilation.[15]

Climate change may also lead to a sharp rise in mental health concerns across the MENA region.[16] Mental health is an important aspect of emotional, social, and psychological well-being. There is strong evidence that climate change is associated with mental health issues.[17] Distressing change in the climate or environmental catastrophes can trigger post-traumatic stress disorder (PTSD), anxiety, depression, suffering from complicated and unexpected grief, surviving guiltiness, and recovery fatigue. If these signs and symptoms are neglected, they may lead to suicidal ideation. Extreme climate change conditions, such as rising temperatures, sea levels changes, drought, reduction in agriculture yield, lack of food security, economic weakness, and infrastructure demolition, can lead to increased stress as well as trigger violence and aggression.[18]

The continued emission of greenhouse gases (GHGs) that cause climate change can result in significant health impacts. Industrial activities associated with climate change include, but are not limited to, pollution, burning fuel, and gas emissions. The resultant toxic gases, such as carbon dioxide (CO_2), build up in the atmosphere and eventually contribute to global warming. The suspended allergens are thought to upsurge during warmer seasons, when the air temperature is raised due to increased CO_2.[19] Additionally, extreme weather events linked to climate change, such as extreme rainfalls, can contribute to indoor pollution by supporting fungi and mold growth, as well as vector, water-, or foodborne diseases.[20]

Food security is another important aspect. Climate change plays a major role in the quality and quantity of available food and in the livestock industries.[21] Climate change impacts the availability, affordability, and accessibility of nutritious food, which could trigger public health challenges such as malnutrition, stunted growth, and the spread of foodborne diseases.[22] Climate change could also impact the availability, affordability, and accessibility of water, which could result in water scarcity and the spread of waterborne diseases.[23] For example, water shortage in Egypt, Iraq, East Africa, Yemen, and Jordan has led some farmers to lose their hope in agriculture.[24] The extreme heat and temperature in the Gulf region, especially in UAE and Saudi Arabia is an additional factor.[25] Many water springs have entirely dried in south Jordan,[26] and the Khabour River in Syria has completely dried up.[27] Between 2006 and 2010, devastating drought atrophied areas of Jordan, Syria, Iraq, and Turkey, leading to the migration of millions of residents and cessation of livestock and crops.[28]

While Africa is naturally suffering from harsh environmental conditions due to the shortage in clean water resources, the Gaza Strip also suffers harsh conditions driven by the Israeli conflict and the blockade. Gaza residents seek water through drilling prospectus wells to reach groundwater, inadvertently hitting the sewage infrastructure system, and in turn contaminating the drinking water. The Gaza Strip is increasingly suffering from several waterborne diseases side by side with the drought.[29]

Fecal-oral infectious disease in East and North Africa are very common; *enteropathogenic Escherichia coli* (EPEC), *enterotoxigenic Escherichia coli* (ETEC), *enteroaggregative Escherichia coli* (EAEC), and *diffusely adherent Escherichia coli* (DAEC) have been documented as the most prevalent diseases of fecal source.[30] Floods collapse the sewage system and allow the contamination of water sources; this is typically coupled with the poor sanitation conditions that sustain the disease transmission.[31] Another example is cholera in Yemen. The disease transmits through a fecal-oral route, caused by *Vibrio cholerae* bacteria, which enter the human body through ingesting contaminated water or food. In 2016, Yemen had one of the worst cholera outbreaks in the 21st century, with approximately 15,000 cases reported and more than 2,500 deaths. The poor socioeconomic conditions along with the climate and environmental factors exacerbated the epidemic.[32] Sandstorms are also a constant concern in the MENA, triggering several diseases, such as respiratory and cardiovascular diseases.[33]

Given the growing evidence that climate change in the MENA has been associated with several diseases, as well as increased morbidity and mortality, it is important for MENA countries to develop integrated law and policy responses that address health impacts of climate change. The following section examines the need for integrated public health responses to climate change. However, despite the clear impetus and obligations in international environmental law that aim to protect the environment through global environmental frameworks, integrating public health into climate change policies and planning is still hindered across the MENA. A number of MENA countries still do not have a clear vision regarding how to address the long-term effects of climate change on public health. Although comprehensive laws and policies have been developed to tackle the climate change factors, there is a need for greater harmonization and integration between public health and environment agencies and institutions to ensure more effective implementation, monitoring, and surveillance.

This chapter discusses how MENA countries can better integrate public health into climate change law and policies in order to ensure coherent and holistic implementation. The chapter is divided into five sections. After this introduction, the second section discusses the need for an integrated public health response to climate change in the MENA region. It discusses international and regional efforts in the MENA region that aim to integrate public health into climate policies. The third section highlights the gaps in integrating the climate change policies into public health practice at domestic levels. The fourth section offers recommendations on how to address implementation gaps and challenges to the effective integration of public health and climate change planning in MENA. The chapter concludes in the fifth section.

The need for an integrated public health response to climate change in the MENA region

As discussed earlier, climate change is associated with increased disease. Further, the agriculture, water, and food security nexus is expected to be severely affected over the forthcoming decades by climate change. About one-third of the region's population work in agriculture, which contributes 13 per cent to the region's GDP, versus 3.2 per cent globally; over 60 per cent of the population live in high to very-high water stressed areas with reduced availability to water for agriculture or drinking.[34] It is estimated, that by 2050, climate change will account for 22 per cent of water shortage, with socioeconomic factors accounting for the rest of the water shortage in the region.[35] Food security is also impacted by climate change; crop yield loss can be due to intolerable temperature, insufficient rain or water scarcity, and dry soil.[36] MENA is expected to have the greatest economic loss among other regions, with a reduction of six per cent of GDP by 2050, due to climate change-related water scarcity and its adverse effects on agriculture, health, and incomes.[37]

Given the interconnectedness and multi-scale nature of climate change-related risks, such as economic crisis, and food and water insecurity associated with political instability, conflict-induced migration, and poverty, climate change poses a large threat to public health and needs to be addressed in an integrated "One Health approach".[38] This is a multi-sectorial, collaborative effort that identifies and assesses the climate change impacts on health through impactful research findings and surveillance reports that rely on documented data and governance system. This approach functions on a broad scale through stating shared goals that are met through comprehensive teamwork in order to achieve maximum efficiency.[39] An essential aspect of the integrated public health approach is policy development. It is not only establishing governmental laws and regulations, but also public engagement and partnership building, based on a public health approach. Accordingly, effective climate change policies should be based on three elements: informing, educating, and empowering populations toward climate resilience through emphasizing health behaviour and communication; mobilizing partnerships through active collaboration with non-health sectors; and development of health policies that support climate mitigation and adaptation.[40] However, these elements are not adequately implemented in the MENA, and these gaps will be discussed later. The Intergovernmental Panel on Climate Change (IPCC) assessment indicates that policies and regulations in MENA regarding climate change need to include public health dimensions in order to achieve the sought sustainable development stated goals.[41]

Efforts to promote an integrated approach to addressing the public health impacts of climate change have increased at international, regional, and domestic levels across the region. Internationally, the World Health Organization (WHO) has called for mitigation of the effect of climate change and its increasing deaths toll.[42] The WHO has taken several initiatives to work collaboratively with supporting states to help building climate-resilient health systems; especially in the developing countries.[43] The United States Center of Disease Control (CDC) has also played an essential role in assessing impacts of climate change and embracing several projects on health, such as the Climate-Ready States and Cities Initiative; however, MENA countries do not have members among the 16 cities included.[44]

International and regional initiatives to promote an integrated approach to public health and climate change

Many countries have signed their Intended Nationally Determined Contributions (INDCs) to the Paris Agreement[45] including 16 countries in the MENA region.[46] The WHO is working with these signatories to assess health gains upon implementation of this agreement.[47] The WHO has also embraced the Health and Climate Change Country Profile Project to monitor the national and global progress on climate change.[48] This indeed supports the United Nations Framework Convention on Climate Change (UNFCCC), and promotes actions that improve health while reducing gas emissions. MENA has also contributed to the five-step framework for Building Resilience Against Climate Effects (BRACE). This initiative identifies the likely climate impacts on MENA communities, including the potential health effects, as well as populations and locations at risk. The BRACE framework assists states to develop and implement health adaptation plans and address malfunctions in critical public health issues.[49]

The WHO-affiliated Regional Centre for Environmental Health Action (CEHA) provides technical support to countries of the Eastern Mediterranean region to reduce morbidity and premature mortality caused by modifiable environmental risk factors. The CEHA provides technical guidance, programmatic support, advisory consultations, and related services to member states and partner agencies. Its ultimate goal is to heighten the leadership role of the public health sector through regulation, surveillance, and management of environmental risks in the fields of water and sanitation, waste management, air quality, food and chemical safety, vector control, and environmental health emergency management.[50]

Further, the WHO's Global Arabic Programme actively contributes to maintaining and enhancing people's health in the region. This initiative provides accurate, high-quality, and up-to-date health-related information, including climate and environmental aspects of health, in Arabic, among other languages, through printed and electronic media.[51]

One of the most prominent initiatives in the Middle East that tackles climate change in the scope of public health is the Masdar project of UAE.[52] The project focuses on enhancing renewable energy or clean energy capacities, which greatly benefit human health and survival, in contrast to the toxic emissions of industrial production. The project addresses climate change issues by conducting energy audits of buildings and embracing strategic frameworks that serve as templates for low-carbon designs, such as installing rooftop solar panels on buildings. The initiative established several partners including some of the world's largest energy companies and elite institutions. The Masdar initiative has four key elements: an innovation centre to support the demonstration and adoption of sustainable energy technologies; the Masdar Institute of Science and Technology with graduate programmes in renewable energy and sustainability to enhance the national capacities and impose education; a development company focused on the commercialization of emissions reduction and clean development mechanism solutions as provided by the Kyoto Protocol; and special economic zone to host institutions investing in renewable energy technology.[53] The Gulf Cooperation Council (GCC) also plays an important role in promoting sustainable development projects that mitigate the adverse health effects of GHGs to enhance health and well-being of the Arabian society.[54]

The League of Arab States (LAS) is a strategic regional partner to the United Nations Development Programme (UNDP) created to build a resilient society, based on the goals addressed in the sustainable development plans, and to implement the new Paris Agreement on Climate Change. [55] Additionally, LAS has signed the UN Environment Programme (UNEP) memorandum, which provides the basis for enhanced cooperation on climate change projects, such as the SDG Climate Nexus Facility.[56] LAS and the Arab Water Council (AWC) are regional partners with the facility; they play an essential role in securing clean water resources for drinking and agriculture. The AWC is important for tackling droughts and preventing malnutrition and famines.[57] In 2010, the Arab Ministerial Water Council (AMWC) launched the Arab Water Security Strategy 2010–2030.[58] It addresses future water challenges, such as floods, waterborne diseases, drinking water scarcity, and agricultural water.[59] Later, the Arab Strategic Framework for Sustainable Development was developed to address the key challenges faced by Arab countries in achieving sustainable development for water resources during the period 2015–2025.[60] In 2016, the GCC developed the Water Strategy and Implementation Plan 2016–2035.[61] Another initiative that plays a role in climate change mitigation is the Islamic Development Bank Group, which supports vulnerable countries against poverty, malnutrition, and food and water insecurity.[62]

The Regional Center for Renewable Energy and Energy Efficiency aims to enable and increase the adoption of renewable energy practices in the Arab region.[63] The GCC is the main promoter of this initiative.[64] This collaboration strengthens stakeholder involvement and efforts with the public health sector, leading to enforcement of the green energy policies and projects and improvement in the health and well-being of the regional society.[65] The natural sun in all the GCC countries is an indispensable resource for alternative energy. All GCC countries have the political will, vision, and funds to pursue renewable and alternative energy.[66] All the GCC countries, prominently UAE, have clear renewable energy policies.[67]

The Arab Framework Action Plan on Climate Change (AFAPCC) is a strategic regional climate policy framework developed by the CAMRE, with a goal to coordinate climate action in the Arab region for the maximum attainable health benefits.[68] The overall objective of AFAPCC is to strengthen the capacities of LAS member states to address the urgent challenge of climate change and ensure a climate-resilient region with reduced climate risks and better healthy living standards. The AFAPCC also highlights the Arab Strategy for Disaster Risk Reduction (ASDRR) through mapping climate risks in the region, conducting integrated vulnerability assessments with the public health sectors, and developing cross-sectoral adaptation strategies to minimize adverse health effect.[69] The Arab Climate Resilience Initiative (ACRI) is also a regional initiative of the Regional Bureau for Arab States that has responded to the growing climate risks in the Arab region. The ACRI increased awareness of regional countries on potential climate impacts, while promoting integrated and cross-sectoral public health approaches for climate-resilient pathways. This initiative involved implementing strategic policies around priority areas, such as water security, drought, and access to sustainable energy.[70]

Evidently, the integrated approach to climate change and public health has gained increased recognition at international and regional levels.[71] The next section discusses its application at domestic and country levels across the region.

Domestic efforts in MENA countries to integrate public health into climate policies: trends and limitations

This section highlights ongoing efforts at the domestic level to integrate public health and climate change laws and policies. Several MENA countries have passed laws and policies that introduce institutional obligations to mitigate health impacts of climate change. For example, Morocco passed the law on the Agency for Development of Renewable Energy and Energy Efficiency, the renewable energy law 13-09.[72] Tunisia's constitution (passed in 2014) states: "The state guarantees the right to a healthy and balanced environment and the right to participate in the protection of the climate."[73] Algeria law focuses also on renewable energy.[74] In Saudi Arabia, the Kingdom introduced climate change within its Vision 2030 highlighting renewable energy as a better energy source for heath.[75] The "King Salman Renewable Energy Initiative" is also to be launched to support reaching the 9.5GW goal.[76] Similarly, the Qatar National Vision 2030 states that Qatar aims to take a leading, international role in the mitigation of the impacts of climate change.[77] Qatar also has passed the Environmental Protection Law to protect biodiversity, population and human health, and the environment from the harmful impact of abroad activities.[78] Qatar's mitigation efforts, including carbon capture and storage to minimize the air pollution and adverse health impacts, are regulated by this law as well.[79] The UAE mandated the National Climate Change Plan 2050 to reduce greenhouse emissions.[80] The UAE Green Growth Strategy 2015 anticipates a production industry that is environmentally friendly; it comprises the Green Agenda 2015–2030 which focuses on reforestation as having a vital role in enhancing air quality and improving environmental health.[81] Lebanon, Morocco, and Syria have also adopted the reforestation strategy.[82] In Bahrain, the National Strategy for the Environment was approved in 2006 to recognize and mitigate various climate change aspects.[83]

In Kuwait, the Energy Conservation Code of Practice (R-6) to efficient energy use is the main environmental law that operates several projects to maintain a climate-resilient environment.[84] Jordan mandated the National Climate Change Policy of the Hashemite Kingdom of Jordan 2013–2020 to ensure a climate risk resilient Jordan.[85] The law also mandates special attention to the vulnerable groups that suffer from the negative effects of climate change.[86] The Egypt Renewable Energy Law has been passed to encourage the private sector to produce electricity from renewable sources.[87] Oman has passed the Ministerial Decision for the Management of Climate Affairs, which states that GHG emitting projects require the permission of the Ministry of Environment and Climate Affairs.[88] Oman's environmental laws also include the Royal Decree No. 8/2011 (Oil and Gas Law) for analyzing climate change in environmental and health impact assessments,[89] and the Royal Decree No. 90 of 2007 for establishing the Ministry of Environment and Climate Affairs.[90] Yemen has passed the Presidential Decree No. 101 of 2005 on the establishment of the Public Environmental Protection Authority,[91] the National Strategy for Environmental Sustainability 2005–2015,[92] and the National Environmental Action Plan 2005–2010 that focuses on increasing sustainability in four key health-related areas: water, land resources, biological diversity and coastal and the marine environment, and waste management.[93] Syria has laws on forestry and renewable energy.[94] The law seeks a rationalization of energy consumption, energy efficiency in all fields, and adoption of renewable energy resources.[95] Palestine has passed the legislative decree No. 14/2015 on renewable energy that encourages alternative and clean energy resources,[96] Law No. 7/1999 which aims to protect the environment and health from pollution in all its forms,[97] and Law No. 12/1995 on the establishment of the Palestinian Energy Authority including research and development for all types of renewable energy.[98] Libya has passed law No. 426 establishing the Renewable Energy Authority of Libya to work towards the governmental target of a ten per cent share of the total energy mix coming from renewable energy by 2020.[99] Finally, Iraq has passed the Act for the Protection and Improvement of the Environment that refers to the importance of renewable energy with no specific reference to climate change and GHGs.[100]

Gaps in integrating climate change policies into the public health practices

Despite the varied levels of advancement in the climate change initiatives across MENA, they are still far from meeting expectations. Unfortunately, the mitigation and adaptation policies are at the disposal of national and economic interests. The MENA region is considered especially vulnerable to climate change due to its political instability, mass migration, resource scarcity, and economic dilemmas.

Climate change priorities are affected by economic, social, and political factors

Although the GCC countries strive to comply with several international and regional organizations that facilitate the frameworks of environmental protection, ministries and agencies are still struggling to play a significant role in addressing climate change. Moreover, the political conflicts create a burden on policy and strategy implementation, as the humanitarian disasters deprioritize climate change issues.[101] The ongoing conflicts in the region, including the Arab Uprisings, have created obstacles for effective governance for climate change.[102] There is also limited human expertise and technical resources which weaken adaptation measures that are essential for survival, such as dealing with climate change-related drought, flood preparedness plans, risk assessments, and management of natural disasters. For example, the droughts in Syria, Jordan, and Africa are still progressing, indicating poor water and agricultural management, as well as deficient coordination between authorities, public health, and other sectors, including the economic, political, and social sectors.[103]

Climate change awareness and education strategies are not well-implemented at individual and institutional levels. The pan-Arab survey revealed that education and awareness was the most important measure, after energy-use reduction, to combat climate change effect.[104] Human activity is a major contributor to climate change; however, education about individual risk behaviour, such as smoking, is not incorporated within the context of climate change prevention.[105] Further, climate change as a health risk is not well communicated to the public by leaders or to patients by healthcare providers in MENA.[106] At institutional and individual levels, there is a lack of policies to enhance education and awareness on adaptation technologies of climate change, such as rainwater harvesting, genetically modified plants that handle heat, soil erosion, and desertification.[107]

Limited partnership with other sectors

Developing cross-sectoral national and regional partnerships is essential for adaptation strategies of climate resilience. The ideal climate resilience partnership mandates active collaboration across nine sectors: water, land and biodiversity, agriculture and forests, human settlements, seas and coastal areas, health, energy, transport, and industry.[108] However, MENA is still behind due to its weak institutional capacities to integrate climate change in its policies, involving multiple sectors. The current climate change initiatives in MENA are generally limited to one ministry per country; there is inadequate partnership with other ministries. This limitation adversely affects the climate change development plans, annual budgets, and implementation policies. Other adverse impacts include lack of prioritization and risk assessment for climate projections, ineffective decision-making with regards to implementing adaptation interventions, and lack of knowledge transfer mechanisms to improve awareness of adaptation technologies among the relevant sectors.[109]

Limited laws and policies focus more on adaptation rather than mitigation

MENA countries have focused more on adaptation plans than mitigation for its subsistence; especially in developing petroleum-rich countries, such as the GCC region, as they largely depend on oil and gas for their economical resilience. The INDCs reveal that many of the MENA countries bounded the climate mitigation policies to their economic growth. In addition, the implementation and surveillance of current mitigation policies is poor. This might be subject to bureaucracy and conflict of interest between the decision makers and stakeholders.[110]

Despite the significant laws and policies in MENA to build climate resilience that enhances societies' health, there is a gap in adopting laws and policies that penalize the intended harmful actions to climate, i.e. adversely impact health. The current laws and policies lack specific straightforward terms that penalize the institutional actions that harm the climate. They also lack clear mandates that enforce an interconnected surveillance and assessment approach between the relevant ministries and public health sector.

Limited up-to-date databases and infrastructure

Though the Arabian region in MENA is interested in climate change on several levels, and there is a growing awareness in the area, the infrastructure is underdeveloped.[111] Limited availability of data could be linked to the on-going conflicts (i.e. resource consumption and lost reports). The actions taken by MENA countries depend largely on data and findings that are western-derived, thus the system lacks the legitimate assessment and evidence of the region's climate change aspects and their direct impact on human health.[112]

Addressing the barriers to an integrated response: Recommendations

Education and increased population awareness

Education is a key element of the global response to climate change. It is important to expand knowledge and information amongst the public, so they understand the health and non-health impacts of climate change. Educating the public and increasing the level of awareness reassures positive changes in health behaviours and public attitude. Education and awareness are two anchors for the institutional and individual response, where everyone should be accountable for contribution to climate change. Education is essential to augment the active involvement of the social, political, financial, and health sectors in climate change-related issues. Climate change education is part of UNESCO's Education for Sustainable Development (ESD) programme.[113] UNESCO aims to make climate change education more accessible to enhance the international response to climate change. Resources and expertise are available to support all countries to integrate climate change into their education systems and facilitate knowledge exchange of experiences.[114] All MENA countries can benefit from the available resources designed to achieve the climate literacy between communities.

Initiate integrated multi-sectorial collaboration

The climate change mitigation and adaptation models in MENA should incorporate dimensions other than health, such as social, economic, and political dimensions. Multi-sectorial collaboration should take place to deepen the understanding of climate related stressors that worsen health outcomes. The MENA region should also focus on improving institutional, forecasting, and technical capacities for managing multi-dimensional risks, such as natural disasters and industrial impacts on climates. Improving resources and capacity requires a coherent multi-sectorial system that designs feasible monitoring and assessment frameworks. Likewise, the climate change law and regulations in MENA must be supported with surveillance to ensure their implementation, as well as expanding the law to include other aspects of climate change, such as food security, resource equity, disaster management, and water resources.[115]

Enforce lawful penalties and synergic emphasis on mitigation and adaptation strategies

Environmental liability is important to legalize the actions that affect the climate and enforce penalties on those who cause any harm to the climate. Climate change law can be included within business, institutional, economic, environmental, or even the civil law. Standards that oblige the polluter to repair the harm caused to the environment are also recommended. These standards allow the authorities to proceed against the polluter and to order reparation for any harm. Such regulations can be contained under the environmental damages legislation.[116]

MENA countries should focus on the threats of mitigation practices at their national level in a collaborative approach. This requires re-prioritization of their economic resilience that lasts beyond gas and oil. Concrete actions should be embraced towards low-carbon development and green growth. Such synergistic economic and environmental resilience requires regional collaboration. Upgraded INDCs could state clear goals regarding the national energy plans. The region may also utilize the available resources under the Paris Agreement mandates, which support MENA with the necessary financial and technical resources to help them mitigate climate change and adapt to its effects.[117]

Emphasis on accessible databases

The Arab region needs to utilize available resources from the UNDP and UNFCCC as per the Paris Agreement. These include expertise, consultation, and funds for disaster risk reduction and to maximize the region's recovery.[118] The activation of documented early warning systems for climate change gives enough time for the authorities to establish their plans and seek support from international entities.[119] Participation in data collection and establishing a system of accessible databases is vital for decision-makers.[120]

Conclusion

The MENA region has moved forward to accommodate various international treaties to mitigate and adapt to the adverse effects of climate change in the context of global environmental laws. Although several policies, laws, and legislative instruments have been passed by several MENA countries to protect the environment and combat the effects of climate change, integrating public health into climate change policies and planning is still limited in the scope of application and implementation across MENA. Multi-sectorial collaborations, establishing proper monitoring and surveillance systems, resolving the political conflicts, and enhancing technological capacities are key factors to combatting the effects of climate change on human health. Further studies are needed to draw climate-resilient models that can be utilized in the mitigation practices in the GCC countries. Other potential areas of study include in-depth exploration into the climate and environmental law in MENA countries.

Notes

1 D. Olawuyi, *Principles of Nigerian Environmental Law* (Ado-Ekiti, Nigeria: Afe Babalola University Press 2015).
2 J. Patz, D. Campbell-Lendrum, T. Holloway, and J. Foley, "Impact of Regional Climate Change on Human Health" (2005) 438 *Nature* 310.
3 D. Gollin and C. Zimmermann, "Global Climate Change, the Economy, and the Resurgence of Tropical Disease" (2012) 19 *International Journal of Mathematical Demography* 51.
4 A. Toumi and others, "Temporal Dynamics and Impact of Climate Factors on the Incidence of Zoonotic Cutaneous Leishmaniasis in Central Tunisia" (2012) 6 *NaturePLoS Neglected Tropical Diseases* 1–8. *Mathematical Population Studies*.
5 K. Hayes and others, "Climate Change and Mental Health: Risks, Impacts and Priority Actions" (2018) 12 *International Journal of Mental Health Systems* https://doi.org/10.1186/s13033-018-0210-6.
6 R. Hanna, "Drivers and Challenges for Transnational Land–Water–Food Investments by the Middle East and North Africa Region" (2020) 7 *WIRES Water* 1–15.
7 Ibid.
8 G. Cissé, "Food-Borne and Water-Borne Diseases under Climate Change in Low- and Middle-Income Countries: Further Efforts Needed for Reducing Environmental Health Exposure Risks" (2019) 194 *Acta Tropica* 181; F. Fouque and J. Reeder, "Impact of Past and On-going Changes on Climate and Weather on Vector-Borne Diseases Transmission: A Look at the Evidence" (2019) 8 (51) *Infectious Diseases of Poverty* https://doi.org/10.1186/s40249-019-0565-1
9 M. Cooper, "How Climate Change Affects Allergies: Quality, Quantity of Allergens Are Shifting Worldwide, so Geographic and Family History Are Key" (2019) 11 *Optometry Times* 10; A. Damialis, C. Traidl-Hoffmann, and R. Treudler, "Climate

Change and Pollen Allergies" in M. R. Marselle, J. Stadler, H. Korn, K. N. Irvine, and A. Bonn (eds), *Biodiversity and Health in the Face of Climate Change* (New York: Springer 2019).

10 K. Reinmuth-Selzle and others, "Air Pollution and Climate Change Effects on Allergies in the Anthropocene: Abundance, Interaction and Modification of Allergens and Adjuvants" (2017) 51 *Environmental Science and Technology* 4119.

11 Ibid.

12 P. Sheffield, K. Weinberger, and P. Kinney, "Climate Change, Aeroallergens, and Pediatric Allergic Disease" (2011) 78 *Mount Sinai Journal of Medicine* 78.

13 See n 10.

14 Ibid.

15 Ibid.

16 See n 9.

17 N. Obradovich, R. Migliorini, P. Martin, and I. Rahwan, "Empirical Evidence of Mental Health Risks Posed by Climate Change" (2018) 115 *Proceedings of the National Academy of Sciences of the United States of America* 10953.

18 Ibid.

19 See n 9.

20 Y. Li and L. Wadsö, "Fungal Activities of Indoor Moulds on Wood as a Function of Relative Humidity during Desorption and Adsorption Processes" (2013) 13 *Engineering in Life Sciences* 528; E. PieckovÁ, "Adverse Health Effects of Indoor Moulds" (2012) 63 *ŠTETNI ZDRAVSTVENI UČINCI PLIJESNI UNUTARNJIH PROSTORA* 545.

21 "Challenge of Desertification: A Looming Threat to Food Security", *Asianet-Pakistan, 2019*; W. Quaye, R. Yawson, E. Ayeh, and I. Yawson, "Climate Change and Food Security: The Role of Biotechnology" (2012) 12 *African Journal of Food, Agriculture, Nutrition and Development* 6354.

22 T. Paramasilvam, "Report: Emerging Issues Facing the Water–Energy–Food Nexus in the Middle East and Asia" (2016) 32 *International Journal for Water Resources Development* 1016.

23 F. Dureab, K. Shibib, R. Al-Yousufi, and A. Jahn, "Yemen: Cholera Outbreak and the Ongoing Armed Conflict" (2018) 12 *Journal of Infection in Developing Countries* 397.

24 A. Ahmadalipour and H. Moradkhani, "Escalating Heat-Stress Mortality Risk Due to Global Warming in the Middle East and North Africa (MENA)" (2018) 117 *Environment International* 215.

25 M. Al-Bouwarthan, M. Quinn, D. Kriebel, and D. Wegman, "Assessment of Heat Stress Exposure among Construction Workers in the Hot Desert Climate of Saudi Arabia" (2019) 63 *Annals of Work Exposures and Health* 505.

26 D. Kaniewski, E. Paulissen, E. Campo, M. Ai-Maqdissi, J. Bretschneider, and K. Lerberghe, "Middle East Coastal Ecosystem Response to Middle-to-Late Holocene Abrupt Climate Changes" (2008) 105 *Proceedings of the National Academy of Sciences of the United States of America* 13941.

27 A. George, "Syria Builds One of the Middle East's Biggest Dams" (1999) 288 *Middle East* 34.

28 Cissé, n 8; M. Lange, "Impacts of Climate Change on the Eastern Mediterranean and the Middle East and North Africa Region and the Water–Energy Nexus" (2019) 10 *Atmosphere* 455.

29 M. Salem, S. Baidoun, and R. Almuzaini, "Water Consumption Demarketing Strategies with Reference to the Gaza Strip, Palestine" (2018) 27 *Polish Journal of Environmental Studies* 297; H. Legge, A. Shaheen, G. Shakhshir, and A. Milojevic, "Access to Water and Morbidity in Children in the Occupied Palestinian Territory, 2000–2014: A Repeated Cross-Sectional Study" (2018) 391 *Lancet* S8–S8.

30 M. Aijuka, A. Santiago, J. Girón, J. Nataro, and E. Buys, "Enteroaggregative *Escherichia coli* is the Predominant Diarrheagenic *E. coli* Pathotype among Irrigation Water

and Food Sources in South Africa" (2018) 278 *International Journal of Food Microbiology* 44.

31 G. Cissé, "Food-Borne and Water-Borne Diseases under Climate Change in Low-and Middle-Income Countries: Further Efforts Needed for Reducing Environmental Health Exposure Risks" (2019) 194 *Acta Tropica* 181.

32 F. Dureab, K. Shibib, R. Al-Yousufi, and A. Jahn, "Yemen: Cholera Outbreak and the Ongoing Armed Conflict" (2018) 12 *Journal of Infection in Developing Countries* 397; "Cholera Cases in Yemen Spike Again" (2019) 364 *Science* 8–9; D. He, X. Wang, D. Gao, and J. Wang, "Modeling the 2016–2017 Yemen Cholera Outbreak with the Impact of Limited Medical Resources" (2018) 451 *Journal of Theoretical Biology* 80.

33 L. Gonzalez and X. Briottet, "North Africa and Saudi Arabia Day/Night Sandstorm Survey (NASCube)" (2017) 9 *Remote Sensing* 896; C. Dang, G. Song, J. Jiang, T. Hou, and Y. Ren, "Comparative Analysis of Pollutant Characteristics during a Fog, Haze and Sandstorm" (2019) 10 *Meteorological and Environmental Research* 18.

34 Ibid.

35 M. Sarraf and M. Heger, "Why is #COP21 Important for the Middle East and North Africa Region?" (The World Bank Blogs, 30 November 2015) <https://blogs.worldbank.org/arabvoices/why-cop21-important-middle-east-and-north-africa-region>.

36 Ibid.

37 Ibid.

38 K. Ebi, J. Smith, and I. Burton, *Integration of Public Health with Adaptation to Climate Change, Lessons Learned and New Directions* (Boca Raton, FL: CRC Press 2005) 260.

39 T. Nyatanyi and others, "Implementing One Health as an Integrated Approach to Health in Rwanda" (2017) 2 *British Medical Journal Global Health* e000121; D. Olawuyi, "Sustainable Development and the Water-Energy-Food Nexus: Legal Challenges and Emerging Solutions" (2020) 103 *Environmental Science and Policy* 1.

40 M. Fox and others, "Integrating Public Health into Climate Change Policy and Planning: State of Practice Update" (2019) 16 *International Journal of Environmental Research and Public Health* 3232.

41 A. Spiess, "Developing Adaptive Capacity for Responding to Environmental Change in the Arab Gulf States: Uncertainties to Linking Ecosystem Conservation, Sustainable Development and Society in Authoritarian Rentier Economies" (2008) 64 *Global and Planetary Change* 244.

42 See, for example, World Health Organization, "Strengthening Health Resilience to Climate Change" (Technical Briefing for the World Health Organization, Conference on Health and Climate 2015) <www.who.int/globalchange/publications/briefing-health-resilience/en/>.

43 World Health Organization, "Climate Change and Human Health: Health and Climate Change Toolkit for Project Managers" <www.who.int/globalchange/resources/toolkit/en/>.

44 M. Sheehan, M. Fox, C. Kaye, and B. Resnick, "Integrating Health into Local Climate Response: Lessons from the U.S. CDC Climate-Ready States and Cities Initiative" (2017) 125 *Environmental Health Perspectives* 1.

45 United Nations Framework Convention on Climate Change, *Adoption of the Paris Agreement*, 12 December 2015, UN Doc. FCCC/CP/2015/L.9 [Paris Agreement].

46 Sarraf and Heger, n 35.

47 World Health Organization, "Climate Change and Human Health: Health Events in the Paris Climate Conference" <www.who.int/globalchange/global-campaign/en/>.

48 World Health Organization, "Climate Change and Human Health: WHO UNFCCC Health and Climate Change Country Profile Project" <www.who.int/globalchange/resources/countries/en/>.

49 J. Hess, J. McDowell, and G. Luber, "Integrating Climate Change Adaptation into Public Health Practice: Using Adaptive Management to Increase Adaptive Capac-

ity and Build Resilience" (2012) 120 <https://www.who.int/globalchange/resources/countries/en/171-79>; Centers for Disease Control and Prevention, "Climate and Health: BRACE Framework" <www.cdc.gov/climateandhealth/BRACE.htm>.

50 C. Dang and others, "Comparative Analysis of Pollutant Characteristics during a Fog, Haze and Sandstorm" (2019) 10 *Meteorological and Environmental Research* 18; "Cholera Cases in Yemen Spike Again," n 32; World Health Organization, Regional Office for the Eastern Mediterranean, "Regional Centre for Environmental Health Action" <www.emro.who.int/entity/ceha/index.html>.

51 World Health Organization, Regional Office for the Eastern Mediterranean, "WHO Global Arabic Programme" <www.emro.who.int/entity/global-arabic-programme/index.html>.

52 "Masdar" <fqhttps://masdar.ae/>.

53 Ibid.

54 IRENA, "Renewable Energy Market Analysis: GCC 2019", January 2019 <www.irena.org/publications/2019/Jan/Renewable-Energy-Market-Analysis-GCC-2019>.

55 Paris Agreement, n 46; UNFCCC, "Assessing Climate Change Impacts in the Arab Region", February 12–13, 2020 <https://unfccc.int/sites/default/files/resource/TechnicalWorkshop_Session1_Majdalani.pdf>; United Nations Economic and Social Commission for Western Asia (ESCWA), "Arab Climate Change Assessment Report – Main Report. Beirut, E/ESCWA/SDPD/2017/RICCAR/Report" <www.unescwa.org/sites/www.unescwa.org/files/events/files/riccar_main_report_2017.pdf>; UNDP, "Climate Change Adaptation in the Arab States," 2018 <www.undp.org/content/undp/en/home/librarypage/climate-and-disaster-resilience-/climate-change-adaptation-in-the-arab-states.html>.

56 K. Khoday and G. Haddad, "The SDG-Climate Nexus: UN Partnerships in the Arab Region," (UNDP, 20 November 2017) <www.undp.org/content/undp/en/home/blog/2017/the-sdg-climate-nexus--un-partnerships-in-the-arab-region.html>;   IISD, "UNP Report Discusses Climate-SDG Nexus in Arab Region," July 31, 2018 <https://sdg.iisd.org/news/undp-report-discusses-climate-sdg-nexus-in-arab-region/>; UNDP Arab States, "The SDG-Climate Facility Project" <www.arabstates.undp.org/content/rbas/en/home/climate-and-disaster-reslience/the-sdg-climate-facility-project.html>.

57 Arab Water Council <www.arabwatercouncil.org/index.php?lang=en>.

58 General Secretariat, Economic Department, Technical Secretariat of the Arab Ministerial Water Council, "Arab Strategy for Water Security in the Arab Region: to Meet the Challenges and Future Needs for Suitable Development (2010–2030)", 2012 <www.unescwa.org/sites/www.unescwa.org/files/events/files/arab_strategy_for_water_securityenglish_translation2012_0.pdf>.

59 Ibid.

60 UNESCWA, "Arab Strategic Framework for Sustainable Development", 2013 <www.unescwa.org/arab-strategic-framework-sustainable-development>.

61 W. Al-Zubaria, A. Al-Turbakb, W. Zahidb, K. Al-Ruwisb, A. Al-Tkhaisc, I. Al-Muatazb, A. Abdelwahabd, A. Murade, M. Al-Harbif, and Z. Al-Sulaymani, "An Overview of the GCC Unified Water Strategy (2016–2035)" (2017) 81 *Desalination and Water Treatment* 1–18.

62 Islamic Development Group, "Climate Change Policy", (ISDB February 2018) <www.isdb.org/sites/default/files/media/documents/2019-04/IsDB%20Climate%20Change%20Policy.pdf>.

63 Regional Center for Renewable Energy and Energy Efficiency (RECREEE) <www.rcreee.org/>.

64 Ibid.

65 M. Khan, "Global Renewable/Alternative Energy Trends and Associates Policies for the UAE" (2011) 15 *Defence Journal* 38.

66 Ibid.
67 Ibid.
68 UNDP, n 56, 32.
69 Ibid, 33.
70 Ibid.
71 See n 39.
72 M. Hochberg, "Renewable Energy Growth in Morocco: An Example for the Region" (*MEI Policy Focus* 2016-26 December 2016) <www.mei.edu/sites/default/files/publications/PF26_Hochberg_Moroccorenewables_web.pdf>.
73 Tunisia's Constitution of 2014 <www.constituteproject.org/constitution/Tunisia_2014.pdf>, Article 45.
74 M. Bouznit, M. del P. Pablo-Romero, and A. Sanchez-Braza, "Measures to Promote Renewable Energy for Electricity Generation in Algeria" (2020) 12 *Sustainability* 1468.
75 Saudi Arabia, "Vision 2030" <https://vision2030.gov.sa/en>.
76 Ibid.
77 State of Qatar, Government Communications Office, "Qatar National Vision 2030" <www.gco.gov.qa/en/about-qatar/national-vision2030/>.
78 Law No. 30 of 2002 Promulgating the Law of the Environment Protection 30/2002 (Qatar).
79 Ibid.
80 UAE, "National Climate Change Plan of the UAE 2017–2050" <https://u.ae/en/about-the-uae/strategies-initiatives-and-awards/federal-governments-strategies-and-plans/national-climate-change-plan-of-the-uae>.
81 Ibid.
82 Lebanon Ministry of Agriculture, "2015–2025 Lebanon National Forest Program" <http://extwprlegs1.fao.org/docs/pdf/leb163865.pdf>; Reforest'Action, "Reforestation in Morocco" <www.reforestaction.com/en/reforestation-morocco>; Syrian Arab Republic, "National Forest Program" <www.fao.org/forestry/14895-0b6ee182942e634eeab4a8e39e22afb26.pdf>.
83 Kingdom of Bahrain, "Bahrain National Environment Strategy", 8 October 2006 <www.sce.gov.bh/en/NationalStrategyforEnvironment?cms=iQRpheuphYtJ6pyXUGiNqkP7woZPUrlc>.
84 State of Kuwait, Energy Conservation Code of Practice (R-6) (1983), revised 2010. Information available at <www.unece.org/fileadmin/DAM/energy/se/pdfs/gee21/projects/others/Kuwait.pdf>.
85 The Hashemite Kingdom of Jordan, Ministry of Environment, "The National Climate Change Policy of the Hashemite Kingdom of Jordan 2013–2020", 2013 <www.jo.undp.org/content/dam/jordan/docs/Publications/Climate%20change%20policy_JO.pdf>.
86 Ibid, 46.
87 Egypt, Renewable Energy Law (Decree No 203/2014) (2014).
88 Oman, Ministerial Decision No. 20 of 2016 Regulations for the Management of Climate Affairs (2016).
89 Oman, Royal Decree No. 8/2011 (Oil and Gas Law) (2011).
90 Oman, Royal Decree No. 90 of 2007 Establishing the Ministry of Environment and Climate Affairs (2007).
91 Yemen, Republican Decree No.101 of 2005 on the establishment of the Public Environmental Protection Authority (2005).
92 Yemen, "The National Strategy for Environmental Sustainability 2005–2015", November 25, 2014 <http://ye.chm-cbd.net/management-and-conservation/strategies-action-plans-and-programs/NSES.pdf>.
93 Yemen, "National Action Plan 2005–2010", November 25, 2014 <http://ye.chm-cbd.net/management-and-conservation/strategies-action-plans-and-programs/NSES.pdf>.

94 Syrian Arab Republic, "Nationally Determined Contributions under Paris Agreement on Climate", November 2018 <www4.unfccc.int/sites/ndcstaging/Published-Documents/Syrian%20Arabic%20Republic%20First/FirstNDC-Eng-Syrian%20Arab%20Republic.pdf> 4–5.

95 Ibid.

96 Palestine, "Legislative Decree no 14/2015 on Renewable Energy and Energy Efficiency", December 2015 <https://climate-laws.org/legislation_and_policies?geography%5B%5D=136>.

97 Palestine, "Law No 7/1999 on the Environment", July 1999 <https://climate laws.org/geographies/palestine/laws/law-no-7-1999-on-the-environment>.

98 Palestine, "Law No 12/1995 on the Establishment of the Palestinian Energy Authority" <https://climate-laws.org/geographies/palestine/laws/law-no-12-1995-on-the-establishment-of-the-palestinian-energy-authority>.

99 Libya, Law No. 426 establishing the Renewable Energy Authority of Libya (REAOL) (2007).

100 Iraq, Law No. 27 of 2009 for Protection and Improvement of Environment (2009).

101 A. Ahmadalipour and H. Moradkhani, "Escalating Heat-Stress Mortality Risk due to Global Warming in the Middle East and North Africa (MENA)" (2018) 117 *Environment International* 215; Lange, n 28.

102 UNDP, n 56, 28.

103 A. Ahmadalipour and H. Moradkhani, "Escalating Heat-Stress Mortality Risk due to Global Warming in the Middle East and North Africa (MENA)" (2018) 117 *Environment International* 215–225; M. Lange, "Impacts of Climate Change on the Eastern Mediterranean and the Middle East and North Africa Region and the Water–Energy Nexus" (2019) 10 *Atmosphere* 455.

104 M. Tolba and N. Saab; "Arab Environment: Impact of Climate Change on Arab Countries" (Report 2009) <https://www.droughtmanagement.info/literature/AFED_climate_change_arab_countries_2009.pdf> accessed 27 July 2020.

105 Ibid.

106 Ibid.

107 M. Tolba and N. Saab, "Arab Environment: Impact of Climate Change on Arab Countries," 2009 Report of the Arab Forum for Environment and Development, 2009 <www.droughtmanagement.info/literature/AFED_climate_change_arab_countries_2009.pdf>.

108 T. Twinning-Ward, "Climate Change Adaptation in the Arab States, Best Practices and Lessons Learned" (UNDP, July 2018).

109 UNDP, n 56, 38–39.

110 S. Greenwood, "Water Insecurity, Climate Change and Governance in the Arab World" (2014) 21 *Middle East Policy* 140–156; J. Powell, "War or Peace? Water Conflicts in the Middle East" (1995) 8 *Geodate* 1.

111 H. Al-Maamary, H. Kazem, and M. Chaichan, "Climate Change: The Game Changer in the Gulf Cooperation Council Region" (2017) 76 *Renewable and Sustainable Energy Reviews* 555.

112 A. Spiess, "Developing Adaptive Capacity for Responding to Environmental Change in the Arab Gulf States: Uncertainties to Linking Ecosystem Conservation, Sustainable Development and Society in Authoritarian Rentier Economies" (2008) 64 *Global and Planetary Change* 244.

113 UNESCO, "Climate Change Education and Awareness" <https://en.unesco.org/themes/addressing-climate-change/climate-change-education-and-awareness>.

114 Ibid.

115 S. Sellers, K. Ebi, and J. Hess, "Climate Change, Human Health, and Social Stability: Addressing Interlinkages" (2019) 128 *Environmental Health Perspectives* 1.

116 J. Sowers, E. Weinthal, and N. Zawahri, "Targeting Environmental Infrastructures, International Law, and Civilians in the New Middle Eastern Wars" (2017) 48 *Security Dialogue* 410; See also "Laws of Environmental Liability, Harm and Crim-

inal Actions" <https://www.umweltbundesamt.de/en/environm-liability-a-environm
-damage-law-envir#the-laws-of-environmental-liability-and-environmental-harm>
accessed 18 July 2020.

117 N. Shafi, "Can Fighting Climate Change Bring the Arab World Closer Together?"
(World Economic Forum, 1 April 2019) <www.weforum.org/agenda/2019/04/to-f
ight-climate-change-the-arab-world-needs-to-come-together/>.

118 H. Al-Maamary, H. Kazem, and M. Chaichan, "Climate Change: The Game Changer in the Gulf Cooperation Council Region" (2017) 76 *Renewable and Sustainable
Energy Reviews* 555.

119 A. Spiess, "Developing Adaptive Capacity for Responding to Environmental
Change in the Arab Gulf States: Uncertainties to Linking Ecosystem Conservation,
Sustainable Development and Society in Authoritarian Rentier Economies" (2008)
64 *Global and Planetary Change* 244.

120 Ibid.

11 Carbon taxation as a tool for sustainable development in the MENA region

Potentials and future directions

Alexander Ezenagu

Introduction

This chapter analyzes the roles that carbon taxation can play in supporting climate change mitigation efforts in the Middle East and North African (MENA) region. After discussing the nature, scope, and elements of carbon taxation as a tool for combating climate change, it examines the legal and policy frameworks for promoting the progressive design and implementation of carbon taxation in the MENA region.

Over the last decade, MENA countries have accelerated efforts aimed at achieving sustainable development, especially through significant infrastructure investments.[1] This is largely observed in the increased international sporting, entertainment, clean technology, and geo-political activities now taking place in MENA countries.[2] However, the rapid infrastructural development in MENA states also means that they have become some of the world's biggest emitters of greenhouse gas (GHG) emissions that contribute to climate change on a per capita basis.[3] As MENA countries race to meet their infrastructural demands, it is expected that GHG emissions will increase significantly.[4] With increased global pressure for countries to reduce their GHG emissions,[5] MENA countries will need to consider their emission practices, with potential implications for their development, in the short, medium, and long term. Moreover, MENA countries already face harsh weather conditions and deteriorating air quality, due to desertification and intensive consumption of petrochemical energy for industries and households.[6] To reverse the current state of the environment and avoid long-term negative outcomes as forecast by experts, MENA countries must embrace policies, clean energy technologies, and practices that reduce carbon pollution and promote low carbon development.[7]

Carbon pricing has become increasingly recognized worldwide as a policy option that allows countries to promote a gradual shift to low carbon development.[8] Carbon pricing is a system that aims to promote GHG emission reduction by putting a price on GHG emission following the polluter-pays principle.[9] By putting a price on GHG emission, carbon pricing causes polluters to bear the burden of pollution and decide whether to seek cleaner options or continue to bear the brunt of emission. As described by the World Bank, "carbon pricing is an instrument that captures the external costs of GHG emissions—the costs of emissions that the public pays for, such as damage to crops, health care costs from heat waves and droughts, and loss of property from flooding and sea level rise—and ties them to their sources through a price, usually in the form of a price on the carbon dioxide (CO_2) emitted".[10] The two main carbon pricing approaches are the cap and trade system (otherwise known as emissions trading systems (ETS)) and carbon taxation. The chapter focuses on carbon taxation. It evaluates the utility, desirability, and potential of achieving low carbon transition and sustainable development in MENA countries through carbon taxation.

This chapter is divided into five sections. After this introduction, the next section examines the nature and scope of the two carbon pricing approaches, with focus on their pros and cons. The third section discusses the design of a carbon tax regime. The fourth section evaluates legal and policy barriers to implementing carbon taxation in the MENA region. The final section is the concluding section.

Market-based approaches to addressing carbon emissions

There are two main carbon pricing approaches, namely: the cap-and-trade system and carbon taxation. This section discusses the nature and scope of both.

Cap and trade system

In a cap and trade system, the government sets emission limits for emission entities, who must abide by the set limits.[11] Emission entities that exceed their emission limits may buy from those under-utilizing their emission quotas. Governments reduce the emission caps each year to set a new pollution target and allocate new emission limits to industries. This way, companies are forced or incentivized to seek other ways to meet their energy needs, thereby potentially embracing the use of green technology. Companies may also explore the option of buying allowances from other emission entities that have not surpassed their limits, which may lead to an increase in their cost of production.[12]

The cap and trade system allows emission entities to reduce emissions based on the cost of reductions, as opposed to requiring all industries to meet the same emission requirements, regardless of cost.[13] Anderson and Sullivan opine that a cap and trade system:

> can help energy intensive industries remain competitive with companies in regions that have no emission caps; eliminates windfall profits, provides money for clean energy development, and helps low-income ratepayers; can be distributed only to major greenhouse gas emitters and suppliers of fossil fuel, simplifying the system; guarantees that specific emissions reduction targets will be met, given that the caps shrink over time; brings in revenues that can be used to ease the burden on those with lower incomes; and assures price stability through the banking, auctioning and safety valves put in place.[14]

Experts who support the cap and trade system argue that environmental issues are best dealt with through the instrumentality of the market.[15] However, there are a number of counterarguments to this claim.[16] First, agreeing on the nature of markets that should be set up and their forms of socio-technical organization could be difficult, time-wasting, and expensive.[17] Second, ensuring effective functioning of the market demands a dedicated process, which many MENA countries may not have the facilities for.[18] Third, there are potential challenges present in the identification of buyers and sellers in the market. Moreover, rivalry among key existing firms and structural barriers to entry for new entrants may mean that carbon emissions are not adequately accounted for and traded properly. Also, potential manipulation of the market and pricing by key players exists.[19] In addition, a cap and trade system has the potential to reward polluters instead of punishing them.[20] Lastly, creating a profit-driven market out of an issue that threatens the existence and survival of the planet earth and of human beings has been described as immoral and counterproductive, and may not be the right approach to take.[21]

Anderson and Sullivan further highlight the practical limitations of a cap and trade system as: large fluctuations and unpredictability in allowance prices to make investment decisions difficult; complexity of a trading regime could foster delay and be difficult to enforce; firms operating under cap and trade are at a competitive disadvantage; free allocation of allowances provides windfall profits for polluters; a comprehensive programme must encompass many sectors, creating complications for trading and enforcement; and increased energy prices will burden low-income families.[22] These demerits have accentuated calls for carbon tax as a more viable approach.

Carbon tax

Meaning, nature, and scope of carbon taxation

Carbon tax is tax paid on the emission of carbon into the atmosphere. It refers to a form of explicit carbon pricing; a tax directly linked to the level of CO_2 emissions, often expressed as a value per tonne CO_2 equivalent (per tCO_2e).[23] Carbon tax is "a fee placed on greenhouse gas pollution mainly from burning fossil fuels. This can be done by placing a surcharge on carbon-based fuels and other sources of pollution such as industrial processes."[24]

A carbon tax puts a price on the real costs of GHG emissions on the economy, the environment, and on the pockets of governments. Kaufman opines that the fuel-specific charges that would be imposed by a carbon tax are a popular policy option because many believe that a carbon tax will reduce emissions of CO_2 in an economically efficient manner.[25] That is, a carbon tax will reduce the use of fossil fuels by spurring technical change and inducing the substitution of capital, labor, and non-energy materials. As of 2018, 26 national and subnational jurisdictions have introduced a direct carbon tax. These include British Columbia, Canada;

Chile; Costa Rica; Denmark; Finland; France; Iceland; Ireland; Japan; Mexico; Norway; South Africa;[26] Sweden; Switzerland; and the United Kingdom.[27]

Rationale for carbon tax

Countries impose taxes on their citizens and residents for four main reasons. First, to raise revenue to provide infrastructure.[28] Second, to re-price commodities to influence behaviour.[29] A third reason is to achieve representation by taxpayers in the affairs of the state by demanding accountability.[30] The fourth is to redistribute wealth.[31] The adoption of a carbon tax regime by MENA countries will be discussed in this chapter, in light of these reasons and others.

One reason for introducing carbon tax is to raise revenue needed by governments to provide infrastructure. State governments are duty bound to provide social amenities and infrastructure for their citizens and residents. Governments, in turn, share the burden with citizens and residents through the instrument of taxation. The basis for this relationship has been defined by scholars as deriving from a "social contract"[32] or "economic allegiance".[33] Carbon tax can be used by governments to raise significant revenue, while at the same time achieving reduction of carbon emissions by polluters—a "double dividend" benefit. States with carbon tax regimes observe significant increased revenue in their tax collection. For instance, British Columbia, a province in Canada which introduced carbon tax in 2008, saw additional yearly revenue contribution of above US$ 1 billion since year 2013[34] and its real GDP grew more than 17 per cent between 2007 and 2015.[35] Furthermore, experts in the wake of the coronavirus pandemic have mulled the idea of introducing carbon tax regimes by countries as a way of increasing government revenue and contributing to economic recoveries.[36]

To cushion the potential effect of carbon tax on taxpayers, some countries adopt a revenue-neutral approach.[37] In these countries, governments have either reduced the tax rates in other areas, devised a pay-back scheme, or expended the revenue from carbon tax on reliefs such as subsidized transportation, health care, clean parks, and clean water, especially for the poor.[38] In other instances, revenue from carbon tax has been used to create new employment opportunities and fund new businesses in clean technology.[39] Thus, unlike the cap and trade system, which prioritizes the carbon market and industry players, carbon tax works for the greater good of society.[40] This way, carbon tax can be used to redistribute wealth to the less privileged.

The other advantage of carbon tax is that it could foster behavioural change in fossil fuel consumption, thus, leading to a reduction in environmental pollution. In the long term, a well-priced carbon tax will cause emitting entities to seek cleaner energy supplies, therefore reducing the combustion of fossil fuel. For instance, in British Columbia, carbon emission saw a 12.9 per cent decrease in per capita emissions from 2008 to 2013.[41] Consumption of fossil fuels was reduced as consumers sought cleaner, cheaper alternatives as a result of the additional carbon tax burden.[42]

Secondly, the cost of administering, complying with, and enforcing carbon tax is low.[43] This low compliance cost results from the recognition that applying carbon tax to all polluters will cause varying rates of abatement, determined by individual marginal cost of pollution abatement.[44] Polluters with high marginal cost of abatement will likely elect to stick with paying the carbon tax, against abating, thus, concentrating abatement measures on low carbon polluters. Similarly, implementation of carbon tax has a low cost of enforcement. Where a tax structure already exists in the country, a carbon tax can be applied in the same way as the value-added tax (VAT) system in some industries, and as a licensing regime in other industries or emission entities.[45] This is particularly relevant for Gulf Cooperation Council (GCC) countries, who recently introduced the Unified VAT Agreement for the Cooperation Council for the Arab States of the Gulf (GCC Treaty).[46] Therefore, existing tax and licensing regimes may be immediately employed to implement carbon tax regimes in MENA states.

Thirdly, introduction of carbon tax can lead to innovation in green technology. The global pressure to reduce carbon emission has witnessed increased investment and adoption of clean energy, such as solar energy and wind energy, causing departure from fossil-fuelled energy supplies.[47] Innovation in clean energy is fast expanding: electric cars are increasingly more accessible and affordable to many and will soon be commonplace in most countries.[48] MENA countries, in developing their states, will have no choice but to adopt innovation in clean energy and green technology.[49] International agreements, such as the Paris Agreement,[50] limit the choices for countries in their developmental process.[51] Introducing carbon tax will cause emission entities to seek cleaner ways of generating and using energy, thus opening up the market for clean energy and green technology in the MENA region.

However, the introduction of carbon tax has its opposition. Some experts argue that a carbon tax does not guarantee decreased carbon emissions.[52] They claim that unlike a cap and trade system, which sets emission limits, a carbon tax focuses on revenue generation.[53] While there is some truth to the claim, taxpayers are generally not receptive to the idea of paying more taxes to the government. Thus, this tax apathy will lead taxpayers to embrace green technology, especially where affordable.[54] This is particularly true for MENA countries, where citizens will reduce fossil fuel combustion if there is a high cost attached to the use of fossil fuels.[55] In some MENA countries, taxpayers may refuse to support the introduction of carbon tax, on the conviction that putting more money in the hands of the governments of their countries, suspected of corruption and misappropriation, will be a bad idea.[56]

Furthermore, there are concerns that a carbon tax may distort the economy, by increasing the cost of business, adding an additional compliance burden, and increasing cost of living.[57] However, evidence from Scandinavian countries shows otherwise.[58] Countries are able to set off the burden of carbon tax through the introduction of palliative measures.[59] As countries consider introducing carbon tax, the focus should not only be on the revenue generation attribute of a carbon tax but also on the overall impact a carbon tax has on the environment, in terms of behavioural change and encouraging the promotion and adoption of clean technology.

Finally, an additional challenge to introducing carbon tax in MENA countries is the lack of adequate information on the volume and sources of carbon emissions, emission entities, and determination of the carbon tax rate. Given that no MENA country presently has a carbon tax regime, generating data on the volume of carbon emissions with high level of accuracy for different activities and industries is relevant for designing an effective tax regime.[60] Also, management of the system poses a challenge, given the dearth of administrative capacities of tax administrators in MENA countries.[61] Another challenge is the conflicting presence of other environmental policy measures or taxes, such as excise taxes, levies for flaring gases, and environmental levies.[62]

Designing a carbon tax regime: legal and institutional prerequisites

The factors that influence the economic returns of carbon tax in any country include the pricing of the carbon, the carbon tax base, and the administration of the carbon tax regime.[63] These factors are best provided in an Act of Parliament. This section explores some of the steps required for an effective carbon tax regime.

Act of Parliament

While MENA countries have committed to reducing their carbon emission footprints, there exists no carbon tax regime in the region. One of the key reasons is the absence of clear and comprehensive laws that establish the legal framework for carbon taxation.[64] Given the rising calls for governments to intensify their climate change action, carbon tax legislation provides the right instrument for achieving the climate change commitments. Carbon tax legislation enacts a charging power, enabling the state to levy the tax on taxable persons. For instance, s 2 of the Carbon Tax Act of South Africa provides that there must be levied and collected for the benefit of the National Revenue Fund, a tax to be known as the carbon tax.[65] The carbon tax law further contains provisions on the tax rate and the tax base, among other provisions.

On tax rate, a carbon tax law must expressly state the rate of carbon taxation as it not only helps with decision-making for consumers but also helps the government in ascertaining its revenue potential. For example, British Columbia has set a price of CA\$30 per tCO_2e,[66] Chile has a pricing of US\$5 per tCO_2e,[67] Sweden charges at a rate of US\$168 per tCO_2e,[68] and South Africa at a price of R120 per tCO_2e.[69]

An optimal tax rate is one that reflects the specific cost that results from the actions of the polluter for the society as a whole.[70] This approach known as the cost-benefit analysis allows the marginal benefit that society acquires from the reduction of damage to be compared with marginal cost to society of cleaning up the pollution.[71] Another approach is the cost-effectiveness analysis, which involves choosing the amount of tax such that it will achieve a previously set emissions reduction goal.[72]

In relation to the tax base, the broader the carbon tax base, the higher the revenues generated by the imposition of carbon tax on emission entities.[73] For instance, s 4 of the Carbon Tax Act of South Africa provides that the carbon tax must be levied in respect of the sum of the GHG emissions of a taxpayer in respect of a tax period expressed as the carbon dioxide equivalent of those greenhouse gas emissions resulting from fuel combustion and industrial processes, and fugitive emissions in accordance with the emissions factors determined in accordance with a reporting methodology approved by the Department of Environmental Affairs.[74]

However, countries tend to exempt particular industries from the payment of carbon tax in their tax laws, thus narrowing the tax base. For instance, Denmark does not tax fuels used for electricity production,[75] Ireland excludes most emissions from farming,[76] and Mexico exempts natural gas from carbon tax.[77] These exemptions are policy considerations aimed at encouraging industrial development. This chapter recommends that an optimal tax base for revenue generation is one that has a Pigouvian effect. A Pigouvian tax[78] is a tax levied on any market activity that generates negative externalities, in this instance, a tax levied on all emission entities generating negative externalities.[79] Thus, from households using fossil fuels to cook and for electricity, to the car owner whose automobile causes damage to the environment, to the cigarette smoker, to the factory owner, or company operating on fossil fuels, all these emission entities must be captured within the tax base. Setting an adequately captured rate will balance off the cost of emission to the emission entities, with the biggest polluters paying the highest cost for emissions.[80] MENA countries, being largely extractive industries, will be positioned to generate high revenue from the exploration and production of their natural resources. Carbon taxes from the oil, gas, and mining industries, if adequately implemented, can generate significant revenue to the governments of these countries.

Administration of the carbon tax regime

Efficient implementation of carbon tax is at the heart of the realization of any meaningful revenue contribution to the economy of any MENA country.[81] This is one area to which MENA countries must pay more attention. Potential revenue generation from carbon tax should encourage MENA countries to seriously consider the introduction of carbon tax.[82] Countries that impose carbon tax report significant revenue generation from the tax: Ireland's carbon tax generates about 400 million euros annually;[83] British Columbia generated CA$6.1 billion between 2008 and 2015;[84] Finland posts revenue of US$750 million yearly;[85] the Netherlands' annual revenue is US$4.819 billion.[86] Sweden generates an annual revenue of US$3.665 billion;[87] the United Kingdom an annual revenue of US$905 million;[88] and Denmark, an annual revenue of US$905 million.[89]

To achieve efficient administration of the carbon tax regime, inter-agency cooperation is especially necessary. Revenue authorities must cooperate with environmental agencies in data collection, monitoring, and evaluation and adjustment of the carbon tax rate. This is clearly seen in the Carbon Tax Act of South Africa, which provides for inter-agency cooperation between the Department of Environmental Affairs and the South African Revenue Service (SARS).[90]

Furthermore, determining who pays the tax is probably a better consideration than who bears the brunt. Granted, the brunt will almost certainly be borne by the end-consumers; however, the question of whether the tax will be paid upstream, downstream, or mid-stream becomes a policy decision for governments. The Center for Climate and Energy Solutions recommends that, for administrative simplicity, carbon tax should be levied at the point where there are relatively few entities subject to the tax, though admits that achieving this will depend on the fuel type.[91] It further opines that wherever the tax is imposed, the price signal it creates will theoretically be passed backwards and forwards through the energy chain in the same way, and that the price signal should, in principle, bring about the same behavioural response and result in the same economic burden to firms and consumers.[92]

A downstream tax will have to capture millions of emission entities, thus making the administration expensive compared to an upstream tax, which captures few emitters and is relatively less expensive to administer. Thus, an effective design will be to adopt the VAT system, where manufacturers are taxed, and the costs spread to end-consumers.[93] For example, a company producing X amount of oil will pay the carbon tax on the expected emissions from the use of the oil by end-users and the carbon tax paid will be factored into the retail price of oil to end-users. Thus, liability to pay the tax rests with the upstream operator, while the operator ensures the tax costs are passed to their consumers. This design is efficient for MENA countries, given the capacities of tax authorities.

This upstream tax system may not capture all energy suppliers given the diversity of sources of energy in most MENA countries. This is where the licensing regime has an important part to play. Carbon tax can be built into the award and renewal of licences and could be deployed to capture major emission entities, such as independent power plants or factories with independent power from fossil fuel combustion. The licensing regime could also be used to capture small emission entities, such as vehicle operators not captured under the VAT system.

In summary, while significant revenue may be realizable from carbon tax for MENA states, the reduction in carbon emission should remain the core purpose for introducing carbon tax in any country. For instance, Murray and Rivers report that British Columbia experienced a 5–15 per cent reduction in carbon emission since the introduction of carbon tax;[94] Sweden's total GHG emissions fell 16 per cent;[95] and in Norway, the carbon tax reduced emissions by about 1.5 to 2.3 per cent.[96]

Legal and policy barriers to implementing carbon taxation in the MENA region

Though the percentage contributions of MENA countries to global emissions is historically and comparably low, notwithstanding the region's emission per capita, the region is not spared from the effects of climate change on its environment.[97] As MENA countries seek to attract foreign direct investment and put infrastructural facilities in place, it is important they set out to curtail, early on, the pollution from building infrastructural facilities. Carbon tax provides an effective tool for encouraging behavioural change and adoption of green technology, alongside securing a significant revenue stream for MENA governments. This section discusses some of the legal and policy barriers to implementing carbon tax.

Absence of comprehensive carbon tax legislation

As stated above, there is an absence of a clear and comprehensive legal regime on carbon tax in MENA countries. The first step towards implementing a carbon tax regime is to enact enabling legislation to that effect. The ease or difficulty of achieving this will be determined by the system of government in place. In GCC states, with a progressive legislative body, enacting a carbon tax law should be easily accomplished. States with bicameral legislature may witness a more difficult path.[98]

However, it should be expected that the introduction of a carbon tax regime will face stiff opposition by taxpayers in MENA countries. This is attributable to the level of tax morale in those countries. The tax to GDP ratio of most of the MENA countries falls below 28 per cent, compared to their European country counterparts averaging 44 per cent.[99] Data from the International Monetary Fund (IMF) reveal the tax to GDP ratio in GCC countries to be as low as 3.8 in Qatar, 0.8 in Bahrain, 2.2 in Kuwait, 6.0 in Oman, 2.3 in Saudi Arabia, and 3.5 in the UAE.[100] This low tax to GDP ratio is caused by a culture of tax apathy, inefficient tax administration, and lack of transparency in many MENA countries.[101] The tax base in many MENA countries is narrow as the informal sector is hardly captured and there is an over-emphasis on the natural resource sectors.[102] Also, income, profits, and consumption are not taxed in some of the countries.[103]

For any meaningful returns from carbon tax to occur, the law must provide for a tax base broad enough to capture all emission entities, save for those expressly exempted by legislation or policy. The tax base must capture emissions from car exhaust, wood and garbage burning, use of fuel stoves, use of diesel electricity generators and petrochemical plants, gas flaring, oil spillages, refining activities, etc. Also, efficient and capable tax authorities with in-depth training on mobilization and collection of domestic revenue are essential to providing these countries with predictable revenue, thereby reducing the reliance on natural resources, and dependence on development aids. In addition, the carbon tax rate must be high enough to contribute significantly to government revenue, while at the same time influencing behaviour. Having predictable domestic revenue sources will go a long way in budget planning and implementation, and will reduce reliance on foreign currencies, thus protecting the local currencies.[104]

Limited availability of clean technology

Another barrier to the introduction of carbon tax in MENA countries is the absence of clean alternatives to fossil fuels. Alternatives, such as biofuels, solar energy, and water-driven hydro projects, are yet to be fully developed and utilized in commercial quantities. Until green technology is developed to replace fossil fuels, the goal of reduction of carbon emissions will be impossible to achieve and imposition of carbon tax will have no significant influence on reduction of carbon emissions, which is the ultimate goal of carbon tax.[105]

Political will questions

Given the high dependence of MENA economies on hydrocarbon industries, the implementation of a carbon tax system may be met with strong opposition by industry stakeholders. For instance, the countries in the GCC hold almost a third of proven crude oil reserves and about a fifth of the world's gas reserves. This natural resource wealth accounts for the bulk of exports from these countries and a significant part of their revenue.[106] Given the hydrocarbon-dependent nature of MENA economies, any policy or discourse which affects or may affect the economy of these natural resource-rich countries may face practical challenges and opposition. Thus, implementing carbon tax, especially accessing relevant data and information on levels of emission, distribution of emission, and emission verification, amongst others, may be difficult without comprehensive legislative and policy backing. This is not an attempt to discredit the clean development mechanisms put in place by MENA governments and commitments to reduce GHG emissions in their countries.[107] However, export-driven combustion of fuels and combustion of fuels for local consumption are approached differently.[108] Strong political will and support is required to develop, implement, and monitor carbon tax policies. Also, given that tax avoidance by major corporations and industry stakeholders is a major problem facing regular tax regimes in many countries across the world, effective monitoring systems backed by government support will be required to avoid similar avoidance problems from major polluters, with respect to carbon taxes.[109]

Fear of competitiveness and introduction of border tax adjustments

There are general concerns that introducing carbon taxes by states may cause goods produced in-state to be non-competitive against foreign products, thus, potentially making the state a dumping ground.[110] One way to address this possibility is through the introduction of a border tax adjustment (BTA). Panezi has defined BTAs as "optional taxes or duties imposed on imports in order to ensure similar market conditions for similar domestic and imported products, when the domestic products are already taxed nationally".[111] BTAs are expected to "level the playing field between domestic and foreign products".[112] However, there is the concern that BTAs may violate a state's international commitments owed to the World Trade Organization (WTO). The trade implication of BTAs is beyond the scope of this paper; however, suffice it to say that the exceptions to the non-discrimination rules—to protect human, animal, and plant health and life, public morals, or exhaustible natural resources—could justify the introduction of a BTA by a state.[113]

Advancing carbon tax systems in MENA countries: concluding recommendations

The advantages of environmental effectiveness, economic efficiency, increased public revenue, and carbon emissions reductions attributable to carbon tax provide strong arguments in favor of introducing carbon tax systems.[114] Carbon taxation provides a tool for MENA countries to fight climate change and its adverse effects on the economy and lives of all. As argued, carbon tax offers "double dividend"—revenue generation and behavioural change—which ultimately reduces carbon emissions.

However, the absence of robust legal and institutional frameworks on climate change across the region is exacerbated by a lack of enabling carbon tax laws, a low tax morale among taxpayers, and the absence of institutional capacity. Legal barriers that stifle the development of coherent carbon taxation systems in MENA countries must be addressed in order to leverage the potential of carbon taxation as a tool for sustainable development in the region. The first step towards achieving this is by enacting a carbon tax law with the right tax rates, tax bases, and institutions to enforce the law.

On setting the carbon tax rate, the rate must reflect an empirical valuation of the cost of carbon emissions to the environment, the economy, and the lives of inhabitants, and must be set at a price that discourages fossil fuels combustions. The emissions limits or goal must be ascertained prior to setting a price and the price must reflect commitment to reduce carbon emissions. The carbon tax rate should rise and fall over time with the growth rate of the marginal damages from emissions. MENA governments should be dissuaded from seeing carbon tax solely as another revenue stream but must pursue a revenue-neutral carbon tax rate. This can be achieved by reducing other taxes or levies paid by inhabitants, and by putting up palliatives. Also, tax bases should be broad enough to capture all emission entities or emissions behaviours, with few (if any) exceptions.

Palliatives must be put in place for two reasons. First, to ensure revenue-neutrality of carbon tax; and second, to cushion the effects of an additional tax burden on the poor. Such palliatives can be a redistribution of revenue from carbon tax to the most vulnerable in the society or those directly impacted by environmental pollution; reduction of other taxes paid by taxpayers such as VAT, income taxes, thus resulting in no effective net tax increase; an massive investment in infrastructure, such as clean parks, better transport system, free transport for taxpayers, education scholarships, planting of trees, etc.

The success of carbon tax depends primarily on the tax authorities of the countries in question. The tax authorities must be equipped and trained to address the challenges of carbon tax. They must interact with other agencies, such as environmental bodies, research institutes, and other stakeholders in creating an effective carbon tax regime. Their activities and approaches must be scientific and transparent, and they must always command confidence from the taxpayers by acting in good faith.

Notes

1 See D. Olawuyi, "Advancing Innovations in Renewable Energy Technologies as Alternatives to Fossil Fuel Use in the Middle East: Trends, Limitations, and Ways Forward" in D. Zillman, M. Roggenkamp, L. Paddock, and L. Godden (eds), *Innovation in Energy Law and Technology: Dynamic Solutions for Energy Transitions* (Oxford University Press 2018).
2 Ibid.
3 World Bank Group, "State and Trends of Carbon Pricing 2019" Washington, DC, 2019 <https://openknowledge.worldbank.org/handle/10986/31755>; A. Al-Sarihi, "How Can Carbon Pricing Support Economic Diversification in the Arab Gulf States?" (Carbon Pricing Leadership Coalition, 13 December 2018) <www.carbon pricingleadership.org/blogs/2018/12/13/how-can-carbon-pricing-support-economic -diversification-in-the-arab-gulf-states> accessed 7 January 2021.
4 M. Daneshvar, M. Ebrahimi, and H. Nejadsoleymani, "An Overview of Climate Change in Iran: Facts and Statistics" (2019) 8 *Environmental Systems Research*, <https ://doi.org/10.1186/s40068-019-0135-3>
5 At the 21st Conference in Parties (COP 21) held in Paris in December 2015, parties to the United Nations Framework Convention on Climate Change (UNFCCC) agreed to limit the global average temperature to well below 2°C more than pre-industrial levels and to further pursue the reduction of temperature increase to only 1.5°C above pre-industrial levels (UNFCCC, Paris Agreement, 13 December 2015, UN Doc. FCCC/CP/2015/10/Add.1, Art. 2(1)(a)).
6 A. Al-Sarihi, "Prospects for Climate Change Integration into the GCC Economic Diversification Strategies" (LSE Middle East Centre Paper Series, No. 20, February 2018).
7 M. Babiker and M. Fehaid, "Climate Change Policy in the MENA Region: Prospects, Challenges and the Implications of Market Instruments" (Economic Research Forum, Working Paper 588, May 2011) <https://ideas.repec.org/p/erg/wpaper/588 .html>; D. Olawuyi, "Can MENA Extractive Industries Support the Global Energy Transition? Current Opportunities and Future Directions" (2020) *The Extractive Industries and Society* <https://doi.org/10.1016/j.exis.2020.02.003>.
8 A. Koniuszewski, "Taxing to Promote Public Goods: Carbon Pricing" in B. Alepin, B. Moreno-Dodson, and L. Otis (eds), *Winning the Tax Wars: Tax Competition and Cooperation* (Alphen aan den Rijnm the Netherlands: Wolters Kluwer 2018).
9 D. Olawuyi, *Principles of Nigerian Environmental Law* (Afe Babalola University Press 2015).
10 The World Bank, Carbon Pricing Dashboard, "What is Carbon Pricing?" <https://ca rbonpricingdashboard.worldbank.org/what-carbon-pricing>.
11 World Bank Group, "A Guide to Greenhouse Gas Benchmarking for Climate Policy Instruments", Partnerships for Market Readiness, Technical Note 14, April 2017; D.

Olawuyi, *The Human Rights Based Approach to Carbon Finance* (Cambridge University Press 2016).

12 L. Raymond and G. Shively, "Market-Based Approaches to CO$_2$ Emissions Reductions" 2008 <https://www.choicesmagazine.org/2008-1/articles/2008-1-01.htm>.

13 G. Andersen and D. Sullivan, "Reducing Greenhouse Gas Emissions: Carbon Cap and Trade and the Carbon Tax," National Conference of State Legislatures, July 2009, www.ncsl.org/documents/environ/Captrade.pdf; The European Commission, "EU Emissions Trading System (EU ETS)", <http://ec.europa.eu/clima/policies/ets/index_en.html>.

14 Ibid.

15 M. Betsill and M. Hoffmann, "The Contours of 'Cap and Trade': The Evolution of Emissions Trading Systems for Greenhouse Gases" (2011) 28 *Review of Policy Research* 83.

16 D. Olawuyi, "From Kyoto to Copenhagen: Rethinking the Place of Flexible Mechanisms in Kyoto's Post 2012 Commitments" (2010) 6 *Journal of Law, Environment and Development* 23.

17 Betsill and Hoffman, n 15

18 Ibid.

19 Ibid.

20 Olawuyi, n 16.

21 Ibid.

22 Andersen and Sullivan, n 13; M. Callon, "Civilizing Markets: Carbon Trading between *in vitro* and *in vivo* Experiments" (2009) 34 *Accounting, Organizations and Society* 535.

23 OECD, "Climate and Carbon: Aligning Prices and Policies," OECD Environment Policy Paper, No. 1, October 2013 <www.oecd-ilibrary.org/environment-and-sustainable-development/climate-andcarbon_5k3z11hjg6r7-en>.

24 David Suzuki Foundation, "Carbon Tax or Cap-and-Trade?" <www.davidsuzuki.org/issues/climate-change/science/climate-solutions/carbon-tax-or-cap-and-trade/>.

25 R. K. Kaufmann, "Limits on the Economic Effectiveness of a Carbon Tax" (1991) 12 *The Energy Journal* 12 International Association for Energy Economics 139.

26 South Africa's carbon tax covers GHG emissions arising from energy use (fuel combustion and gasification) and non-energy industrial processes (World Bank Group, n 3).

27 World Bank Group, n 3.

28 M. Jakob, C. Chen, S. Fuss, A. Marxen, N. Rao, and O. Edenhofe. "Carbon Pricing Revenues Could Close Infrastructure Access Gaps." (2016) 84 *World Development* 254–265.

29 N. Butler, "We Need a Carbon Tax to Change Consumer Behaviour" *Financial Times* (2019) <https://www.ft.com/content/2c5f19d6-7245-11e9-bf5c-6eeb837566c5>.

30 S. Carattini, M. Carvalho, and S. Fankhauser, "How to Make Carbon Taxes More Acceptable" (2017) *The Grantham Research Institute on Climate Change and the Environment* <http://www.lse.ac.uk/GranthamInstitute/wp-content/uploads/2017/12/How-to-make-carbon-taxes-more-acceptable.pdf>.

31 J. Villarreal, "To Stop Climate Change Don't Just Cut Carbon. Redistribute Wealth" (2016) *Foreign Policy in Focus* <https://fpif.org/stop-climate-change-dont-just-cut-carbon-redistribute-wealth/>.

32 J. Wolff, "Hobbes and the Motivations of Social Contract Theory" (1994) 2 *International Journal of Philosophical Studies* 271.

33 L. van Apeldoorn, "BEPS, Tax Sovereignty and Global Justice" (2018) 21 *Critical Review of International Social and Political Philosophy* 478.

34 C. Lammam and T. Jackson, "Examining the Revenue Neutrality of British Columbia's Carbon Tax," Fraser Institute, February 2017 <www.fraserinstitute.org/sites/default/files/examining-the-revenue-neutrality-of-bcs-carbon-tax.pdf>.

35 J. P. Tasker, "What You Need to Know: Federal Carbon Tax Takes Effect in Ont., Manitoba, Sask., and N. B. Today", April 1, 2019 <www.cbc.ca/news/politics/tasker -federal-carbon-tax-explainer-1.5077445>.

36 K. Dervis and S. Strauss, "The Carbon Tax Opportunity", May 6, 2020 <www.brooki ngs.edu/opinions/the-carbon-tax-opportunity/>; K. Bond, "Carbon Tax – An Idea Whose Time Has Come," April 6, 2020 <https://carbontracker.org/carbon-tax-an-idea-whose-time-has-come/>; M. Moore and W. Prichard, "How Should We Tax after the Pandemic?" <www.ictd.ac/blog-author/mick-moore/>.

37 J. Carl and D. Fedor, "Revenue-Neutral Carbon Taxes in the Real-World" (2012) Hoover Institution, Stanford University <www.hoover.org/taskforces/energy-policy>.

38 D. B. Marron and A. C. Morris, "How to Use Carbon Tax Revenues" (2016) Urban Institute & Brookings Institution <https://www.brookings.edu/wp-content/uploads/ 2016/07/howtousecarbontaxrevenuemarronmorris.pdf>.

39 A. Bowen, "Carbon Pricing: How Best to Use the Revenue?" (2015) *Policy Brief* <http://www.lse.ac.uk/granthaminstitute/wp-content/uploads/2015/11/Bowen-pol icy-brief-2015.pdf>.

40 Lammam and Jackson, n 34.

41 C. Komanoff and M. Gordon, "British Columbia's Carbon Tax: By the Numbers," A Carbon Tax Centre Report, December 2015 <https://www.carbontax.org/wp-conten t/uploads/CTC_British_Columbia's_Carbon_Tax_By_The_Numbers.pdf>; S. Elgie, "British Columbia's Carbon Tax Shift: An Environmental and Economic Success," September 10, 2014 <https://blogs.worldbank.org/climatechange/british-columbia-s -carbon-tax-shift-environmental-and-economic-success>.

42 Ibid.

43 H. Zhang, X. Xu, J. Jiang, and M. Zhang, "Carbon Tax and Trading Price on Power Plant with Carbon Capture and Storage under Incentive Regulation Theory" (2020) 2020 *Discrete Dynamics in Nature and Society* 8509834

44 Ibid.

45 I. Parry, R. Ploeg, and R. Williams, "How to Design a Carbon Tax" (2012) OxCarre Policy Paper 14.

46 R. Scalia, "VAT in United Arab Emirates, Saudi Arabia and Bahrain – Transitional Rules" (2019) 30 (1) *International VAT Monitor* <https://www.ibfd.org/sites/ibfd.org/file s/content/pdf/ivm_2019_01_gcc_1.pdf?utm_source=twitter&utm_medium=social-me dia&utm_campaign=tweet-week-25&utm_content=pdf/ivm_2019_01_gcc_1.pdf>

47 Olawuyi, n 1.

48 Ibid.

49 R. Poudineh, A. Sen, and B. Fattouh, "Advancing Renewable Energy in Resource-Rich Economies of the MENA" *Oxford Institute for Energy Studies* (September 2016) <https://www.oxfordenergy.org/wpcms/wp-content/uploads/2016/10/Advancing -Renewable-Energy-in-Resource-Rich-Economies-of-the-MENA-MEP-15.pdf>.

50 F. Cahill-Webb, "International Environmental Governance and the Paris Agreement on Climate Change: The Adoption of the 'Pledge and Review' Governance Approach" (2018) Working Paper, No. 99/2018, Hochschule für Wirtschaft und Recht Berlin, Institute for International Political Economy (IPE), Berlin <https://www .econstor.eu/handle/10419/175427>.

51 Ibid.

52 Andersen and Sullivan, n 13.

53 Ibid.

54 M. Mansour, "Tax Policy in MENA Countries: Looking Back and Forward" (2015) IMF Working Paper 15/98.

55 A. Jewell, M. Mansour, P. Mitra, and C. Sdralevich, "Fair Taxation in the Middle East and North Africa" (2015) IMF Staff Discussion Note.

56 A. Jewell, M. Mansour, P. Mitra, and C. Sdralevich, "Fair Taxation in the Middle East and Northern Africa" International Monetary Fund (2015) SDN/15/16 <https:/ /www.imf.org/external/pubs/ft/sdn/2015/sdn1516.pdf>.

57 Andersen and Sullivan, n 13.
58 P. Criqui, M. Jaccard, and T. Sterner, "Carbon Taxation: A Tale of Three Countries" (2019) 11 *Sustainability* 6280. doi: 10.3390/su11226280 <www.mdpi.com/journal/sustainability>.
59 C. G. Fernández, "Carbon Tax and Economic Crisis: The Need to Change the Current Production Model" *La balsa de piedra, n° 10, enero-marzo* (2015) p. 2. ISSN: 2255-047X <https://dialnet.unirioja.es/descarga/articulo/4905315.pdf>.
60 J. Meltzer, N. Hultman, and C. Langley, "Low-Carbon Energy Transitions in Qatar and the Gulf Cooperation Council Region" (2014) Global Economy and Development Program.
61 Jewell et al., n 56.
62 Meltzer et al., n 60.
63 J. Carl and D. Fedor, "Tracking Global Carbon Revenues: A Survey of Carbon Taxes Versus Cap-and-Trade in the Real World" *Energy Policy* (September 2016), pp 50–77 <https://www.sciencedirect.com/science/article/pii/S0301421516302531>.
64 N. Fujiwara, M. Alessi, and A. Georgiev, "Carbon Market Opportunities in the Middle East and North Africa," MEDPRO Technical Report No. 8, Centre for European Policy Studies, March 2012 <https://www.files.ethz.ch/isn/141241/MEDPRO%20TR%20No%208%20Fujiwara%20et%20al%20MENA%20Carbon%20Markets.pdf>.
65 Carbon Tax Act of South Africa Act No. 15 of 2019, s 2 <www.gov.za/sites/default/files/gcis_document/201905/4248323-5act15of2019carbontaxact.pdf>.
66 M. Beck, N. Rivers, R. Wigle, and H. Yonezawa, "Carbon Tax and Revenue Recycling: Impacts on Households in British Columbia" (2015) 41 *Resource and Energy Economics* 40–69.
67 World Bank and Ecofys, 2018, "State and Trends of Carbon Pricing" (May 2018), World Bank, Washington, DC <https://openknowledge.worldbank.org/bitstream/handle/10986/29687/9781464812927.pdf?sequence=5&isAllowed=y.>
68 Partnership for Market Readiness (PMR). "Carbon Tax Guide: A Handbook for Policy Makers." World Bank, Washington, DC. (2017) License: Creative Commons Attribution CC BY 3.0 IGO https://openknowledge.worldbank.org/bitstream/handle/10986/26300/Carbon%20Tax%20Guide%20-%20Main%20Report%20web%20FINAL.pdf?sequence=1&isAllowed=y.
69 Carbon Tax Act of South Africa, n 65, s 5.
70 OECD, "Effective Carbon Rates: Pricing CO$_2$ through Taxes and Emissions Trading Systems," September 26, 2016 <https://www.oecd-ilibrary.org/taxation/effective-carbon-rates_9789264260115-en>.
71 K. Coplan, "Carbon Tax Based on Social Cost of Carbon: Cost Benefit Analysis in Disguise?" *Green Law Blog* (28 April 2016) <https://greenlaw.blogs.pace.edu/2016/04/28/carbon-tax-based-on-social-cost-of-carbon-cost-benefit-analysis-in-disguise/#:~:text=Under%20cost%20benefit%20analysis%2C%20the,harms%20avoided%20by%20the%20regulation.&text=So%20here's%20the%20problem%20with%20a%20social%2Dcost%20based%20carbon%20tax>.
72 For more information on carbon pricing, see J. Elbeze and C. De Perthuis, "Twenty Years of Carbon Taxation in Europe: Some Lessons Learned", Les Cahiers de la Chaire Economie du Climat, Information and Debates Series, No. 9, April 2011.
73 W. Gale, S. Brown, and F. Saltiel, "Carbon Taxes as Part of the Fiscal Solution" (2013) Urban-Brookings Tax Policy Center.
74 Carbon Tax Act of South Africa, n 65, s 4.
75 OECD, "Towards Green Growth" OECD Environmental Performance Reviews: Denmark (2019) <https://www.oecd-ilibrary.org/sites/3093cb23-en/index.html?itemId=/content/component/3093cb23-en>.
76 Teagasc, "Irish Agriculture, Greenhouse Gas Emissions and Climate Change: Opportunities, Obstacles and Proposed Solutions" Teagasc Working Group on Greenhouse Gas Emissions (28 January 2011) <https://www.teagasc.ie/media/website/publications/2011/61_ClimateBillSubmission.pdf>.

77 J. Altamirano and J. Martínez, "Mexico's 3 Big Steps towards Comprehensive Carbon Pricing" (14 April 2017) <https://www.wri.org/blog/2017/04/mexicos-3-big-steps-towards-comprehensive-carbon-pricing>.
78 Named after Arthur Cecil Pigou (see, for example, P. Agarwal, "The Pigou Effect" 4 August 2017, <https://www.intelligenteconomist.com/pigou-effect/>).
79 T. Jiang, "Earmarking of Pollution Charges and the Sub-Optimality of the Pigouvian Tax" (2001) 45 *The Australian Journal of Agricultural and Resource Economics* 623.
80 World Bank Group, "The FASTER Principles for Successful Carbon Pricing: An Approach Based on Initial Experience", OECD, September 2015 <www.oecd.org/environment/tools-evaluation/FASTER-carbon-pricing.pdf>.
81 J. Aldy, "Design and Administration of a Carbon Tax," Harvard Kennedy School, Resources for the Future, National Bureau of Economic Research, 18 October 2011 <https://media.rff.org/archive/files/sharepoint/Documents/Events/111018_Fiscal_Reform/111018-Aldy.pdf>.
82 N. Saidi, "Carbon Tax can Fund Energy Transition: There is No Trade-Off between Economic Growth and Decarbonised Economies", January 4, 2017 <https://gulfnews.com/opinion/op-eds/carbon-tax-can-fund-clean-energy-transition-1.1956326>.
83 J. Carl and D. Fedor, "Tracking Global Carbon Revenues: A Survey of Carbon Taxes Versus Cap-and-Trade in the Real World" (2016) 96 *Energy Policy* 50.
84 B. C. Murray and N. Rivers, "British Columbia's Revenue- Neutral Carbon Tax: A Review of the Latest 'Grand Experiment' in Environmental Policy", Nicholas Institute for Environment for Environmental Policy Solutions, Duke University, University of Ottawa, Working Paper NI WP 15-04, May 2015 <https://nicholasinstitute.duke.edu/sites/default/files/publications/ni_wp_15-04_full.pdf>.
85 Ibid.
86 J. Sumner, L. Bird, and H. Dobos, "Carbon Taxes: A Review of Experience and Policy Design Considerations" (2011) 11(2) *Climate Policy* 922.
87 Ibid.
88 Ibid.
89 J. Sumner, L. Bird, and H. Smith, "Carbon Taxes: A Review of Experience and Policy Design Considerations", National Renewable Energy Laboratory Technical Report NREL/TP-6A2-47312, December 2009 <https://www.nrel.gov/docs/fy10osti/47312.pdf>.
90 Carbon Tax Act of South Africa, n 69, s 4.
91 Center for Climate and Energy Solutions, "Options and Considerations for a Federal Carbon Tax", February 28, 2013 <www.c2es.org/publications/options-considerations-federal-carbon-tax>.
92 Ibid.
93 Ibid.
94 Murray and Rivers, n 84.
95 J. J. Andersson, "Carbon Taxes and CO_2 Emissions: Sweden as a Case Study" (2019) 11(4) *American Economic Journal: Economic Policy* 1.
96 A. Bruvoll and B. M. Larsen, "Greenhouse Gas Emissions in Norway: Do Carbon Taxes Work?" (2004) 32 *Energy Policy* 493.
97 M. Sarraf and M. Heger, "Why is COP21 Important for the Middle East and North Africa Region?" (World Bank Blogs, 2015) <https://blogs.worldbank.org/arabvoices/why-cop21-important-middle-east-and-north-africa-region>.
98 Library of Congress, "National Parliaments: Gulf Cooperation Council Countries" (2020) <https://www.loc.gov/law/help/national-parliaments/gulf.php>.
99 H. Almutairi, "Competitive Advantage through Taxation in GCC Countries" (2014) 13 *International Business & Economics Research Journal* 769.
100 A. Alreshan, N. Ltaifa, T. Callen, Z. Liu, V. Stepanyan, and M. Mansour, "Diversifying Government Revenue in the GCC: Next Steps," GCC Tax Policy Paper, Annual

Meeting of Ministers of Finance and Central Bank Governors, October 26, 2016 <www.imf.org/external/np/pp/eng/2016/102616.pdf>.

101 A. Jewell, M. Mansour, P. Mitra, and C. Sdralevich, "Fair Taxation in the Middle East and Northern Africa" International Monetary Fund (2015) SDN/15/16 <https://www.imf.org/external/pubs/ft/sdn/2015/sdn1516.pdf>.

102 R. Bird, "Tax Challenges Facing Developing Countries," Institute for International Business Working Paper No. 9, March 31, 2008 <https://papers.ssrn.com/sol3/papers.cfm?abstract_id=1114084>.

103 Ibid.

104 The effects of over-reliance on foreign currencies on the economy are adverse as currently being experienced by most African countries, with the devaluation of their local currencies. See, B. Massell, S. Pearson, and J. Fitch, "Foreign Exchange and Economic Development: An Empirical Study of Selected Latin American Countries" (1972) 54(2) *The Review of Economics and Statistics* 208.

105 J. Meltzer, "Financing Low Carbon, Climate Resilient Infrastructure: The Role of Climate Finance and Green Financial Systems", Global Economy and Development Working Paper 96, September 2016 <https://www.brookings.edu/wp-content/uploads/2016/09/global_20160920_climate_finance.pdf>.

106 A. Al-Badia and I. Al Mubarak, "Growing Energy Demand in the GCC Countries" (2019) 26 *Arab Journal of Basic and Applied Sciences* 488.

107 See Qatar's National Vision 2030, which seeks to manage the environment such that there is harmony between economic growth, social development, and environmental protection ("Qatar National Vision 2030", General Secretariat for Development Planning, July 2008 <www.gco.gov.qa/wp-content/uploads/2016/09/GCO-QNV-English.pdf>).

108 H. Sun, S. Clottey, Y. Geng, K. Fang, and J. Amissah, "Trade Openness and Carbon Emissions: Evidence from Belt and Road Countries" (2019) 11 *Sustainability* 2682.

109 R. Bird and K. Davis-Nozemack, "Tax Avoidance as a Sustainability Problem" (2018) 151 *Journal of Business Ethics* 1009.

110 R. Martin, L. de Preux, and U. Wagner, "The Impact of a Carbon Tax on Manufacturing: Evidence from Microdata" (2014) 117 *Journal of Public Economics* 1.

111 M. Panezi, "When CO_2 Goes to Geneva: Taking Carbon across Borders without Violating WTO Obligations", Centre for International Governance Innovation Papers No. 83, November 2015 <www.cigionline.org/sites/default/files/cigi_paper_no.83_web.pdf>.

112 Ibid.

113 L. Mittal, "A Carbon Border Tax Is the Best Answer on Climate Change" *Financial Times* (12 February 2017) <https://www.ft.com/content/8341b644-ef95-11e6-ba01-119a44939bb6>.

114 World Bank Group, "Carbon Tax Guide: A Handbook for Policy Makers", Partnership for Market Readiness, March 2017 <https://openknowledge.worldbank.org/handle/10986/26300>.

12 Mobilizing and leveraging Islamic climate finance in the MENA region

The potential role of national green participative banks

Dalal Aassouli

Introduction

This chapter examines the roles of National Green Participative Banks (NGPB) in supporting the transition to a green economy and the mobilization of Islamic climate finance in Middle East and North Africa (MENA) countries.

MENA countries face several climate change challenges.[1] With increasing water supply shortage and frequent droughts resulting in adverse impacts, climate change could significantly slow down the achievement of sustainable development objectives in MENA countries.[2] For example, the Intergovernmental Panel on Climate Change (IPCC) predicts that climate change will rapidly reduce precipitation in the region.[3] The resulting hydrological changes could reduce water availability per person by 30–70 per cent by 2025, diminish agricultural productivity, and heighten the risk of flooding in highly populated urban coastal areas.[4] Therefore, large-scale mitigation and adaptation investments are needed to reduce the adverse impact of climate change. Efficient climate finance mobilization is critical to achieve these objectives. The United Nations Framework Convention on Climate Change (UNFCCC) defines climate finance as "local, national or transnational financing–drawn from public, private and alternative sources of financing—that seeks to support mitigation and adaptation actions that will address climate change".[5]

Today, the rising awareness of climate change and sustainability issues influences the global climate finance landscape, requiring the consideration of new approaches in redirecting financing to climate-friendly projects, along with enabling policies that promote inclusive and equitable financing solutions.[6] While Islamic finance is gaining ground in the financial landscape of the MENA region,[7] its engagement in climate adaptation and mitigation projects with environmental and social impacts is still insignificant.[8] This is despite the fact that Islamic finance is rooted in developmental aims.[9]

Climate finance projects often have two characteristics: they require a large amount of capital and they come with significant investment risks.[10] Few individual investors have the capital to finance these projects. In addition, even those with the required capital may be reluctant to invest a significant portion of their wealth in a single, risky venture.[11] These projects, therefore, require enormous planning work by governments to ensure their viability, to attract private investors and large partnerships, and to necessitate smart planning, technology, and adequate funding sources.[12]

This chapter explores the potential of Islamic finance in mobilizing and leveraging private climate finance in the MENA region. It examines ethical proposals for an effective mobilization of climate finance at the national level. It then discusses the roles of NGPBs in supporting the transition to a green economy and the mobilization of climate finance in MENA countries. The NGPB can use several financial instruments and strategies to mobilize private sector financing. These include participative microfinance, social finance, crowdfunding, as well as green *sukuk*. Green *sukuk* is a new asset class that targets both Islamic and socially responsible investors (SRI). SRI is considered the umbrella term for sustainable investing, responsible investing, ethical investing, and impact investing.[13] Other mechanisms, such as those supporting household investment in climate-friendly projects, are also discussed.

This chapter proceeds in five sections. After this introductory section, the second section discusses the current climate finance landscape in the MENA region and the key stakeholders. The third section suggests innovative financial mechanisms together with the enabling environment to promote the effective mobilization of private climate finance. The fourth section analyzes how green finance guidelines and regulations can support climate finance initiatives and, therefore, the transition to a green economy. The fifth section is the conclusion. It suggests that the scaling-up of green projects is not only dependent on financial resources but also on a range of enabling factors, including the provision of the necessary infrastructure, and the appropriate incentives for investors and partnerships between governments, the private sector, development institutions, social finance institutions, and civil society.

The current climate finance landscape in the MENA region

Countries in the MENA region are highly vulnerable to climate change, which is likely to impact the achievement of sustainable development objectives.[14] The MENA region is already the most water-scarce region in the world and has to import more than half of its food.[15] Moreover, per capita emissions in many MENA countries are 60 per cent higher than the average among developing countries.[16] As governments in the region have limited funds, attracting private climate finance with the support of multilateral development banks (MDB), in the form of investment guarantees, for example, is critical to mitigate these challenges.[17] MDBs are among the providers of climate financing. The MDB's latest joint report on climate financing said US$ 27.9 billion, or 79 per cent of the 2017 total, was devoted to climate mitigation projects that aim to reduce harmful emissions and slow down global warming. The remaining 21 per cent, or US\$ 7.4 billion, of financing for emerging and developing nations was invested in climate adaptation projects that help economies deal with the effects of climate change such as unusual levels of rain, worsening droughts, and extreme weather events.[18]

As shown by the UNFCCC definiton,[19] the term "climate finance" is closely associated with related concepts, such as green finance and sustainable finance.[20] Climate finance in the MENA region is largely concentrated on a small number of large projects in the form of loans or concessional loans, funded by the Clean Technology Fund (CTF).[21] The total amount of finance approved between 2003 and 2016 amounted to US\$1.2 billion for 94 projects, whereas the actual increased to US\$1.4 billion for 103 projects between 2003 and 2018, mainly targeting mitigation projects despite pressing adaptation needs in the region, especially for water conservation and food security measures.[22] Table 12.1 summarizes the approved funding per theme between 2003 and 2018:

Table 12.2 below summarizes the main funding sources in the MENA region, as well as the projects and amounts approved between 2003 and 2018.

Climate Finance Update (CFU) data shows that of the 21 MENA countries, only 15 countries between 2003 and 2018 were recipients of climate finance. The top two recipients, Morocco and Egypt, received respectively and 54 per cent and 27 per cent of total approved climate finance in the region. The majority of funding is allocated to adaptation projects.[23]

On the other hand, the European Union (EU) committed €48 billion in June 2018 for the two regions of Southern and Eastern Mediterranean and sub-Saharan Africa (SSA) under the European External Investment Plan (EIP). The EIP targets climate-related investment projects by the private sector.[24]

Among the MDBs active in climate finance is the Islamic Development Bank (IsDB), which joined the MDB climate finance tracking groups in October 2017.[25] IsDB's financing is structured based on Islamic finance principles. Its climate finance initiatives were undertaken to support programs relating to renewable energy and energy efficiency in the MENA region, with the Government of Netherlands and the Dutch Development Bank (FMO), for potentially setting up a climate change financing facility along with the EU, in cooperation with the Regional Centre for Renewable Energy and Energy Efficiency (RCREEE).[26]

Table 12.1 Approved funding per theme between 2003 and 2018

Theme	Approved amount (USD millions)	Projects approved
Adaptation	255.8	37
Mitigation	1139.7	49
Multiple	53.7	17

Source: Watson and Schalatek, 2019.
The table is adapted from The World Bank, p. 86.

Table 12.2 Climate finance funding sources in MENA region (2003–2018)

Fund	Funder	Amount Approved (USD millions) 2003–2018	Projects approved 2003–2018
Clean Technology Fund (CTF)	The World Bank Group	864.8	10
Global Environment Facility (GEF)	Global Environment Facility, World Bank Group (Trustee)	108.6	47
Special Climate Change Fund (SCCF)	Global Environment Facility, World Bank Group (Trustee)	43.6	8
Adaptation Fund (AF)	2% share of proceeds of the Certified Emission Reductions, Multiple countries	48.7	10
Germany's International Climate Initiative	Federal Ministry for the Environment, Nature Conservation and Nuclear Safety, Germany		
Least Developed Countries Fund (LDCF)	United Nations	35.1	8
Green Climate Fund (GCF)	Majorly developed countries, some developing countries	287.8	6
Adaptation for Smallholder Agriculture Programme (ASAP)	International Fund for Agricultural Development, United Nations	23.0	4
Pilot Program for Climate Resilience (PPCR)	African Development Bank's Climate Investment Funds		
MDG Achievement Fund	United Nations Development Programme	7.60	2
Strategic Priority on Adaptation (SPA) (from GEF4)	Global Environment Facility, World Bank Group (Trustee)		
Partnership for Market Readiness	World Bank Group	11.0	6
Global Climate Change Alliance (GCCA)	European Union	3.4	1
Global Energy Efficiency and Renewable Energy Fund (GEEREF)	European Union, Germany, Norway, and private sector investors	16.6	1

Source: Watson and Schalatek, 2019.
The table is adapted from The World Bank, p. 86.

Financial innovation to promote the effective mobilization of private climate finance

Climate change mitigation requires an effective mobilization of the necessary financial resources from several stakeholders in order to drive climate action.[27] These key stakeholders include: domestic, bilateral, and multilateral development finance institutions (DFIs), corporations, and private commercial institutions.[28] However, the problem in many countries, especially developing ones, is the inefficient use of funding sources.[29] Therefore, the establishment of an NGPB could enable a broader and more efficient mobilization of resources through a blended finance mechanism as well as the guarantee of an inclusive financing for projects targeting rural populations. An NGPB is a financial institution that uses innovative blended financing strategies to raise and attract needed capital required for climate change, low carbon, and clean technology projects.[30] The World Economic Forum and the OECD define blended finance as "the strategic use of development finance and philanthropic funds to mobilize private capital flows to emerging and frontier markets".[31] Blended finance enables the blending of various financings (development finance, social finance, Islamic finance, green finance, etc.) using different structuring approaches. The purpose is to attract additional private capital while overcoming investment barriers.[32]

In addition, the NGPB will allow the use of public and concessional funds, which are limited[33] to support climate change mitigation and adaptation projects, in order to mobilize private sector financing. The model borrows its characteristics from participation investment banks, national development banks (NDB), and National Climate Funds (NCFs). Participation, or participative finance, commonly known as Islamic finance, is based on five key principles. Three exclusionary principles, i.e. the prohibition of usury (*riba*), uncertainty (*gharar*), and speculation (*mayssir*), and two positive principles, i.e. real assets backing and profit and loss sharing.[34] The NGPB will have two other distinguishing features, which are the promotion of ethical and green investment culture as well as support for research in climate and participative financing sectors.

NDBs are institutions owned by, or affiliated with, the government which have a specific mandate to promote economic development.[35] NDBs do not have a standard operating model. They have various ownership structures and differ in their political and financial objectives, their governance mechanisms, and the financial instruments they use.[36] Development banks have historically been established for economic and social reasons.[37] They are generally in a privileged position in their local markets because of their unique characteristics and can play a crucial role in the development and promotion of climate projects. These characteristics allow them to efficiently mobilize financing for green projects. For example, during the pre-investment phase, NDBs can promote demand for green investments and financing by fostering the development of an enabling framework for green projects. Then, during the investment phase, NDBs can mobilize capital, both in the form of equity, debt, and incentives to the private sector.

Financial innovation consists in improving the efficiency of financial flows by diversifying funding sources, enhancing resources allocation on the basis of result-based financing,[38] and adapting financing strategies to the nature and size of projects. The choice of financial instruments should depend on the capacity of local agencies to manage them effectively and efficiently, and local financial markets to understand and use them.[39] Good governance can be achieved through transparency and impact measurement.

Considering the specific case of renewable energy, several MENA countries have set ambitious targets to promote energy transition.[40] These strategies can be adapted according to the economic status and energy access of the country. For low-income countries, such as those in sub-Saharan Africa, that suffer from energy poverty, access to energy is a major challenge that has a negative impact on education, health, agriculture, and the well-being of the population.[41] In middle-income countries, which already have high electricity access rates, such as Morocco, efforts must focus on energy independence and efficiency in order to reduce greenhouse gas emissions. Moreover, and given the income group and the credit rating of these countries, isolated financing can be very expensive.[42] The pooling of financial resources through an NGPB can decrease the average cost of funding through partnerships with, for example, development finance institutions, such as MDBs and climate funds. These institutions can improve the credit quality of sovereign issuances.[43] Another area of MDBs support is technical assistance.[44] A study of energy sector initiatives and programs in Africa by the Africa-EU Energy Partnership (AEEP), which covers 58 programs and initiatives on the promotion of renewable energies, highlights five categories of technical assistance for the energy sector: information and analysis, dialogue and networking, project preparation, policy support, and skills development.[45]

The NGPB uses existing capital dedicated to development projects by public and philanthropic institutions, DFIs, and green funds to leverage additional private sector investment. The resources mobilized can be used to finance green projects in the following sectors.

In this chapter, we will focus on energy efficiency and renewable energies given their rise in national visions of many MENA countries as tools for combating climate change. There are three main focus areas for widespread and inclusive access to clean energy: (i) electricity generation in urban and rural areas, (ii) green buildings including schools, administrations, hospitals, mosques, and housing, and (iii) agriculture through clean energy solutions for irrigation.[46] For the success of these strategies, the implementation of a feed-in tariffs (FIT) policy

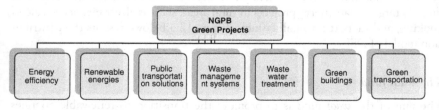

Figure 12.1 Categories of green projects to be financed by the NGPB. Source: Author's own.

is important.[47] Also, partnerships with the various real estate stakeholders are necessary for an-effective deployment of these strategies (administrations, real estate developers, etc.).[48]

There are two categories of projects:

a. Large-scale green projects: These projects require long-term financing and national and international public-private partnerships given the size of the required investments. For these projects, the NGPB can use green *sukuk*, project co-financing, and public-private partnerships. Public-private, national-international partnerships promote the exchange of expertise and experience and enable the effective implementation of strategic objectives.[49]
b. Small-scale green projects: Small projects are usually localized and involve local populations. They target bottom of the pyramid (BOP) populations mainly in the rural areas. These populations generally do not have access to finance and the small size of these projects creates significant problems in obtaining private financing. The objective is to increase (thanks to the economies of scale) access to inclusive and impact financing that is intended for micro-renewable energy projects, such as biomass and access to clean energy for cooking. In this case, the mobilization and solidarity of local populations is important. The institutions of *zakat* (almsgiving), *waqf* (endowment) associated with participative microfinance, and *qard hasan* (interest-free loans) can be used. The establishment of a crowdfunding platform can also promote the banking of projects. This assistance can be done in collaboration with platforms, such as the Sustainable Energy for All (SE4All) African Platform, which is mandated to implement a regional support program for "green" mini-grids, in collaboration with the Sustainable Energy Fund for Africa (SEFA).[50]

The resources provided by DFIs can be structured based on a *mudaraba* (partnership) contract, whereby DFIs represent *rab al mal* (fund provider) and the NGPB the *mudarib* (agent).[51] The contract will also integrate the principles of results-based financing, which consist in setting up key performance indicators KPIs (economic, social, and environmental) to evaluate the performance of funded projects and determine the financing terms in the *mudaraba* agreement. This technique is not widely developed in participative financing modes. Resources from philanthropic institutions, such as the Bill and Melinda Gates Foundation (Gates Foundation), can be structured as social impact bonds (SIB).[52] They can directly target clean energy projects in rural areas, such as electricity access, clean cooking, and support for small farmers. Figure 12.3 below presents the contribution model of DFIs.

The waqf fund

The aim of the *waqf* fund is to promote the transition to renewable energies and responsible investment through three action areas: technical assistance,

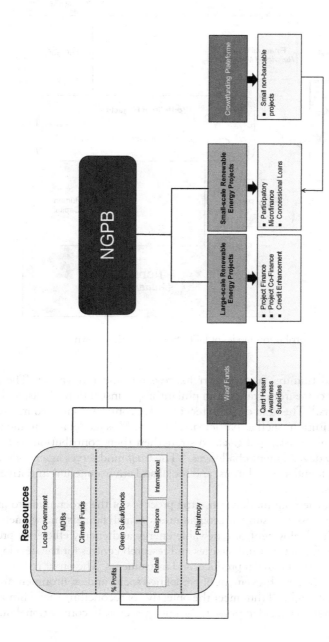

Figure 12.2 The NGPB model. Source: Author's own.

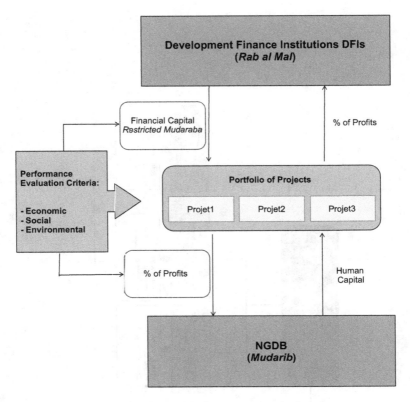

Figure 12.3 The contribution model of DFIs. Source: Author's own

awareness and training, and support for research and innovation. The fund's resources will come from grants from philanthropic institutions, DFIs, and green *sukuk* investors.[53] The fund can also receive *zakat* contributions from individuals, as well as institutions such as participative banks. This model aims to modernize the institutions of *waqf* and *zakat* to strengthen their contribution to tackling contemporary development challenges.[54] The *waqf* fund serves as a conduit linking investors, foundations, donors, and social and environmental initiatives with funding needs.

The issuer, the *waqf* fund, can use the proceeds of the programme to finance training, awareness, and support programmes for small farmers and women cooperatives for the deployment of renewable energy and energy efficiency projects. The fund can also support universities and research centres for the development of innovative green financial products and instruments. The profit to investors is contingent upon the achievement of predefined social and environmental objectives. The *waqf* fund will thus meet the objective of reconciling social innovation and environmental considerations in the energy sector. In conventional markets, several foundations are active in this area. Among them, the Rexel Foundation and Energy Access Practitioner Network.

The *waqf* fund can be structured in two ways (Figures 12.4 and 12.5). In the first structure, periodic profits are paid by the *waqf* fund. In the second structure, a third party "the donor" pays the profits to the investors. The donor can be a charity or a philanthropic structure.

Key financial instruments used by the NGDB

Among the key innovative instruments that can be used by the NGPB are green *sukuk*, participative microfinance, and crowdfunding.

Green sukuk

The *sukuk* market is one of the fast-growing segments in the Islamic financial industry with 24.2 per cent of Islamic finance global assets in 2018 and a compound annual growth rate (CAGR) of 30.6 per cent between 2003 and 2018.[55] The Accounting and Auditing Organization for Islamic Financial Institutions (AAOIFI) defines *sukuk* as certificates of equal value representing undivided shares in ownership of tangible assets, usufructs, and services or (in the ownership of) the assets of particular projects or special investment activity.[56] Since its inception in 2001, the *sukuk* market has witnessed new entrants from non-Islamic sovereigns and corporates, an increase in cross-border transactions, and has become an important source of funding for several countries in Asia, Europe, and Africa. As a result, global *sukuk* issuance is expected to reach $783 billion by 2023.[57]

Sukuk, often qualified as Islamic bonds, present two key features that position them as a viable option for financing green projects. First, the asset-backing requirement[58] facilitates their link to the real economy and therefore widens the scope of sectors that can be financed. These could include financing projects that target climate change, agriculture, education, poverty alleviation, etc. Second, *sukuk* can be structured in various ways using single or hybrid Islamic contracts such as *wakala, musharaka, mudaraba, ijara*, etc. This flexibility offers tremendous opportunities for innovation and for addressing specific financing needs, even though this complexity may lead to a lack of standardization and therefore higher transaction costs. The proposed green *sukuk* model borrows its principles from the traditional structures of *sukuk*, social impact bonds (SIBs), green bonds, and SRI. The model targets retail and diaspora investors in order to widen the investor base and mobilize public savings to finance green projects.

POSSIBLE STRUCTURES OF GREEN *SUKUK*

Istisna'a ijara for individual project financing   *Ijara* (sale and lease-back) structure has been used since 2002 in several local and international sovereign and corporate issuances.[59] The *ijara* requires a tangible underlying asset that can be transferred and leased. The value of the underlying asset must be equal or greater

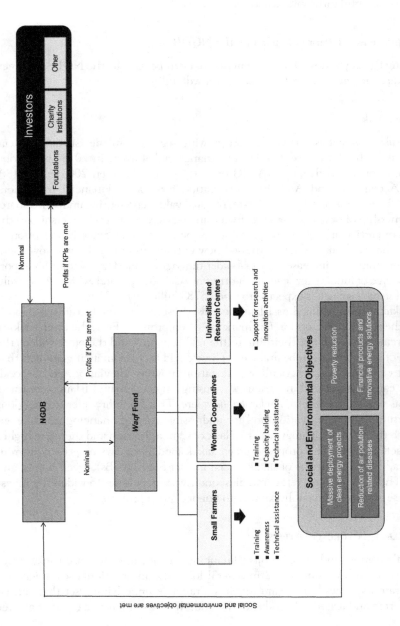

Figure 12.4 First structure of the waqf fund. Source: Author's own.

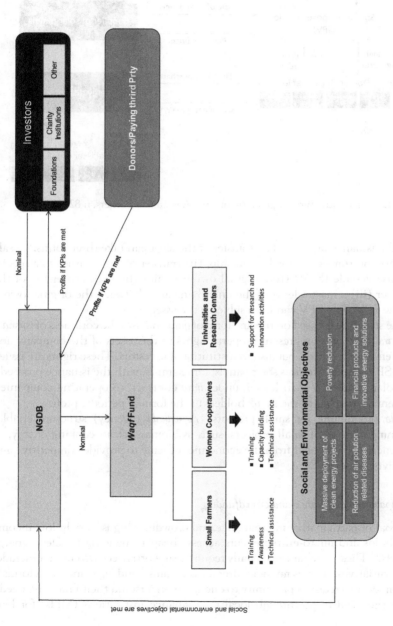

Figure 12.5 Second structure of the waqf fund. Source: Author's own.

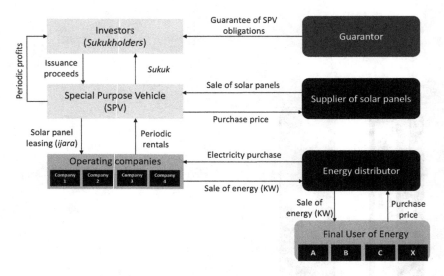

Figure 12.6 Ijara Structure for green sukuk. See Aassouli and others, n 82.

than the issuance amount. The transfer of the asset can take three forms: (i) sale of assets and transfer of legal ownership, (ii) transfer of the usufruct to a special purchase vehicle (SPV) (the original owner retains the legal ownership of the asset), or (iii) a master lease of the underlying asset between the original owner and the SPV. The SPV can then sublease the asset.

The structure is based on the first form of *ijara* and uses the contracts of *istisna'a* and *ijara*. The *sukuk* represent property rights in the assets of the company and are open to individuals, but also to institutional investors. These rights are issued by an SPV, which acquires solar panels, for example, with the issuance proceeds. The solar panels are then leased, under *ijara* contract, to operating companies. The rentals are paid to the *sukuk* holders in the form of periodic profits.

Wakala programme for small projects   A *wakala* (agency) structure enables governments to pool small projects, such as biomass, clean cooking energy, or solar energy, and benefit from the economies of scale to provide competitive and inclusive financing.[60]

Participative microfinance and crowdfunding

The goal of participative microfinance and crowdfunding is to help low-income households and micro-entrepreneurs access cheaper financing for clean energy projects.[61] These mechanisms usually require a supportive ecosystem that includes other initiatives, such as financial literacy programs[62] and legal and tax reforms.[63] This mode also requires proximity to clients to understand their financing needs, hence the need to go through local microfinance institutions (MFIs) for loan

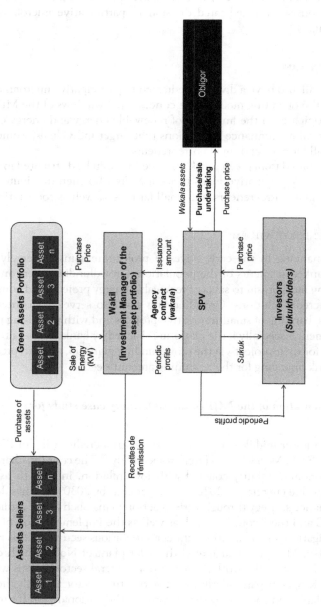

Figure 12.7 Wakala programme for green sukuk. See Aassouli and others, n 82.

distribution. In addition, the NGPB will have the institutional capacity to support microfinance loan beneficiaries and the crowdfunding platform through technical assistance, advisory services, and public authorities' relationship management. The achievement of these objectives can be supported by the *waqf* fund in collaboration with the local MFIs.

Figure 12.8 summarizes the operating model of the NGPB alternative financing, which consist in a dedicated division to participative microfinance and crowdfunding.

ISLAMIC MICROFINANCE

The NGPB will also have a division dedicated to participative microfinance that is similar to the operating mode of the concessional windows of the MDBs. The division participates in the financing of renewable energy and energy efficiency projects of local microfinance institutions that target individuals, women cooperatives, small farmers, and small entrepreneurs.

The funds raised from green *sukuk* issuance can then be distributed in the form of microfinance participative loans to local MFIs who then distribute them to individuals, small entrepreneurs, and small farmers, as well as cooperatives.

THE CROWDFUNDING PLATFORM

Access to financing in the early stages of projects remains extremely difficult for the unbanked segments of the population. Crowdfunding offers an alternative financing mechanism to support renewable energy projects and complement green microcredit solutions. Crowdfunding can also serve as an intermediate phase for the banking of small innovative projects and with promising prospects for development, especially if the platform is backed by banks. Thus, after having raised funds for their project via the crowdfunding platform, an entrepreneur can apply for bank financing for the remaining amount needed.

The implementation of the NGPB: an exploratory case study for Morocco

Morocco is a lower middle-income country[64] with a credit rating of BBB-, Ba1, and BBB- by S&P, Moody's, and Fitch, respectively.[65] The country, characterized by macroeconomic stability coupled with low inflation, intends to use 42 per cent of renewable energies by 2020 and 52 per cent by 2030[66] and has developed a climate finance strategy through various action plans, such as the ratification of the UNFCC and the Kyoto Protocol, as well as the implementation of national climate mitigation and adaptation strategies in various sectors. In this regard, in February 2016, Morocco inaugurated the solar plant of Noor, which claims to become the largest in the world.[67] In addition, several sectoral plans were initiated by the Kingdom particularly in the agriculture sector through the "Green Morocco" plan[68] and the energy sector through the national plan for renewable

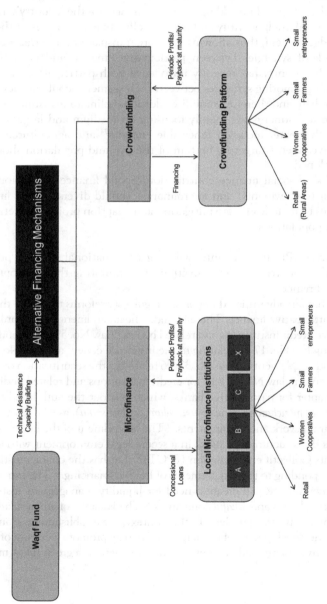

Figure 12.8 The alternative mechanisms of green financing of the NGPB. Source: Author's own.

energies.[69] Morocco has also joined Africa50, an investment fund dedicated to financing infrastructure in the African continent with a subscription of 10 million dollars.[70] However, the Kingdom faces several challenges that limit the effective mobilization of private climate finance. These include:

a. The country does not identify climate change as a category within its national budget.[71] Embedding climate change in the country's national budgets would help identify and monitor climate expenditures in the country's budget system, thus allowing for more transparency and traceability.
b. The lack of synergies between participative finance and green finance. While the country has recently introduced both participative[72] and green finance,[73] promoting synergies between both segments could attract a wider investor base and therefore mobilize additional climate investments.
c. Climate investments are mainly focusing on medium and large-scale projects with greater focus on renewable energies[74] and agriculture.[75] Small-scale investments targeting bottom of the pyramid population should also be developed.
d. The lack of social finance contribution. Social finance institutions, such as *awqaf* (endowments) and microfinance, should diversify their financing activities to include climate mitigation and adaption projects targeting low-income populations.

Setting up an NGPB in the country will target two national strategic priorities: supporting the country's energy transition and promoting the development of participative finance.

Morocco has already initiated important regulatory reforms to enable the introduction of participative finance. These include: the amendment of the banking law No. 103-12 for credit institutions and related bodies,[76] law No. 59-13 amending and supplementing law No. 17-99 relating to the insurance code,[77] and law No. 119-12 modifying and supplementing law No. 33-06 related to the securitization of receivables.[78] The banking law No. 103-12 for credit institutions and related bodies has a dedicated chapter for participative banks, which sets out the authorized products (*murabaha, ijara, moucharaka, moudaraba, salam*, and *istisna'a*), as well as the *shari'ah* compliance framework and requirements.[79] The amendment of the law on securitization allows the structuring of *sukuk* in a secure legal environment with an Fond de Placements Collectif en Titrisation (FPCT)[80] that has the characteristics of an SPV while responding to the requirements of Islamic financing.[81] The introduction of participative finance will create a need for liquidity management instruments for participating banks and *takaful* companies.[82] The issuance of green *sukuk* by the NGPB can help solve this problem. In this context, the establishment of an NGPB will support the development of participative finance, promote a culture of participative green investment, and facilitate the development of a green *sukuk* market.

The potential for retail and institutional investors

The *sukuk* issued by the NGPB can facilitate liquidity management for participative banks while providing investment opportunities for retail investors. A study conducted by Islamic Finance Advisory and Assurance Services (IFAAS)[83] to analyze and evaluate the potential of participative finance in Morocco, confirms the potential of retail investors and their interest in *shari'ah* compliant saving products.[84] An investigation conducted by the Conseil Déontologique des Valeurs Mobilières (CDVM) in 2012 among institutional investors in Morocco reveals that the factors motivating any investment decision are liquidity (100 per cent), the yield of the instrument (92 per cent), investment security (92 per cent), and the need for diversification (62 per cent). The decision-making process is also influenced by the tenor of the instrument, the investment strategy, and the risk-return appetite. Ninety-two per cent of the surveyed management companies indicated the importance of the marketability of the product because of the liquidity and the preference of their customers.[85]

Energy transition in the country

The state energy investment company (SIE) is the country's financial vehicle facilitating investments in the energy sector, including renewable energy and energy efficiency. Its role is to contribute to achieving the objectives set by the national strategy promulgated by the Ministry of Energy, Mines, Water and Environment under the leadership of His Majesty King Mohammed VI. In this regard, SIE contributed to the capitalization of MASEN, the Moroccan agency for solar energy, with a 25 per cent equity participation.[86]

The financing of the energy transition in Morocco is supported by various mechanisms, which include public funds, commercial banks, the European Investment Bank (EIB), as well as international donors like Kfw, the World Bank, French Development Agency (Afd), African Development Bank (AfDB), etc.[87] Table 12.3 presents the technical data sheet of the first three phases of the Noor solar project.

Table 12.3 shows how the pooling of all these sources of funding under an NGPB can increase their effectiveness. This requires revising the mandate of the SIE either to integrate it into the NGPB or to use it as a separate platform within the framework of the NGPB.

The NGPB can also support the development of small farmers, women cooperatives, and small entrepreneurs by targeting two objectives that improve their quality of life and well-being: clean energy that facilitates their operational activities, as energy is vital for entrepreneurship and agriculture; and the achievement of environmental objectives. In addition, the NGPB can facilitate the extension of the green mosques program to the rest of the Kingdom.[88]

Table 12.3 Data sheet of the first three phases of the Noor project

Noor Project			
Projects	Noor-I	Noor-II	Noor-III
Sponsor	Masen	Masen	Masen
Developer	Acwa Power	Acwa Power	Acwa Power
Gross Capacity	160 MW	200 MW	150 MW
Thermal storage	3 hours	8 hours	8 hours
Construction start date	2013	2015	2015
Developers	Acciona, Sener, TSK	Sener, Power China, Sepco	Sener, Power China, Sepco
Operations start date	2015	2017	2017
Concessional financing mobilized	~730 million €	~810 million €	~810 million €
Funding sources	Agence Française de Développement (AFD), African Development Bank (AfDB), European Bank for Reconstruction and Development (EBRD), World Bank, Clean Technology Fund, KfW Bankengruppe, European Union	African Development Bank (AfDB), European Bank for Reconstruction and Development (EBRD), World Bank, Clean Technology Fund, KfW Bankengruppe, European Union	Agence Française de Développement (AFD), African Development Bank (AfDB), European Bank for Reconstruction and Development (EBRD), World Bank, Clean Technology Fund, KfW Bankengruppe, European Union

<http://masen2.emagin.ma/noor/>.

The agricultural sector

Morocco's economic growth is highly dependent on the agricultural sector, which contributes nearly 15 per cent of GDP[89] and employs nearly 40 per cent of the active population.[90] Morocco has set up an ambitious strategy for the agricultural sector through the Green Morocco Plan,[91] as well as a national program to promote clean irrigation.[92] The water program aims to enable small and medium-sized farmers to equip, with the help of a grant from the Agricultural Development Fund, water pumps powered by electricity produced from solar panels, with the aim of improving their efficiency and production, while promoting water and energy savings.[93]

Green finance guidelines and regulations and the transition to a green economy

Green guidelines and regulations promote standardization and the harmonization of taxonomies and definitions. Standardization is important to promote the development of a new instrument. For example, in the case of green bonds and *sukuk*, standard terminology on the "green" qualification of investments facilitates the understanding of both instruments among investors and issuers and helps issuers claim the "green" status of their issuances and therefore attract environment-friendly investors.[94] Among these guidelines is the Climate Bonds Initiative (CBI) standard taxonomy for qualifying climate bonds, which includes energy, energy efficiency, transport, water, waste management, land use, and adaptation infrastructure. Each category comprises several subcategories. The energy category includes: renewables, distribution/management, energy storage, and products and technology that support smart grid and data centres using renewable energy.[95]

In addition to the taxonomies, green bonds, and *sukuk* principles issued by MDBs, the governments of Malaysia, Indonesia, Morocco, and China have enhanced transparency and the understanding of the instrument. These frameworks usually distinguish five criteria that should be considered to qualify as a green bond:[96]

a. Categories of eligible projects;
b. Selection process of eligible projects;
c. Rules governing the management and allocation of the issuance proceeds whereby issuance proceeds are ring-fenced in a sub-portfolio to be allocated to green eligible projects;
d. Communication and impact measurement;
e. Conformity assessment through third party certification.

Moreover, additional clauses are integrated in the legal documentation of green bonds and *sukuk* to ensure the traceability of the proceeds allocation.

On the other hand, green finance regulations enable policymakers to leverage on regulatory intervention to accelerate the transition to a green economy.[97] In this regard, in May 2018, the EU Commission adopted a package of measures implementing several key actions announced in its action plan on sustainable finance. The package includes four legislative proposals: taxonomy, disclosure and duties, benchmarks, and sustainability preferences.[98] Similarly, China has employed various economic instruments to promote green growth and investments in clean-technologies. These include taxes and subsidies, tradable permits, and public finance mechanisms.[99]

Conclusion and policy recommendations

The successful implementation of an NGPB requires an action plan based on five key areas.

First, the action plan should include financial innovation by promoting national and international public-private partnerships, as well as an efficient mobilization of the resources needed to mitigate climate change. This can be facilitated by the use of blended finance with synergies between Islamic, green, climate, and social finance.

Second, there should be government support through climate legislation, strategies, and policies, as well as tax and regulatory incentives. For example, strategies and policies for low-emission development include, *inter alia*, measures such as reforms of fossil fuel subsidies, renewable energy feed-in tariffs, and energy efficiency programmes.[100] In addition, climate policy instruments can support private investments that support reducing emissions.[101]

Third, the action plan requires market education and awareness. To ensure an ethical deployment of its objectives, considerable communication and awareness efforts are needed to attract investors seeking social and environmental impacts of their investments. Effective promotion demands consistant targeting of a wider investor base of both SRIs and Islamic investors. In a study conducted by Sustainable Square targeting 1,500 organizations across 18 countries in the MENA region covering 16 sectors, 21 per cent of the surveyed organizations in the region confirmed that the lack of awareness about climate-related risks is their main challenge related to climate change mitigation.[102]

Fourth, NGPB implementation requires transparency and impact assessments. This may be achieved through the development of frameworks for climate finance reporting that enhance traceability and accountability.

Fifth, climate finance must be integrated within national development plans and budgets. This requires a strategic approach to resource mobilization that is integrated with national development plans.[103]

The success of the NGPB depends largely on good governance and the quality, expertise, and awareness of internal teams when it comes to ethical issues and sustainability. We propose this model for the MENA region countries that have established participative finance activities and climate change mitigation and adaptation strategies. However, the model can also contribute to the mobilization efforts of other countries globally. Renewable energies were selected solely for the purpose of illustrating how climate finance can be mobilized. However, these strategies can very well be applied to other sectors, such as agriculture, transportation, real estate, amongst others.

Notes

1　See D. Olawuyi, "Advancing Innovations in Renewable Energy Technologies as Alternatives to Fossil Fuel Use in the Middle East: Trends, Limitations, and Ways Forward" in Donald Zillman, Martha Roggenkamp, LeRoy Paddock, and Lee Godden (eds), *Innovation in Energy Law and Technology: Dynamic Solutions for Energy Transitions* (Oxford University Press 2018) 354.

2　J. Devlin, "Is Water Scarcity Dampening Growth Prospects in the Middle East and North Africa?" (Brookings, 24 June 2014) <www.brookings.edu/opinions/is-water-scarcity-dampening-growth-prospects-in-the-middle-east-and-north-africa/>.

3　S. Patel, C. Watson, and L. Schalatek, "Climate Finance Regional Briefing: Middle East and North Africa", (November 2016) <www.odi.org/sites/odi.org.uk/files/resource-documents/11042.pdf>.

4　Ibid.

5　United Nations Framework Convention on Climate Change (UNFCCC), "Introduction to Climate Finance" <https://unfccc.int/topics/climate-finance/the-big-picture/introduction-to-climate-finance>.

6 D. Olawuyi, 'Financing Low-Emission and Climate-Resilient Infrastructure in the Arab Region: Potentials and Limitations of Public-Private Partnership Contracts' in W. Filho and A. Meguid (eds), *Climate Change Research at Universities: Addressing the Mitigation and Adaptation Challenges* (Springer 2017) 533–547.

7 As of 2018, the share of MENA (excluding GCC) in global Islamic finance assets is estimated at 25.1%. (see Islamic Financial Services Board, "Islamic Financial Services Industry Stability Report 2019" July 2019).

8 H. Ahmed, M. Mohieldin, J. Verbeek, and F. W. Aboulmagd, "On the Sustainable Development Goals and the Role of Islamic Finance" May 2015, World Bank Group Policy Research Working Paper No. WPS 7266.

9 M. Asutay, "Islamic Moral Economy as the Foundation of Islamic Finance" in Valentino Cattelan (ed.), *Islamic Finance in Europe* (Edward Elgar Publishing 2013) 55.

10 D. Olawuyi, *The Human Rights Based Approach to Carbon Finance* (Cambridge University Press 2016) 1-25.

11 N. Lindenberg, "Public Instruments to Leverage Private Capital for Green Investments in Developing Countries" (German Development Institute and Deutsches Institut für Entwicklungspolitik Discussion Paper 4/2014).

12 See S. A. Ngum and D. Elhassan, "Inclusive Climate-Smart Finance" (February 2018) African Development Bank Group <https://www.afdb.org/fileadmin/uploads/afdb/Documents/Publications/AfDB_Inclusive_Climate-Smart_Finance.pdf>.

13 M. Kerste, N. Rosenbook, B. Jan Sikken, and J. Weda, *Financing Sustainability: Insights for Investors, Corporate Executives, and Policymakers* (Amsterdam: VU University Press 2011) 158.

14 K. Waha and others, "Climate Change Impacts in the Middle East and Northern Africa (MENA) Region and Their Implications for Vulnerable Population Groups" (2017) 17 *Reg Environ Change* 1623.

15 Patel and others, n 3.

16 Ibid.

17 The World Bank, "A New Plan to Support Action on Climate Change in the Arab World" (November 2016) <www.worldbank.org/en/news/feature/2016/11/15/a-new-plan-to-support-action-on-climate-change-in-the-arab-world>.

18 Inter-American Development Bank, "Joint Report on Multilateral Development Banks' Climate Finance," African Development Bank, Asian Development Bank, European Bank for Reconstruction and Development, European Investment Bank, IDB Invest, Islamic Development Bank, World Bank Group (2017).

19 UNFCCC, n 5.

20 The Organization for Economic Co-operation and Development (OECD) defines green finance as finance for "achieving economic growth while reducing pollution and greenhouse gas emissions, minimizing waste and improving efficiency in the use of natural resources" <https://www.oecd-ilibrary.org/fr/environment/green-finance-and-investment_24090344>. The international capital market association (ICMA) defines sustainable finance as finance that incorporates climate, green, and social finance while also adding wider considerations concerning the longer-term economic sustainability of the organizations that are being funded, as well as the role and stability of the overall financial system in which they operate <https://www.icmagroup.org/assets/documents/Regulatory/Green-Bonds/Sustainable-Finance-High-Level-Definitions-May-2020-110520v4.pdf>).

21 Patel and others, n 3.

22 C. Watson and L. Schalatek, "Climate Finance Regional Briefing:Middle East and North Africa" (February 2019) <https://us.boell.org/sites/default/files/cff9_2018_eng-digital.pdf>.

23 Patel and others, n 3.

24 "Joint side event from the Dutch Government, the Islamic Development Bank, and the Regional Center for Renewable Energy and Energy Efficiency for COP24: 'Sus-

tainable Energy Transition in North Africa and the Middle East: Role of Climate Finance'" (Regional Center for Renewable Energy and Energy Efficiency, Islamic Development Bank, and the Kingdom of the Netherlands, December 2018) <https://rc reee.org/sites/default/files/cn_cop24_isdb_fmo_rcreee_v2.pdf>.

25 African Development Bank, Asian Development Bank, European Bank for Reconstruction and Development, European Investment Bank, Inter-American Development Bank, IDB Invest, Islamic Development Bank, World Bank Group, "2017 Joint Report on Multilateral Development Banks' Climate Finance" (June 2018) <https://publications.iadb.org/publications/english/document/2017-Joint-Report-on-Multilateral-Development-Banks-Climate-Finance.pdf>.

26 Regional Center for Renewable Energy and Energy Efficiency (RCREEE), "Sustainable Energy Transition in North Africa and the Middle East: Role of Climate Finance" (RCREEE, December 2018).

27 Olawuyi, n 10.

28 B. Buchner, A. Clark, A. Falconer, R. Macquarie, C. Meattle, and C. Wetherbee, "Global Landscape of Climate Finance 2019" (Climate Policy Initiatiev, November 2019)   <https://climatepolicyinitiative.org/publication/global-landscape-of-climate-finance-2019/>.

29 V. Vandeweerd, Y. Glemarec, and S. Billett, "Readiness for Climate Finance. A Framework for Understanding What it Means to Be Ready to Use Climate Finance" (UNDP, December 2015).

30 D. Smallridge and others, "The Role of National Development Banks in Catalyzing International Climate Finance," (IADB, March 2013).

31 World Economic Forum and the OECD, "Blended Finance Vol. 1: A Primer for Development Finance and Philanthropic Funders" (WEF, September 2015).

32 Ibid.

33 See The World Bank Group, "Strategic Use of Climate Finance to Maximize Climate Action. A Guiding Framework" (The World Bank Group, September 2018).

34 D. Aassouli, "Can the Integration of SRI Principles in Islamic Finance Help Bridge the Gap between Aspirational Islamic Moral Economy and Realistic Islamic Finance?" (Society for the Advancement of Socio-Economics, 24 June 2016) <https://sase.confex.com/sase/2016am/webprogram/Paper4489.html>.

35 Smallridge and others, n 39, 11.

36 J. Martinez, C. L. Vicente, A. bin Arshad, R. Tatucu, and J. Song, "2017 Survey of National Development Banks" (The World Bank, 2017) <http://documents1.worldbank.org/curated/en/977821525438071799/pdf/2017-Survey-of-National-development-banks.pdf>.

37 Ibid.

38 The World Bank defines Result-Based Financing as "any program that rewards the delivery of one or more outputs or outcomes by one or more incentives, financial or otherwise", after the principal has verified that the agent has delivered the agreed-upon results (P. Musgrove, "Rewards for Good Performance or Results: A Short Glossary", The World Bank, March 2011, 3).

39 M. Z. Hussain, "Financing Renewable Energy Options for Developing Financing Instruments Using Public Funds" (The World Bank and Climate Investment Funds, January 2013).

40 The International Renewable Energy Agency (IRENA) highlights that the ambitious targets set by all countries of the region are expected to translate into a combined 80 GW of renewable capacity by 2030 (IRENA, "Middle East & North Africa" <www.irena.org/mena>).

41 See D. Olawuyi, "Energy Poverty in the Middle East and North African (MENA) Region: Divergent Tales and Future Prospects" in I. Del Guayo, L. Godden, D. Zillman, M. Montoya, and J. Gonzalez (eds), *Energy Law and Energy Justice* (Oxford University Press 2020).

42 For the relationship between borrowing costs and the credit rating of sovereign bonds, see D. Ratha, P. De, and S. Mohapatra, "Shadow Sovereign Ratings for Unrated Developing Countries", June 2007, World Bank Policy Research Working Paper 4269 <http://documents1.worldbank.org/curated/en/759061468323725184/pdf/wps4269.pdf>.

43 D. Aassouli, M. Asutay, M. Mohieldin, and T. C. Nwokike, "Green Sukuk, Energy Poverty, and Climate Change: A Roadmap for Sub-Saharan Africa" (World Bank Group Policy Research Working Paper No. WPS8680, December 2018).

44 Ibid.

45 R. Quitzow and others, "Mapping of Energy Initiatives and Programs in Africa" (May 2016) <www.euei-pdf.org/sites/default/files/field_publication_file/mapping_of_init iatives_final_report_may_2016.pdf>.

46 Olawuyi, n 1.

47 Ibid

48 Ibid.

49 Olawuyi, n 6.

50 <https://www.afdb.org/en/topics-and-sectors/initiatives-partnerships/sustainable-ene rgy-for-all-se4all>.

51 Mudaraba is a partnership contract whereby the fund provider (*rabb ul Mal*) contributes money and the manager (*mudarib*) provides work in exchange for a share of the profits.

52 The OECD defines SIBs as innovative financing mechanisms in which governments or donors respectively enter into agreements with service providers and investors to pay for delivery of pre-defined social results (outcomes). See OECD, "Social Impact Bonds-Promises and Pitfalls" (Summary Report of the OECD Expert Seminar, 15 April 2015) <http://www.oecd.org/cfe/leed/SIBsExpertSeminar SummaryReport -FINAL.pdf>.

53 Green *sukuk* investors can choose to forego the profits, the principal, or both of the *sukuk* they hold subject to meeting certain Key Performance Indicators (KPIs). This arrangement was used in the Khazanah SRI *sukuk* issued in 2015 whereby the return to the sukukholders is based on the outcome of the project and a set of KPIs to be achieved (see Khazanah SRI sukuk prospectus).

54 *Waqf* and *zakat* institutions are criticized for their inefficient governance and limited contribution to development. As a result, many scholars called for the modernization of these two institutions. See Ahmed and others, n 6.

55 Islamic Financial Services Board (IFSB), "Islamic Financial Services Industry Stability Report" (IFSB, July 2019).

56 AAOIFI standard FAS 33 "Investments in Sukuk, Shares and Similar Instruments".

57 Thomson Reuters, "Islamic Finance Development Report Building Momentum" (Center for Exclennce in Islamic Finance, 2018) <https://ceif.iba.edu.pk/pdf/Reuters -Islamic-finance-development-report2018.pdf>.

58 Real asset-backing is one of the key principles of Islamic finance whereby financing is based on real economic activities rather than pure financing.

59 Based on International Islamic Financial Market (IIFM) data <www.iifm.net>.

60 Ibid.

61 See Enclude, Abt Associates and USAID, "Clean Energy Lending Toolkit" February 2014, <https://www.climatelinks.org/sites/default/files/asset/document/Clean% 20Energy%20Lending%20Toolkit.pdf>.

62 See D. Karlan and M. Valdivia, "Teaching Entrepreneurship: Impact of Business Training on Microfinance Clients and Institutions" (2011) 93(2) *The Review of Economics and Statistics* 510.

63 See A. F. Cicchiello, F. Battaglia, and S. Monferrà, "Crowdfunding Tax Incentives in Europe: A Comparative Analysis" (2019) 25(18) *The European Journal of Finance* 1856.

64 <https://data.worldbank.org/?locations=MA-XN>.

65 See Morocco credit ratings by Moody's, S&P and Fitch.

66 REN 21, "Renewables 2020 Global Status Report" <https://www.ren21.net/gsr-2020/>.

67 <https://www.afdb.org/fileadmin/uploads/afdb/Documents/Generic-Documents/NOORo_Press_Kit_Eng.pdf>.

68 The Green Morocco Plan (2008–2020) aims to contribute to strengthening the competitiveness of the agricultural sector for inclusive economic growth. The programme is structured around seven major goals: (i) make agriculture the main driver of growth, (ii) adopt aggregation as a model of organization for agriculture, (iii) ensure the development of agriculture as a whole, (iv) promote private investment, (v) adopt a contractual approach to execute this plan, (vi) sustain the development of agriculture, and (vii) prepare for the overhaul of this sector's framework (see African Development Bank, "Country Results Brief 2019- Morocco" (AFDB, November 2018) <https://www.afdb.org/sites/default/files/2019/10/15/crb_morocco_en.pdf>.

69 The energy strategy emphasizes reduction of energy imports through the development of renewable energy by creating an additional 10 GW of installed renewable capacity to meet the target to generate 52% of energy from renewable sources by 2030. See Oxford Business Group country report <https://oxfordbusinessgroup.com/analysis/viable-alternative-plans-add-10-gw-power-renewable-sources-2030>.

70 Africa50 2018 annual report <https://www.africa50.com/fileadmin/uploads/africa50/Documents/Knowledge_Center/AFRICA50_ANNUAL_REPORT_2018_-_ENGLISH.pdf>.

71 P. Grant, "Climate Change Financing and Aid Effectiveness: Morocco Case Study" (OECD, July 2011) <www.oecd.org/environment/environment-development/48458464.pdf> 6.

72 For the country's experience in participative finance see D. Aassouli, "Overview of Sharia'ah Governance in Morocco" in Syed Nazim Ali, Wijdan Tariq, and Bahnaz Al Quradaghi (eds), *The Edinburgh Companion to Shari'ah Governance in Islamic Finance* (Edinburgh University Press 2020) 233.

73 According to Oxford Business Group, as of early 2019 the total amount of green bonds issued in Morocco had reached Dh3.65bn (€328.3m). The main issuers include the Moroccan Agency for Solar Energy, BMCE Bank of Africa, and Banque Centrale Populaire <https://oxfordbusinessgroup.com/analysis/financing-change-green-bonds-are-starting-support-switch-cleaner-energy-attitudes-banking-sector>.

74 D. Ezouine and B. Naima, "Climate Finance Strategy in Morocco" (2019) 8 *American Journal of Climate Change* 482, 488.

75 Ibid.

76 Aassouli, n 129.

77 Ibid.

78 Ibid.

79 Ibid.

80 The FPCT Clifford Chance, "Securitization Law Reform: New Perspectives for Financing the Moroccan and Wider African Economy", April 2013, Client Briefing <https://www.cliffordchance.com/briefings/2013/04/securitisation_lawreformnewperspectivesfo.html>.

81 Ibid.

82 D. Aassouli, and others, 'Green Sukuk, Energy Poverty, and Climate Change: A Roadmap for Sub-Saharan Africa' (2018) World Bank Group Policy Research working paper no. WPS 8680 < http://documents.worldbank.org/curated/en/595861545145005026/Green-Sukuk-Energy-Poverty-and-Climate-Change-A-Roadmap-for-Sub-Saharan-Africa > accessed 16 July 2019.

83 For a summary of the survey findings see Thomson Reuters, IRTI, CIBAFI, Al Baraka (2014), "Morocco Islamic Finance 2014: Unlocking the Kingdom's Potential", 16.

84 Ibid.

85 Conseil Déontologique des Valeurs Mobilières (CDVM), "Sukuk, Quel potentiel de développement au Maroc?" (2012) <www.ammc.ma/sites/default/files/Rapport_Sukuk_07122012_0.pdf>.

86 SIE 2017 Annual Report <http://www.siem.ma/sitesie/images/publications/ra2017.pdf>.

87 See the climate finance page of the climate change online knowledge platform <http://www.4c.ma/fr/projets/liste/financement>.

88 The Energy Efficiency in Mosques initiative was introduced in March 2014 and aims to improve the energy efficiency of mosques by using energy-saving lighting, photovoltaic electricity generation, and solar water heating. See Deutsche Gesellschaft für Internationale Zusammenarbeit (GIZ), "Morocco: Green Mosques" (GIZ, 2017) <https://www.giz.de/en/downloads/giz2017-en-utilisation-en-Allemagne.pdf>.

89 African Development Bank, "Country Results Brief 2019- Morocco" (AFDB, November 2018) <https://www.afdb.org/sites/default/files/2019/10/15/crb_morocco_en.pdf>.

90 International Labor Office, "Morocco: Young Women's Employment and Empowerment in the Rural Economy" (ILO Country Brief, April 2018).

91 See n 125.

92 <https://www.mem.gov.ma/en/Pages/secteur.aspx?e=3&prj=21>.

93 P. Talks and I. Oudra, "FAO/WB Cooperative Programme: Nationally Determined Contribution Support on the Groundwater, Energy and Food Security Nexus in Morocco". World Bank Group <http://documents.worldbank.org/curated/en/353851560191063136/FAO-WB-Cooperative-Programme-Nationally-Determined-Contribution-Support-on-the-Groundwater-Energy-and-Food-Security-Nexus-in-Morocco>.

94 Aassouli and others, n 82.

95 See Climate Bonds Taxonomy <https://www.climatebonds.net/standard/taxonomy>.

96 Ibid.

97 Olawuyi, n 10.

98 UNEP Finance Initiative and the UN Global Compact, "The European Commission Action Plan: Financing Sustainable Growth" (UN Principles for Responsible Investment, July 2018) <www.unpri.org/sustainable-financial-system/explaining-the-eu-action-plan-for-financing-sustainable-growth/3000.article>.

99 A. Baietti, A. Shlyykhtenko, R. La Rocca, and U. D. Patel, "Green Infrastructure Finance: Leading Initiatives and Research" (The World Bank, 2012).

100 C. Lüdemann and O. C. Ruppel, "International Climate Finance: Policies, Structures and Challenges" in O. C. Ruppel, C. Roschmann, and K. Ruppel-Schlichting (eds), *Climate Change: International Law and Global Governance: Volume II: Policy, Diplomacy and Governance in a Changing Environment* (Baden-Baden, Germany: Nomos Verlagsgesellschaft Mbh & Co. 2013) 375.

101 A good example of climate policy instruments is the sector-specific instruments implemented in Sweden. These include energy performance certificates and building norms in the energy and housing sectors, emission performance standards for new vehicles in the transport sector, etc. See Ministry of the Environment and Energy Sweden, "Sweden's Seventh National Communication on Climate Change" (Government of Sweden, 2018) <https://www.government.se/49c920/contentassets/f947f4f4b7ac4af3baadfc827d97557a/swedens-seventh-national-communication-on-climate-change.pdf>.

102 Sustainable Square, "The 2018 State of Sustainability and CSR in the Mena Region" (Sustainable Square, May 2019) <http://sustainablesquare.com/wp-content/uploads

/2019/05/2018-STATE-OF-SUSTAINABILITY-AND-CSR-IN-THE-MENA-RE-GION.pdf> 14.
103   A good example supporting this objective is the Low Emission Development Strategies Global Partnership (LEDS GP), which provides a platform that supports countries in their strategies to achieve low emission and climate resilient development (LEDS Global Partnership <https://ledsgp.org/about/?loclang=en_gb>).

13 Leveraging renewable energy technologies for climate change mitigation and adaptation in the Middle East

The role of public-private partnerships

Cameron Kelly

Introduction

The chapter examines how the Middle East is beginning to successfully leverage private sector expertise and capital to support the region's climate change mitigation and adaptation objectives, principally through the delivery mechanism of public-private partnership (PPP) projects in the renewable energy sector.

The Middle East has long been associated with energy, possessing just under half of the world's oil and gas reserves.[1] In terms of oil deposits, the Kingdom of Saudi Arabia (KSA) has the region's largest reserves and the second largest in the world.[2] The Middle East is also the world's largest oil-producing region, accounting for more than a third of global oil production and just over a third of global oil exports.[3] The Middle East also possesses the largest natural gas reserves in the world—Iran has the region's largest reserves (second globally), then Qatar (third globally), followed by the KSA and the United Arab Emirates (UAE) (sixth globally).[4]

For decades, much of the Middle East's economic prosperity has been underpinned by its oil and gas reserves. However, the COVID-19 pandemic of 2020 has sent the price of oil tumbling to all-time lows, as lockdowns are imposed and travel restrictions (both domestic and international) enforced. The fall in oil prices means that the commodity's future looks uncertain, depending, as it does, on a gallimaufry of commuting, travel, government intervention, capital spending, and price recovery assumptions.[5] The International Energy Agency (IEA) has attempted to quantify the energy impacts of the pandemic-induced global recession associated with multi-month restrictions on mobility and economic activity.[6] The IEA suggests that the decline in energy demand over the course of 2020 will be the largest contraction in 70 years in percentage terms, the largest ever in absolute terms, and seven times the impact of the global financial crisis.[7] Despite macroeconomic stimulus efforts, recovery remains suppressed and is accompanied by a significant (possibly even permanent) loss in economic activity.[8] Global oil major BP supports the IEA's projections, positing that global oil demand will never regain the levels seen in 2019 with demand falling by ten per cent this decade and by as much as 50 per cent over the next 20 years.[9]

Since 2012, Middle Eastern oil revenues have fallen by almost 50 per cent.[10] Inevitably, the falls in energy demand contemplated by the IEA will exert significant fiscal pressure on the region's oil-reliant economies. Oman needs an oil price of US$87 per barrel to break even whilst the region's wealthiest oil producers—KSA and the UAE—require US$80 and US$68 per barrel, respectively.[11] Among the region's oil and gas exporters, only Qatar can balance its budget at oil's mid-2020 price of less than US$50 a barrel.[12]

Even before the coronavirus pandemic gripped the world, an ever-increasing proportion of cost-competitive renewable energy sources (RES) in global electricity grids—together with a transition away from fossil fuels—meant that the days of high oil prices were numbered. At the United Nations Framework Convention on Climate Change's (UNFCCC) 21st Conference of the Parties (CoP) in Paris in 2015, all Middle Eastern countries attended when RES were universally accepted as a cost-effective tool to combat climate change. The negotiated outcome of the CoP—the Paris Agreement—was signed by 15 Middle East and North African (MENA) countries the following year, establishing the policy infrastructure through which member states could advance towards achieving the UNFCCC's "well-below 2°C" goal.[13]

Renewable power capacity is forecasted to expand by 50 per cent, globally, between 2019 and 2024. Solar photovoltaic (PV) power alone is projected to account for almost 60 per cent of this expansion, with onshore wind representing one quarter.[14] The global trend away from fossil fuels and towards increasingly cost-competitive RES suggests that oil prices may remain depressed for the foreseeable future and lends support to the projections offered by the IEA and BP. If such circumstances end up eventuating, no global region will be more affected than the hydrocarbon reliant economies of the Middle East.[15]

Notwithstanding the global significance of its oil and gas resource, the Middle East has begun to seriously invest in utility-scale renewable energy projects. This investment is supported by the declining demand for oil and gas fuel sources; increasingly competitive levelized costs of energy (LCOE)[16] for mature renewable technologies like solar PV and onshore wind; low operating costs; and, in some cases, preferential access to grids.[17] In 2018, just 867 megawatts (MW)—less than one per cent of the Middle East's power capacity—came from RES, yet an estimated seven gigawatts (GW) of renewable energy projects are proposed,[18] with capacity in the region expected to increase by 85 per cent (23 GW) by 2023.[19]

Over the years, the Middle East's abundant hydrocarbon resources have prompted significant interest in the region's approach to energy policy. More recently, a body of literature has emerged that addresses the legal and governance innovations that are required to effectively integrate electricity that is generated from an ever-increasing share of RES, in Middle Eastern grids. Whilst academic research addressing Middle Eastern PPP models has similarly evolved in recent years, less has been written on how such models have been used to support the region's deployment of large-scale renewable energy projects. This chapter attempts to address this gap, by examining the roles of PPPs in addressing climate change mitigation projects, especially those relating to renewable energy technologies.

As in 2016, the Middle East exhibited the lowest level of PPP deployment in the world, particularly in the power sector.[20] Apart from eliciting reasons behind this observation, this chapter posits that certain jurisdictions in the region—such as the KSA—are making significant progress in reversing this trend. Given its expository nature, the chapter adopts a doctrinal research methodology. This involves the systematic analysis of primary sources of laws that regulate an action or activity—in the present case, being how PPPs may lend support to Middle Eastern renewable energy targets and projects.[21] After this introduction, the second section examines several preconditions for successful implementation of PPPs in support of renewable energy projects, drawing on international best practice, and identifies some key barriers to effective PPP deployment. The third section looks at the opportunities, success stories, and challenges that remain in supporting renewable energy PPP projects through a doctrinal legal analysis of a late but significant entrant to the region's renewable energy PPP market, the KSA. The fourth section discusses how the KSA's approach to integrating PPP and renewable energy objectives has overcome many challenges and how, in so doing, it may offer a template for other jurisdictions—both within and outside the Middle East—to implement their own PPP and decarbonisation objectives. The final section concludes the chapter.

PPPs in support of Middle Eastern renewable energy targets

PPPs are an internationally adopted and well-recognized project delivery model. They allow a governmental authority to initiate a project involving a public sector defined objective—such as a renewable energy target (RET)—whilst a private sector party implements the project in accordance with specified contractual agreements.[22] Apart from the obvious benefit of having suitably experienced private sector developers design, construct, operate, and maintain long-lived assets like a wind or solar farm, PPPs free up capital by allowing governments to pay for their share of the asset over many years.[23] The requirement for project developers to invest in a project company or special purpose vehicle (SPV) in exchange for a return on their equity incentivizes them to ensure that the asset is constructed on time and on budget, and in accordance with contractual performance specifications.[24] Contractual approaches include build-transfer-operate (BTO), build-own-operate (BOO), build-own-operate-transfer (BOOT), and build-operate-transfer (BOT), among others.

Barriers to PPP deployment in the Middle East

While Egypt was the first Middle Eastern jurisdiction to introduce specific PPP legislation following the GFC in 2010,[25] the oil price crash of 2014 preceded a raft of PPP legislative and policy initiatives as the region's governments responded to falling hydrocarbon revenues by increasing collaboration with the private sector to develop and modernise key infrastructure assets.[26] Jordan[27] and Kuwait[28] introduced PPP legislation in 2014, followed by Dubai (2015),[29] KSA (2018), Abu Dhabi[30] and Oman (both 2019),[31] and Qatar (2020).[32] Yet across the region, effective implementation of such laws remains mixed, with incompatible or poorly functioning legal structures (including contract enforceability and governance),[33] poor regulatory frameworks, lack of standardized contracts, inadequate (or absent) supporting institutional frameworks, and local capacity gaps, cited as being some of the key barriers to the effective deployment of PPPs.[34] In terms of legal structures, Middle Eastern jurisdictions which recognise Sharia law concepts may[35] be required to accept an internationally recognized system of law supported by rigorous case law precedent (such as English law) as a project's governing body of law,[36] in order to demonstrate a project's "bankability" to international financiers.[37] International lenders may not be inclined to invest if a PPP project is governed by general laws, like a civil code or investment law, which may lack legally robust mechanisms in support of bankability.[38]

Middle Eastern PPP projects have also been hindered by precise risk allocation between a PPP's public and private sector parties via an internationally recognized and standardized contractual framework.[39] Attempting to document such arrangements in poorly structured or inadequately recognized contracts inevitably increases the risk of disputes; financiers and insurers are also disinclined to invest when faced with unfamiliar or inconsistent terms and conditions.

An absence of, or a poorly functioning, national authority or institution mandated to promote and administer PPP projects (as well as enter into PPP contracts on behalf of government) is a further impediment to effective deployment of Middle Eastern PPPs.[40] Of the region's jurisdictions which have enacted specific PPP legislation, a majority have centralized PPP units, yet success in deploying PPP projects across the region remains mixed. In Egypt, for example, PPP projects may be procured through the government's system of public economic entities and public utilities legislation, as well as through sector-specific or project-specific laws.[41] Accordingly, Egypt's PPP legislation (being Law 67 of 2010) is not the only legal and policy framework governing PPP procurement models, which in turn has hindered ministry and sector-wide support for the effective operation of Egypt's PPP central unit.[42]

Local human resource capacity in key project-related disciplines, such as engineering, finance, and law is a further factor impeding the effective deployment of Middle Eastern PPPs.[43] Such capacity gaps may have knock-on effects to other PPP partcipants (such as international lenders seeking to understand a host government's contracting or procurement approach) or functions (such as coordination of due diligence investigations). Domestic nationalization programmes in many Middle Eastern jurisdictions (such as the UAE's Emiratisation Programme, itself a key performance indicator of the UAE's Vision 2021)[44] attempt to reverse this trend; however, progress is often slow.

Middle Eastern RETs

Most Middle Eastern jurisdictions have published ambitious RETs.[45] Such targets form integral components of the "National Visions" of the KSA, Jordan, the UAE, Bahrain, Kuwait, Oman, and Qatar (amongst others), each of which seeks to achieve diversified, sustainable, and decarbonized economies.[46] Persistent low oil prices restrain effective government-led implementation of National Visions. Decarbonization objectives must therefore be underpinned by significant private sector participation and capital.[47] The technical and project expertise required to foster this type of energy transition has not always been readily or locally available. The success of the National Visions' low carbon components will largely depend on the ability of each country to attract foreign and private sector participation by companies with recognised expertise in (renewable energy) project development.[48]

For decades, governments around the world have understood that private investment—such as through PPP models—can be an effective approach to freeing up capital for other purposes (such as delivering RETs).[49] The Middle East is no exception to this dynamic, where global economic shocks wrought by the GFC of 2008/2009, the oil price collapse of 2014, and the COVID-19 pandemic-induced oil price collapse of 2020 have delivered record low revenues for the region's economies. It is therefore unsurprising that a key pillar of the energy transition goals of many Gulf Cooperation Council (GCC) countries (Bahrain, Kuwait, Oman, Qatar, the KSA, and the UAE) has been to promote private sector participation in RETs via PPP-style projects.[50]

For large greenfield, capex intensive renewable energy projects in the Middle East, a BOT (or BOOT) structure tends to be the most common form of PPP.[51] Under this structure, the private sector entity is granted a concession from the public sector and assumes full responsibility (and risk) for the financing, development, construction, operation, and maintenance of the solar or wind farm over a prescribed period. Such structures have delivered a string of record-breaking solar power tariffs.[52] In 2015 (the same year that Dubai issued its PPP legislation), the UAE set a new record by awarding PV capacity at less than US$0.06 per kilowatt hour (kWh) under the first phase of its 950MW Mohammed bin Rashid Al Maktoum solar PV plus concentrated solar power (CSP) project in Dubai. Additional capacity was awarded a year later at half this price.[53] By 2019, solar projects with a combined capacity of 4.22GW were under construction via a variety of PPP models across the region. These included the 1.8GW Benban solar power complex in Egypt, the 50MW Al Husainiyah solar park in Jordan, the 1GW Miraah CSP plant in Oman, the 300MW Sakaka PV park in KSA, and phase 4 of the UAE's Al Maktoum solar PV project.[54]

Despite the advent of the COVID-19 pandemic, 2020 has seen no diminution in utility-scale renewable energy PPP projects offering regionally (or globally) significant tariffs. Abu Dhabi's 2GW Al Dhafra project—constructed over an area of 20 square kilometres and constituting the largest PV project in the world—set a new record-breaking tariff of $0.0135/kWh.[55] Rounds 2 and 3 of the KSA's Renewable Energy Project Development Office broke existing tariff records while Qatar awarded its first large-scale (800MW) solar PV project at Al-Kharsaah at a record price of US$0.016/kWh.[56] The year 2020 also saw the fifth and final phase of the Al Maktoum solar PV project reach financial close, with power to be sold at just $US16.95 per MWh under a 25-year power purchase agreement (PPA) with the Dubai Electricity and Water Authority.[57]

Legal approaches to delivering Middle Eastern PPPs: lessons from the KSA

Background

The KSA is home to the world's largest oil reserves and enjoys the fourth largest concentration of natural gas resources.[58] With population increases and industrial development, consumption of domestic oil continues is expected to reach 7 million barrels per day by 2030.[59] As the vast majority (around 80 per cent) of the KSA's revenue is reliant on oil exports, increasing domestic consumption means the amount available for export will continue to decline, with resultant impacts on the nation's revenues. Consistent with the intent of Vision 2030, key drivers for RES development in the KSA are analogous to those of other GCC countries with globally significant oil and gas reserves. These drivers include freeing up domestic oil and gas reserves for more profitable (export-oriented) uses, satisfying growing demand for energy, economic diversification, and creating more jobs for the local population, leveraging an advantageous geography and climate,[60] and reducing greenhouse gas (GHG) emissions.[61]

Given the extent and profitability of its hydrocarbon resource, it is perhaps unsurprising that the KSA has had several false starts in implementing its renewable energy PPP ambitions. Established by Royal Decree in 2010, the King Abdullah City for Atomic and Renewable Energy (KACARE) was set up to administer procurement processes, implement policy, develop regulations, and establish a viable legal and commercial structure for the renewable and nuclear energy sectors. To advance these aims, the Sustainable Energy Procurement Company (SEPC) and National Grid Company (NGC) were formed.[62] In 2013, KACARE issued a white paper addressing the procurement methodology for the KSA's newly announced RET of 54GW by 2032, together with a broad outline of the proposed regulatory framework and required infrastructure (White Paper).[63] The White Paper detailed the first three procurement rounds, being an introductory round of 500–800MW, a first and second round of 2,000–3,000MW, and 3,000–4,000MW, respectively, with further rounds to be announced thereafter. The introductory round was to consist of up to seven projects at grid-ready sites pre-supplied with the required real estate rights and consents to design, build, and operate projects on a BOO basis and involving long-term (20-year) offtakes with the SEPC.[64]

Barriers to effective PPP deployment in the KSA

Despite its ambitious approach to procuring 7GW of renewable energy capacity, KACARE involved a number of critical issues that impeded its ability to effectively attract (PPP) investors.[65] As of 2016, the following regulatory, institutional, and capability issues remained outstanding:

- *Unsustainable fuel subsidies.* Fossil fuel subsidies remained high in the KSA (with power being amongst the cheapest globally) which dampened the private sector's appetite to make substantial RES investments.[66]
- *Nascent policies and regulations.* The KSA was still establishing a credible regulatory framework to support power supply from RES and grid connections.[67]
- *Emerging institutional framework.* Institutions capable of centrally administering PPP-style investments in RES were yet to establish the required expertise and partnerships to achieve their stated objectives.[68]
- *Lack of local Research and Development talent.* The nationalization and localization of human resource expertise had been identified as a priority by the KSA government but in terms of development, was still in its infancy.[69]
- *Inadequate supporting infrastructure.* The KSA electricity grid was not set up to accept the additional proposed (distributed) RES generation assets. Significant coordination was required between the entity responsible for the KSA's grid—Saudi Electricity Company (SEC) and KACARE—and effective coordination between these entities was elusive.[70]

By 2017, no real progress had been made in delivering the White Paper's RET, yet market participants maintained the view that the KSA could (at least potentially) deliver significantly on its renewable PPP ambitions. Whilst some market participants believed low oil prices would lessen domestic appetite for renewables (due to reduced funding on offer), others were of the view that lower oil prices would result in the KSA recommitting to a competitive domestic renewables programme due to a requirement to export (rather than consume) oil.[71]

Overcoming KSA's PPP barriers

The National Center for Privatization & PPP (NCP) was established in 2017[72] to coordinate the KSA's PPP (and privatization) programmes, both of which form key components of the Saudi Vision 2030.[73] As a centre of excellence, NCP helps to formulate PPP regulations, create PPP frameworks, and identifies PPP opportunities. The NCP's mandate includes provision of funding for privatization supervisory committees (PSCs); review of sector PPP strategies and proposed regulatory changes; development of PPP directives, standards, guidelines, and key performance indicators (in line with Vision 2030); and development of a national regulatory framework for PPPs.[74] NCP personnel include experts with legal, financial, advisory, strategy, communication, risk management, marketing, privatization, and project management backgrounds.

A long-standing feature of the KSA's PPP market was that, despite the nation's track record of PPP deployment in the utilities sector, the nation lacked a specific PPP law. This changed in 2018 when the NCP issued a draft of the "Private Sector Participation Law" (PSP Law) regulating both PPP projects and privatizations.[75] The PSP Law focuses on delivering the key reform agendas contained in both Vision 2030 as well as the "National Transformation Plan" (NTP), both of which target greater private sector participation in delivering key infrastructure assets.[76] This included a target of 14 PPP investments across ten sectors (including energy) to be delivered by 2020.[77] The PSP Law is supported by Implementing Regulations, which (amongst other matters) support entry into framework agreements and reverse auction tendering via electronic portals.[78]

Institutional and regulatory support: incentives and exemptions under the PSP Law

The PSP Law includes a number of incentives specifically designed to attract and retain international PPP investors. These include:

- *Attracting international sponsors, private equity, and infrastructure investors.* The PSP Law includes several provisions designed to specifically attract international (PPP) investors. These include an ability to remit a project's financial returns, an ability for foreign investors to completely own a project's SPV, as well as rights to recover losses associated with a change in law, unlawful action, or where a public authority acts inappropriately or arbitrarily.[79] PPP contracts can also be in a non-Arabic languages.
- *Protecting foreign investors.* The PSP Law provides that foreign legal entities are to be afforded the same treatment as their Saudi counterparts concerning any "procedures, conditions, rights and obligations arising from this [the PSP] Law and any Contract".[80]
- *Government-backed financial support.* Saudi PPP projects have historically included a form of government-sponsored credit support offered to the private sector entity. The PSP Law continues such support, and includes a range of measures which the government may rely on outside of conventional guarantees. These include concessional loans, revenue guarantees, subsidies, and protections against foreign exchange risk. Availability of alternative forms of credit support under the PSP Law sits alongside the Commercial Pledge Law (CPL).[81] The CPL significantly improved how Saudi security is granted and enforced, giving secured creditors the right to enforce their interests against collateral through perfection of centrally maintained security interests register.
- *Measures to improving dispute resolution.* Instead of defaulting to KSA courts, parties are permitted (subject to approval of the Council of Economic Affairs and Development) to resolve contractual disputes by means of arbitration.[82]
- *Enhance cross-sector application.* Contrary to some GCC states (which have adopted varying approaches to PPP legislation or excluded the utilities sector from PPP laws), the PSP Law enjoys cross-sectoral application, and can involve a wide range of transactions. PPPs are broadly defined, and involve multi-year contractual arrangements between government and a private entity for the purposes of infrastructure development. Arrangements include

provisions for the deployment of public services (such as construction, management, operation, and maintenance of an asset), appropriate risk allocation between parties, and performance-orientated contractual payment mechanisms.[83]

• *Enhancing participation.* Like Jordanian and Kuwaiti PPP legislation, the PSP Law permits the submission of unsolicited (or direct) project proposals. The PSP Law requires unsolicited proposals to be "subject to the provisions of this Law, and the Regulations issued thereunder".[84]

In addition to the incentives outlined above, the PSP Law provides several regulatory exemptions designed to benefit international PPP investors. Aside from various customs duty exemptions and tax breaks, these include:

• *Exemption from the Procurement Law.* The PSP Law supports Royal Decree M/101 which exempted PPP projects from the Government Tenders and Procurement Law 2006 (Procurement Law).[85] The Procurement Law enjoys widespread application to infrastructure projects procured across the KSA. However, its more traditional approach to procurement meant that it was never intended to regulate PPP procurements. The Procurement Law involves several incompatible provisions with PPP project development, resulting in a need to obtain exemptions, with inevitable delays on long-lead items.[86]

• *Exemptions from the Labour Law.* Recent PPP procurements in the KSA have strictly applied local content requirements, in relation to the Labour Law and Saudisation. Against a backdrop of ever-increasing local content requirements, the PSP Law contemplates project-specific changes to (or exemptions from) the Labour Law.[87]

PPPs in support of KSA's RET

In 2016, the KSA's Vision 2030 noted that "the Kingdom's impressive natural potential for solar and wind power generation remains largely untapped" and promised to generate 9.5GW of renewable energy by 2030—this equated to around 30 per cent of the country's power mix to be sourced from RES by 2030.[88] To meet this objective, the Saudi Government announced an intention to attract between US$30 billion and US$50 billion in new renewable investments over the same time frame, with plans to tender the entire 9.5GW of solar and wind capacity by 2023.[89] The year 2016 also saw the establishment of the Renewable Energy Project Development Office (REPDO).[90] This institution was set up to deliver the National Renewable Energy Program (NREP).[91] Working with the KSA's key energy sector stakeholders (including KACARE, the Electricity and Cogeneration Regulatory Authority (ECRA), and the SEC), REPDO was tasked with coordinating the KSA's initiatives in research, measurement, data acquisition, regulation, pre-development, and tendering for renewable energy projects.[92] Cumulatively, these initiatives aimed to maximize the potential of renewable energy project development in the KSA, and sought to diversify local energy sources and stimulate economic development.[93]

In 2017, REPDO announced the launch of Round One of NREP's 9.5 GW target by issuing a request for qualifications (RfQ) for three large-scale green-field renewable energy projects—the 300MW Sakaka solar PV project, the 400MW Midyan wind farm, and the 400MW Dumat Al Jandal wind project.[94] The projects were developed on a PPP (BOO) basis via privately owned SPVs, and offered a number of PPP-style incentives and exemptions consistent with those provided under the PSP Law.[95] Prior to the launch of Round One, SEC acted as offtaker for all generated electricity in the KSA. For the REPDO rounds, NREP established a (dedicated) "Principal Buyer" (known as the Saudi Power Procurement Company (SPPC)) to act as the primary offtaker for all forms of electricity. This greatly enhanced the bankability of related PPAs and was key to attracting foreign lenders.

Apart from the establishment of the SPPC, other incentives under REPDO's Round One included complete (100 per cent) foreign direct ownership of a project's SPV (as opposed to the more traditional limit of 50 per cent), a renewables-specific account manager with the Saudi Arabia General Investment Authority (to assist in procuring required foreign investment licences), availability of attractive concessional debt (for up to three-quarters of total project costs),[96] facilitation of local content (by paying up to half of Saudi nationals' salaries via the Human Resources Development Fund),[97] and free registration of real estate interests and accelerated asset depreciation.[98] Exemptions included tax concessions (such as reduced corporate tax rates, tax exemptions on capital expenditure, and an ability to carry forward taxation losses over an unlimited period), customs duty exemptions for raw materials and spare parts,[99] and no restrictions on repatriation of capital.[100]

In 2019, REPDO announced record-breaking LCOEs of US$0.0199 per MWh for the $500 million Dumat Al Jandal wind PPP and US$0.0234 per MWh for the Sakaka PV project.[101] The Dumat Al Jandal bid was a record low for the Middle East in onshore wind. By financial close, this price was cut by a further 6.5 per cent, a global record price for onshore wind.[102] REPDO launched Round Two of the NREP in mid-2019 by tendering for six projects (with generation capacities of between 20 and 600MW),[103] yielding a combined capacity of almost 1.5GW.[104] Like Round One, the second round of REPDO projects included long-term (20 to 25 year) offtake arrangements with SPPC. Sixty companies were pre-qualified, with 28 being from the KSA.[105] The projects were divided into smaller (up to 100MW) "Category A" projects[106] and larger (over 100MW) "Category B" projects.[107] Round Two projects also required a minimum local content percentage (of 17 per cent).[108] REPDO issued an RfQ for Round Three of the NREP on 9 January 2020. This round attracted a record 83 applications (including 28 Saudi companies).[109] Round Three included four solar PV projects with a total capacity of 1.2GW.[110]

The tendering methodology for Round Three continues the strategy established by REPDO under the previous round, whereby the smaller (Category A) projects have relaxed qualification criteria (with local developers able to participate as lead members) with the two larger (Category B) projects targeting international sponsors and financiers.[111] Like the previous rounds, Round Three projects are backed by long-term (25-year) PPAs, with SPPC as the offtaker.

Leveraging PPPs for climate change mitigation: discussion and recommendations

In 2019, KSA required an oil price greater than US$80 a barrel to balance its budget, yet brent crude has failed to trade over US$50 since early 2020.[112] For Middle Eastern governments struggling to balance coronavirus-affected budgets, PPP's unique pairing of public and private sector expertise can deliver better value for money in project delivery than more traditional procurement models. PPPs have greater potential to deliver projects on time and within budget, and— through international and local financiers conducting rigorous due diligence before committing capital—can offer greater budgetary certainty at the time of contracting for the duration of the project's offtake.[113]

While KSA has sponsored PPP construction models for many years, the use of such models has undoubtedly increased in the wake of persistently low oil revenues and associated increased government spending deficit.[114] This is consistent with the NTP 2020, which calls for a significant increase in private sector participation in the nation's GDP (by assuming almost half of the funding burden) and is further supported by the NCP, an aim of which is to accelerate the deployment of PPP models in key infrastructure assets like utility-scale solar PV and wind farms.[115] Typical PPP models in the KSA include BOOT and BOT models,[116] with project documentation generally involving a government employer and a private sector engineering, procurement, and construction (EPC) contractor using an internationally recognized form of contract such as the International Federation of Consulting Engineers (FIDIC) form of contract.[117] Renewable energy project documents typically also include power purchase agreements (PPAs), operation and maintenance (O&M) agreements, and (in some cases) supply and intellectual property licensing agreements.

The KSA's NREP has tendered over 2GW of renewable energy capacity, of which 700MW was awarded in 2019 with an additional 1,470MW to be awarded by the end of 2020.[118] Coming from a country which (as recently as 2016) possessed negligible renewable capacity and a questionable commitment to meeting its RET, deploying this much renewable capacity in less than three years is undoubtedly a significant achievement, both regionally and globally. As a key component of Vision 2030, the REPDO renewable energy rounds are paving the way for the KSA to catalyze a viable and fully funded renewable energy sector over the next decade, relying heavily on the mechanism of utility-scale power (solar PV and onshore wind) PPP projects.

Over the 2020–2021 period, PV and CSP projects across the Middle East (many of which will involve PPP models) with a total capacity of 4.6GW have been announced, with an additional 7GW to be tendered.[119] Following the

2014–2015 oil price collapse, the string of record-breaking tariffs set for large-scale renewable energy PPP projects awarded in the UAE, the KSA, and Qatar, underscores the ever-increasing competitiveness of key renewable technologies like solar PV. To facilitate the effective deployment of such PPPs in support of Middle Eastern RETs, a number of general recommendations aimed at attracting and retaining private sector expertise and financing have arisen. Such recommendations include updating (and where required, reforming) existing laws which may operate to hinder renewable energy PPP projects, legislating specific renewable energy PPP laws and policies which have regard to (and in fact demonstrate) international best practice, establishing dedicated institutions to centrally promote and administer renewable energy PPPs, and promoting regional cooperation and knowledge sharing.[120]

In terms of best practice, PPP laws should ideally include fiscal and tax incentives to attract and retain financing, streamline conflicting PPP regulatory regimes, and establish specified institution(s) to (centrally) administer PPP investments.[121] The KSA's PSP Law, together with institutions like the NCP and supporting documentation like the NTP 2020, provide an effective template for REPDO to procure large-scale solar PV and onshore wind energy in a cost-effective and competitive manner. The output and performance focus of government-specified PPPs (implicit within the KSA's Vision 2030, as well as other Middle Eastern National Visions) provides greater opportunity for the private sector to competitively price innovative solutions for proposed solar and wind projects.

The KSA has implemented significant regulatory reform by enacting the PSP Law and eliminating inconsistent or conflicting procurement regimes (such as application of the Procurement Law). Establishment of centralized institutions, like the NCP (in support of the PSP Law) together with REPDO (in support of the NREP) has meant that notwithstanding COVID-19-related declines in oil demand and government revenues, the KSA stands a good chance of delivering on its ambitious decarbonization agenda. The PSP Law, together with NREP's competitive procurement of the REPDO-administered renewable energy rounds, address the majority of the aforementioned recommendations. By employing many of the key legislative features of the PSP Law (which was introduced between the first and second of the REPDO rounds), the REPDO projects are expected to attract over US$1.5 billion of private sector investment[122] and should provide a significant contribution to the KSA's RET of 27.3GW of renewable energy capacity by 2024 and 58.7GW by 2030[123] (of which an estimated 40GW is likely to be in the form of solar PPPs).[124] In addition to adopting further measures to reduce unsustainable domestic oil subsidies,[125] upgrading the electricity grid to connect more distributed RES and enshrining its RET into domestic legislation, it is recommended that the KSA consider emerging distributed ledger technologies, like blockchain, which have the potential to dramatically improve processes and systems (such as a project's permitting and approval processes) across a solar or wind PPP's value chain.[126] It is also recommended that the KSA consider sharing its approach for integrating PPP and RET policy objectives with other Middle Eastern jurisdictions with similar climate change mitigation, adaptation, and decarbonization objectives.

Conclusion

In recent years, the Middle East has begun to leverage private sector expertise and finance in support of the region's renewable energy and broader decarbonization objectives through the mechanism of PPP projects in earnest. Declining government revenues wrought by the oil price collapse of 2014–2015, as well as the cornovirus pandemic-induced collapse of 2020, have been key factors in prompting many of the region's governments to re-evaluate their policy and regulatory approaches to implementing the RET, decarbonization, and diversification agendas implicit in their National Visions. Barriers to effective implementation—such as poor legal structures and regulatory frameworks, lack of standardized contracts, inadequate institutional frameworks, local capacity gaps and unsustainable fuel subsidies—have all acted as impediments to the adoption of international best practice PPPs. Yet despite a historic reliance on globally significant hydrocarbon reserves to fund infrastructure and power projects, it is evident that jurisdictions like the KSA are, for the first time, making substantial progress in developing an enduring model which effectively combines ambitious RETs with dedicated PPP legislation to deliver power projects. With oil prices threatening to remain depressed for the foreseeable future, the ability of PPP models to cost-effectively procure power from key renewable technologies, like solar PV, augurs well for the future deployment of such project delivery models across the Middle East. By accelerating PPP-administered investments into utility-scale renewable energy projects, the region is poised to provide an important contribution to the world's attempt to chart a sustainably based recovery from the coronavirus pandemic.

Notes

1 BP, "Statistical Review of World Energy" (68th edition, June 2019) 14 and 30.
2 Ibid, 14.
3 Ibid, 16.
4 Ibid, 14.
5 The Economist, "Arab States Are Embracing Solar Power" (7 May 2020) <www
.economist.com/middle-east-and-africa/2020/05/07/arab-states-are-embracing-solar
-power?utm_campaign=the-climate-issue&utm_medium=newsletter&utm_source
=salesforce-marketing-cloud&utm_term=2020-05-18&utm_content=ed-picks-arti-
cle-link-7>.
6 International Energy Agency (IEA), "Global Energy Review 2020" (2020) <www.iea
.org/reports/global-energy-review-2020>.
7 Ibid, 5.
8 Ibid.
9 BP, "Energy Outlook 2020" (14 September 2020) <www.bp.com/content/dam/bp/
business-sites/en/global/corporate/pdfs/news-and-insights/press-releases/bp-energy-o
utlook-2020-pr.pdf>.

10 The Economist, "Twilight of an Era – The End of the Arab World's Oil Age is Nigh" (18 July 2020) <www.economist.com/middle-east-and-africa/2020/07/18/the-end-of-the-arab-worlds-oil-age-is-nigh>.

11 Ibid.

12 Ibid.

13 M. Wilder, R. Saines, P. Curnow, G.Stuart, F. Ruanova Guinea, and R. Amaral, "The Paris Agreement – Putting the First Universal Climate Change Treaty in Context" Baker & McKenzie (11 January 2016) <www.lexology.com/library/detail.aspx?g=7 f89d079-a97c-4c87-99ba-16f3c9ff5fdf>.

14 IEA, "Renewables 2019 Market Analysis and Forecast from 2019 to 2024" (IEA, October 2019) <www.iea.org/reports/renewables-2019>.

15 Ibid.

16 LCOE is a measure of the average net present cost of electricity generation for a generating plant over its lifetime.

17 T. Snider, J. Rahman, and A. Nair, "Energy Arbitrations in the Middle East" Al Tamimi & Company (20 April 2020) <www.lexology.com/library/detail.aspx?g=9 3b43faa-a81e-4f34-affb-f699c8259f62&l=90JQYW1>.

18 International Renewable Energy Agency, "Renewable Energy Market Analysis: GCC 2019" IRENA Abu Dhabi 2019 <www.irena.org//media/Files/IRENA/Agency/Publication/2019/Jan/IRENA_Market_Analysis_GCC_2019.pdf> 9.

19 International Energy Agency, "Renewables 2018 Analysis and Forecasts to 2023" (2018) <https://webstore.iea.org/download/summary/2312?fileName=English-R enewables-2018-ES.pdf> 95.

20 E. Somma and A. Rubino, "Public-Private Participation in Energy Infrastructure in Middle East and North African Countries: The Role of Institutions for Renewable Energy Sources Diffusion" (2016) 6 *International Journal or Energy Economics and Policy* 621; G. Liddo, A. Rubino, and E. Somma, "Determinants of PPP Infrastructure Investments in MENA Countries: A Focus on Energy" (2019) 46 *Journal of Industrial and Business Economics* 523.

21 T. C. Hutchinson and N. Duncan, "Defining and Describing What We Do: Doctrinal Legal Research" (2012) 17 *Deakin Law Review* 83.

22 M. Sawalha, "Public-Private Partnerships in Jordan," Al Tamimi & Co (31 January 2015).

23 J. Bailey, M. Turrini, and A. Pierson, "An Expanding PPP Framework for Qatar" White & Case (8 September 2016) <www.lexology.com/library/detail.aspx?g=509dbf 0c-8284-44e2-a1b2-93879a0c3d85&l=7SF9XEJ>.

24 Ibid.

25 Law 67 for year 2010 regulating Partnerships with the Private Sector in Infrastructure Projects, Services, and Public Utilities.

26 European Bank for Reconstruction & Development (EBRD) and CMS Law & Tax, "Public-Private Partnership Assessment 2017–2018" (2018) 9.

27 Public-Private Partnership Law No 31 of 2014.

28 Law No. 116 of 2014 Concerning Partnerships between the Public and Private Sectors.

29 Law No. 22 of 2015.

30 Law No. 2 of 2019.

31 Sultani Decree No. 52/2019. Oman published the implementing regulations (Decision No. 4/2020 of the Public Authority for Privatisation and Partnership for Decree No. 51/2019 in October 2020).

32 Law No. 12 of 2020 Organising the Partnership between the Public and Private Sector.

33 Liddo and others, n 20.

34 D. Olawuyi, "Financing Low-Emission and Climate – Resilient Infrastrstructure in the Arab Region: Potentials and Limitations of Public-Private Partnership Contracts" in W. L. Filho (ed), *Climate Change Research at Universities Addressing the Mitigation and Adapatation Challenges* (Hamburg: Springer 2017) 539–541.

35 Islamic finance is concerned with the conduct of commercial and financial activities in accordance with Islamic law, or Sharia. Islamic finance emphasizes productive economic activity over pure speculation, and encourages transaction counterparties to share profits and losses to promote collaborative efforts. See M. Hussain, "Islamic Finance" Milbank LLP (4 September 2020) <www.lexology.com/library/detail.aspx?g=f122a244-8b65-45fe-b4fa-c9baef79488e&utm_source=Lexology+Daily+Newsfeed&utm_medium=HTML+email+-+Body+-+General+section&utm_campaign=Australian+IHL+subscriber+daily+feed&utm_content=Lexology+Daily+Newsfeed+2020-09-24&utm_term=>.

36 EBRD and CMS Law & Tax, n 26, 21.

37 "Bankability" refers to a project's ability to secure finance (debt and/or equity) thereby reaching financial close.

38 Mechanisms in support of bankability may include mechanisms to enable technical and financial due diligence, arrangements for fair compensation on termination, sovereign guarantees to mitigate currency risk, and direct agreements affording lenders' step-in rights, amongst others (EBRD and others, n 26, 21).

39 Olawuyi, n 34, 540.

40 Ibid, 541.

41 European Investment Bank, "Facility for Euro-Mediterranean Investment and Partnership (FEMIP) – Study on PPP Legal & Financial Frameworks in the Mediterranean Partner Countries (Volume 2 – Country Analysis)," May 2011, 8.

42 Ibid, 9.

43 Olawuyi, n 34, 542.

44 UAE, "Vision 2021 and Emiratisation" <https://u.ae/en/information-and-services/jobs/vision-2021-and-emiratisation>.

45 For an analysis of the various RETs across the Middle East, see C. Kelly, "Developing Renewable Energy Projects in the Middle East and North African region" in T. Hunter, I. Herrara, P. Crossley, and G. Alvarez, (eds), *Routledge Handbook of Energy Law* (Milton Park, Abingdon-on-Thames, Oxfordshire: Taylor & Francis 2020) 507.

46 UAE National Vision 2021 <www.vision2021.ae/en>; Bahrain National Vision 2030 <www.bahrainedb.com/en/about/Pages/economic%20vision%202030.aspx>; Saudi Arabia Vision 2030 <http://vision2030.gov.sa/en>; Kuwait National Development Plan <www.newkuwait.gov.kw/en/plan/>; Oman Vision 2040 and Qatar National Vision 2030 <www.gco.gov.qa/en/about-qatar/national-vision2030/>.

47 D. Olawuyi, "Advancing Innovations in Renewable Energy Technologies as Alternatives to Fossil Fuel Use in the Middle East: Trends, Limitations and Ways Forward" in D. Zillman, M. Roggenkamp, L. Paddock, and L. Godden (eds), *Innovation in Energy Law and Technology Dynamic Solutions for Energy Transitions* (University Press 2018) 364.

48 Ibid.

49 M. Brown, "Projects into 2020: PPP Leads the Way", Al Tamimi & Company (31 January 2020) <www.lexology.com/library/detail.aspx?g=9027c13c-27a4-458e-9ba5-ea7705bc59ec>.

50 Olawuyi, n 34, 364.

51 For example, Oman's PPP legislation of 2019 provides that a PPP must adopt a BOOT structure (T. Butcher, and A. Haque, "GCC PPPs 2016" DLA Piper (28 November 2016) <www.lexology.com/library/detail.aspx?g=35839e89-0dd9-48f0-b170-fee88b4b8087&l=7T47YGJ>.

52 Brown, n 49.

53 Hunter and others, n 45.

54 A. Dimitrova, "MENA with 4.2 GW of Solar under Construction in 2019 – MESIA" (Renewables Now, 21 January 2020) <https://renewablesnow.com/news/mena-with-42-gw-of-solar-under-construction-in-2019-mesia-684298/>.

55 J. Hill, "World's Largest Solar Power Plant Moves Forward with World's Lowest Price" (Renew Economy, 28 July 2020) <https://reneweconomy.com.au/worlds-largest-solar-power-plant-moves-forward-with-worlds-lowest-price-61208/>.

56   The Al Kharsaah project is valued at US$462 million (E. Bellini, "Qatar's 800 MW Tender Draws World Record Solar Power Price of $0.01567/kWh" 23 January 2020 <www.pv-magazine.com/2020/01/23/qatars-800-mw-pv-tender-saw-world-record-fin al-price-0-01567-kwh/>.

57   This phase of the Al Maktoum project will be constructed by an SPV in which the ACWA-Gulf Investment Corporation consortium owns 40 per cent, while Dubai Electricity & Water Authority (DEWA) holds the remaining 60 per cent; T. Tsa-nova, "ACWA Announces Fin Close for 900-MW PV Project in Dubai" (Renewa-bles Now, 3 September 2020) <https://renewablesnow.com/news/acwa-announces -fin-close-for-900-mw-pv-project-in-dubai-712262/>.

58   Eversheds Sutherland and Price Waterhouse Coopers, "Developing Renewable Energy Projects – A Guide to Achieving Success in MENA", 4th edition, June 2016. p. 110.

59   Ibid.

60   Saudi Arabia has the greatest potential for renewable energy in the Middle East with direct (normal) irradiation of 2,500 kWh/m2/year, net wind speeds of over seven metres per second, and large tracts of available land (ibid, 111).

61   The KSA is one of six GCC countries that are in the top 15 per capita emitters of carbon dioxide in the world, mostly due to a high reliance on heavy fuel oil to gener-ate domestic electricity (ibid, 111).

62   Ibid, 111.

63   Ibid.

64   Ibid, 112.

65   Ibid, 114.

66   Ibid, 115.

67   Ibid.

68   Ibid.

69   Ibid.

70   Ibid.

71   Ibid, 116.

72   Council of Ministers Resolution No. 665 (dated 8/11/1438H (1/8/2017G)) establish-es the institutional framework under which the NCP operates, pursuant to which the NCP has promoted a system of sectoral/ supervisory committees relating to sectors targeted for privatization, including the energy sector, F. Patalong, "KSA Privatisa-tion, Corporatisation and PPP Schemes in Healthcare", Al Tamimi & Company (26 November 2019) <www.lexology.com/library/detail.aspx?g=614a9017-4160-4a0f-8f b6-e4dded0ed6e8>.

73   <https://www.ncp.gov.sa/en/Pages/NCP-In-Brief.aspx>.

74   National Center for Privatization <www.ncp.gov.sa/en/Pages/Home.aspx>.

75   T. Burbury and T. Smith, "Ten Reasons Why International Investors Should Be Ex-cited about Saudi Arabia's New PSP (PPP) Law" King & Spalding LLP (8 March 2019) <www.lexology.com/library/detail.aspx?g=b0161c0f-4f0e-4f05-881a-27b56255f6da>.

76   As part of delivering Vision 2030, the National Transformation Program was launched as a Vision Realization Program (VRP) involving some 24 government agencies (ibid).

77   Patalong, n 72.

78   Ibid.

79   Ibid.

80   Ibid.

81   Introduced by Royal Decree No. M/86 on 24 April 2018, alongside related Imple-menting Regulations.

82   Burbury and Smith, n 75.

83   Ibid.

84   Ibid.

85 Government Tenders and Procurement Law 2006, as enacted by Royal Decree No. M/58 dated 27 September 2006 and its Implementing Regulations as enacted by Minister of Finance Resolution No. 362 dated 10 March 2007.

86 Burbury and Smith, n 75.

87 Royal Decree M/51 23 Sha'ban 1426/September 27 2005 (as amended) (Burbury and Smith, n 75).

88 KSA <https://vision2030.gov.sa/en/node/385>; M. Brown and F. Patalong, "Time for Renewables in KSA," Al Tamimi & Company (31 January 2020) <www.lexology.co m/library/detail.aspx?g=0600db6a-015e-4a0d-82d5-57387ae7f19a>.

89 Ibid.

90 REPDO was established as a unit within Saudi's Ministry of Energy, Industry and Mineral Resources.

91 Kingdowm of Saudi Arabia, Ministry of Energy, Industry and Mineral Resources. Renewable Energy Project Development Office (REPDO) 2020 <www.powersaudia rabia.com.sa/web/index.html>.

92 Ibid.

93 Ibid.

94 T. A. Thraya, C. Hallab, and R. Goldberg, "Saudi Arabia Announces RFQ for Renewable Energy Projects," Mayer Brown (28 February 2017) <www.lexology.com/lib rary/detail.aspx?g=1a769171-3b52-4dcf-bfd7-2a0e1072a0d1>.

95 Ibid.

96 Subject to a US$320 million cap.

97 Up to two years only.

98 Thraya and others, n 94.

99 Up to two years only.

100 V. Terblanche, E. Hills, and D. McKinley, "Saudi National Renewable Energy Program Round 1 – Scaling Up", Latham & Watkins LLP (24 April 2017) <www.lexolo gy.com/library/detail.aspx?g=a982e233-6d31-452c-b962-b9353ccc656d>.

101 Brown and Patalong, n 88.

102 Ministry of Energy, Industry and Mineral Resources, Kingdom of Saudi Arabia, "Dumat Al Jandal Wind Project Beats Record Low Price for Onshore Wind Power", 8 August 2019 <www.powersaudiarabia.com.sa/>.

103 Round Two involved the following projects: the 50MW Madinah Solar PV IPP, the 45MW Rafha Solar PV IPP, the 200MW Qurayyat Solar PV IPP, the 600MW Al Faisaliah Solar PV IPP, the 300MW Rabigh Solar PV IPP, the 300MW Jeddah Solar PV IPP, and the 20MW Mahd AlDahab Solar PV IPP.

104 A further six projects are due to be tendered in 2020, bringing an additional 1.58 GW of renewable capacity (Brown and Patalong, n 88).

105 Compared to 42 companies which qualified for the Round One projects.

106 The smaller-scale projects are designed to create greater opportunities for local companies to participate.

107 Brown and Patalong, n 88.

108 Ibid.

109 Ministry of Energy, Kingdom of Saudi Arabia, "Saudi Arabia Invites Bids for Round Three of the National Renewable Energy Program" National Renewable Energy Program – Press Release (8 April 2020) <www.powersaudiarabia.com.sa>.

110 Ibid.

111 Ibid.

112 The Economist, "Profits Fall Sharply at Saudi Aramco, the World's Biggest Oil Firm" (12 August 2020) <www.economist.com/business/2020/08/12/profits-fall-sharply-a t-saudi-aramco-the-worlds-biggest-oil-firm?utm_campaign=the-economist-today &utm_medium=newsletter&utm_source=salesforce-marketing-cloud&utm_term =2020-08-12&utm_content=article-link-2>.

113 DLA Piper, "Improving Public Private Partnerships: Lessons from Australia" (29 May 2020) <www.dlapiper.com/en/australia/insights/publications/2020/05/impro ving-public-private-partnerships---lessons-from-australia/>.
114 A. Hammad, "Project and Construction Documentation and Transactional Structures in Saudi Arabia", Hammad & Al-Mehdar Law Firm (23 July 2019) <www.l exology.com/library/detail.aspx?g=c59b0ce7-ae30-4eea-90cf-28c306f16df1>.
115 Patalong, n 72.
116 Ibid.
117 FIDIC forms of contract often used in Saudi PPP (renewable energy) projects include: Conditions of Contract for Construction for Building and Engineering Works Designed by the Employer ("Red book"), Second edition 2017; Conditions of Contract for Plant & Design-Build for Electrical & Mechanical Plant & for Building & Engineering Works Designed by the Contractor ("Yellow book"), Second edition 2017; Conditions of Contract for EPC Turnkey Projects ("Silver book"), Second Edition, 2017); Client/Consultant Model Services Agreement ("White book"), Fifth Edition 2017; Conditions of Contract for Design, Build and Operate Projects ("Gold book") First Edition 2008; The Short Form of Contract ("Green book"), First Edition 1999; and Conditions of Contract for Underground Works ("Emerald book"), 2019 Edition <www.fidic.org>.
118 Ministry of Energy, Kingdom of Saudi Arabia, n 109.
119 Renewables Now, "MENA with 4.2 GW of Solar under Construction in 2019 – MESIA" (21 January 2020) <https://renewablesnow.com/news/mena-with-42-gw-of-so-lar-under-construction-in-2019-mesia-684298/>.
120 Olawuyi, n 34, 540.
121 Olawuyi, n 47, 368–370.
122 Ministry of Energy, Kingdom of Saudi Arabia, n 109.
123 Brown and Patalong, n 88.
124 Ernst & Young, "Renewable Energy Country Attractiveness Index (RECAI)" Issue 55 (May 2020) <www.ey.com/en_gl/recai#:~:text=The%20Renewable%20Energy %20Country%20Attractiveness,attractiveness%20and%20global%20market%20t rends>.
125 J. Blazquez, L. Hunt, and B. Monzano, "Oil Subsidies and Renewable Energy in Saudi Arabia: A General Equilibrium Approach" (2020) 38 *The Energy Journal* 29–32.
126 Organisation for Economic Co-operation and Development (OECD), "Blockchain Technologies as a Digital Enabler for Sustainable Infrastructure," OECD Environment Policy Paper No. 16, 2019 <www.oecd-ilibrary.org/docserver/0ec26947-en.pdf ?expires=1599820684&id=id&accname=guest&checksum=D34C5086D3F8806D DD6087200150483D> 9.

Part III

Lessons learned and future directions

14 Tackling the legally disruptive problem of climate change with disruptive legal education

Hilary Christina Bell

Introduction

This chapter aims to examine innovative pedagogy to educate future climate change actors for evidence-based policymaking in the ever dynamic area of climate change law and policy. It discusses the importance of technology-driven and enquiry-based learning approaches as critical tools to educate future climate change practitioners and policy leaders effectively.

Legal practice relating to climate change has accelerated over the last decade, resulting in climate law emerging from the umbrella of environmental law, as a discrete area demanding action.[1] Complex and rapidly changing scientific understanding, as well as political, moral, and economic considerations, inform legal responses to climate change.[2] The dynamical complexities and the associated social issues make climate change a "super-wicked" problem.[3] Moreover, climate change is also legally disruptive, as a conventional approach and existing legal doctrine are inadequate to address it.[4] For example, the causes and effects of climate change are polycentric and uncertain, and the predicted effects will cause a fundamental upheaval of social and economic orders.[5] Similarly, although climate law practice will continue to grow, it is still a nascent discipline, and codification is at best a moving target.[6] Climate change actors and policymakers must wrestle with a multiplicity of factors to develop holistic and coherent bodies of law and policies that respond to the dynamic nature of climate change and its associated mitigation and adaptation approaches.[7]

Evidence-based policymaking (EBP) is central to establishing a new and effective approach for addressing complex and interdisciplinary problems such as climate change. EBP refers to policymaking based on hard evidence, rigorous research, and dispassionate analysis.[8] Holistic and informed law and policy action is needed at international, national, and local levels to address climate change.[9] By providing a rational, rigorous, and systematic approach to policymaking, EBP offers enormous potential for change in developing countries.[10] EBP connects research with practical societal relevance, grounding policies in available evidence and rational analysis, rather than merely promoting an ideology.[11] EBP requires actors to be adept in adapting quickly to changing environments, understanding evolving systems, and envisioning different solutions and future scenarios.[12]

However, climate change education remains an under-researched area in the MENA region. Notably lacking are recommendations on how to embed the topic in curricula.[13] Conventional approaches to legal education, namely case-based teaching and the lecture and seminar format, are deemed inadequate to prepare future climate lawyers to take on the layered and contentious subject matter of climate change.[14] Instead, a more interactive approach to teaching is recommended to foster critical and independent thinking.[15] Enquiry-based learning engages students in higher-order thinking and develops their critical thinking skills.[16] Education technology (EdTech), or disruptive education, offers an efficient means through which students can undertake comprehensive enquiry-based learning, supported by technology.[17] EdTech can support enquiry-based learning in both traditional face-to-face courses and online courses.

Preparing future climate actors necessitates learning outcomes that go further than merely transferring knowledge of the current, inadequate climate law regimes; educators must develop learners' higher-order thinking to prepare competent, evidence-based policymakers.[18] Integrating enquiry-based learning approaches and technologies into training programs and courses on climate change can prepare future climate actors for the challenges they will face.[19] Educators must focus on developing students' critical reasoning skills, enabling them to synthesize solutions and draft policy responses to novel challenges. In short, climate law courses must be designed to introduce the "wicked" problem of climate change, and simultaneously teach students to be creative, resourceful, and ingenious.

Current literature demonstrates that the exigencies of climate change require regulation in the MENA region,[20] that EBP is the best approach to policymaking in developing countries,[21] and that education in the region is failing to integrate the use of EdTech fully.[22] This chapter fills a gap in the existing literature by examining how MENA educators can adopt disruptive education to prepare climate law actors for these epistemic challenges.

After this introduction, the second section discusses why the breadth and rapidly evolving nature of climate law makes enquiry-based learning preferable to a more passive form of instruction. This discussion demonstrates why using disruptive education technology is the recommended approach to provide a deep understanding of the challenges of climate change, and the necessary skills to be a climate law actor. Additionally, this section considers the learning requirements of utilizing disruptive education to teach climate law through enquiry-based learning. The third section examines the practical challenges MENA educators will face when designing climate courses using enquiry-based learning. The fourth section proposes how educators can navigate these challenges by careful course design and deliberate facilitation. This discussion explores the essential scaffolding requirements of enquiry-based learning and how this optimizes student learning and motivation. Finally, it proposes polycentric educational collaborations to address the particular challenges educators face in the region. The final section is the conclusion.

Developing future climate actors' skills through EdTech supported enquiry-based learning

EdTech supported enquiry-based learning teaches students to design solutions to complex real-world problems, making it the preferable learning device to prepare climate law actors with relevant skills.[23] Disruptive education refers to the practice of breaking from existing education models by using education technology resources (EdTech) to enhance learning and teaching.[24] Disruptive education can revolutionize learning experiences[25] by allowing educators to "harness the potential of digital tools to accelerate the instructional quality and student learning productivity".[26] To properly understand the role of EdTech, it is essential to know that EdTech does not take on any role as an educator. The true potential of disruptive education lies in changing the way learning is achieved.[27] In the context of teaching climate change law and policy, it requires instructors to have students use online education platforms to achieve self-directed learning, rather than take up their traditional role of providing information to passive students.[28] Enquiry-based learning requires students to undertake enquiry or research to solve a problem, gaining a deep understanding of the subject in the process.[29] By using EdTech to conduct enquiry-based learning, the transformative potential of EdTech is realized.

As noted earlier, the rapidly evolving nature of climate law necessitates that climate actors be adaptable in the face of emerging situations. The challenges faced by climate law actors are different from those faced in traditional legal practice, so educators need to develop skills that match the needs of climate actors. Enquiry-based learning is preferable to passive instruction for climate actors because it prepares them to be skilled problem solvers. Enquiry-based learning was developed in the 1960s to replace passive learning with the active engagement of students in their learning process.[30] The process of enquiry-based learning is flexible, incorporating the following broad elements: problem identification; enquiry design; investigation; drawing conclusions; and presenting the findings of the enquiry.[31] A key benefit of enquiry-based learning is the learner can direct the process, empowering students to determine the scope of the problem they work on, and the lines of enquiry they pursue, enhancing student autonomy and motivation.[32] As the lines of enquiry do not follow a pre-determined pathway, they can result in new ways of looking at things. They can potentially identify novel solutions to "wicked" problems, making it the optimal approach to prepare future climate actors for the challenges they will face.[33]

Implementing enquiry-based learning in a climate law course

Enquiry-based learning can be implemented after the students are introduced to the context and pivotal concepts of climate law, including the scientific causes of climate change and the economic and social effects of climate change, regionally. These concepts provide sufficient context for the students to ground their understanding of the objectives, principles, and instruments of climate law.[34] Due to the scale of the area, educators are cautioned against attempting to cover the legal doctrine and technical detail of climate law exhaustively.[35] Arguably, singling out a discrete aspect of climate law avoids the pedagogical challenges of the breadth of the subject and allows a detailed consideration of the selected issue.[36] Educators can target numerous narrow areas of substantive climate law as the focus of enquiry-based learning, e.g., climate mitigation, adaptation strategies, nationally determined contributions, and governance provisions under the Paris Agreement.[37] Alternatively, renewable energy contracting, infrastructure project finance, or corporate climate risk disclosure.[38]

Having selected a specific area of climate law instructors can guide students through guided enquiry exercises in the selected area, familiarizing the students with the process of enquiry. Further, allowing students to contextualize the new information within the general context of their introductory instruction. Moreover, the instructor should assist students' developing understanding at this stage by helping them to contextualize the new information about climate law, within their existing knowledge of related areas of law, such as administrative law and international law. This should give the students enough knowledge to identify a specific problem and conduct further research to design the enquiry they will undertake.

Involving students in designing learning goals, enquiry strategies, and monitoring progress, improves academic achievement and student persistence.[39] Students empowered with autonomy enjoy increased well-being; this is especially important for law students who traditionally experience disproportionately high levels of stress and anxiety, compared to students from other disciplines.[40] Instructors can modify the amount of structure provided to best support each student, from structured enquiry, to guided enquiry, to open enquiry for more proficient learners.[41] Student-paced learning moves the locus of control and the cognitive load to the students, thereby maximizing both their potential and their motivation.[42]

Using EdTech to enhance enquiry-based learning

EdTech provides an optimal platform for implementing enquiry-based learning for three key reasons: the "gatekeeper" role of the instructor is removed as the learner is independent rather than reliant upon the instructor;[43] vast online resources facilitate access to a greater volume of information;[44] and EdTech learning management systems facilitate greater student engagement, enhancing their learning experience and motivation.

Timely, consistent feedback is essential to facilitate the process of enquiry-based learning, helping to develop students' skills.[45] Appropriate feedback reminds the students of their learning objective, indicates how close the work is to the objective, and directs students to tasks and strategies to adopt in order to achieve their learning objective.[46] Feedback needs to be given at each of the key stages of implementation (problem design, enquiry design, investigating solutions, drawing conclusions, and presenting the findings). Feedback is most constructive when it is regular and timely, giving students' opportunities to act upon it while they are working on their project.[47] EdTech allows teachers and learners to engage asynchronously; as a result, instructors can give written feedback without the time demands of arranging regular face-to-face meetings with all students.[48] The time saved can be invested in providing personalized, tailored instruction to students. Students can give feedback to the instructor about their learning activities, through reflection journals that critically assess their enquiry process and their learning progress.[49] Guiding the students in this way creates a deeper understanding of complex problems and allows them to process a greater complexity of information.[50] With the adequate scaffolding of timely feedback and structured self-reflection[51] students develop a metacognitive awareness of their enquiry strategies, embedding their skills.[52]

EdTech can enhance student engagement in the enquiry-based learning process through online discussion forums. Online discussion forums let students demonstrate the progress of their enquiry and benefit from peer learning. Research shows that students engage more in online discussions than in-class discussions; student-to-student participation increases throughout the course, and the majority of contributions are constructive and collaborative.[53] Online discussions held at the outset of the course encourage early peer interaction and foster a collaborative environment. Moreover, early discussions allow the students and instructor to familiarize themselves with the EdTech platform. Online facilitation requires the instructor to move away from personalized communications towards communicating with the group, engaging all the students in communications.[54] Actively encouraging knowledge sharing facilitates understanding of critical concepts, skill development, and peer learning.[55] Active moderations from instructors will include relating the connections between discussion statements and the crucial concepts, enabling students to develop their ideas.[56] By adding alternative points of view and further sources of evidence to the discussion, instructors can make discussions more engaging, and raise the level of learning.[57] Asynchronous discussions mean students have time to reflect; introverts, who may otherwise have remained silent, contribute making for a richer discussion with more points of view.[58] With no verbal and non-verbal clues to test understanding in online discussions, instructors must purposefully engage in communications with students online.[59]

Peer learning results in meaningful learning and involves cooperation, communication, and giving and receiving feedback.[60] EdTech learning platforms allow for students to engage in meaningful peer interactions, helping to develop enquiry skills, through in-depth peer discussions outside of class, and allowing students to evaluate each other's work and construct knowledge collaboratively.[61] Peer review can be facilitated through the EdTech learning platform by requiring learners to critique others' work at assigned stages throughout the course. Ensuring students' understanding of the importance of this task can be done by a formal assessment as students equate formal assessment with importance.[62] Peer assessment and self-reflection enhance students learning, motivation, foster creativity, and place responsibility on the students for their learning.[63]

Option to integrate service learning into enquiry-based learning

Integrating service learning in legal education can profoundly impact students and promote leadership skills.[64] Additionally, the emotional benefits of participating in service-based learning can shape students' political attitudes,[65] potentially instilling a lifelong commitment to climate law practice. Vermont Law School (VLS) has very successfully taught climate law through service-based learning.[66] During the semester-long course, students experience working onsite at the annual Conference of the Parties (COP) of the United Nations Framework Convention on Climate Change (UNFCCC) by attending the negotiation with delegates from their service-learning partner.[67] Uniting formal knowledge of climate law and the experience of its practice, service-based learning benefits students and society by helping partners who cannot otherwise access legal support.[68] A modified programme could be implemented by MENA law schools, and provide the students with a formative experience. In order to incorporate service-based learning, a service partner could be identified as part of the problem identification stage. Thereafter, the enquiry design would be directed by the needs of the service partner.

Leveraging EdTech supported enquiry-based learning for climate education in the MENA region: limitations and challenges

The changes in climate will have extreme consequences for the region. There is, therefore, an urgent need to address this with specific legislation.[69] Appropriate policies and regulatory frameworks are needed to address mitigation (reducing emissions) and adaptation (steps to reduce the impact of temperature increases) through national policies.[70] Implementing corresponding legal frameworks and governance systems are necessary to support climate policies.[71] The region faces extreme exigencies[72] and urgently requires climate lawyers and leaders to manage the multifaceted issues of climate change.

MENA governments, judiciaries, private corporations, interest groups, law firms, and legal consultancies need climate law expertise, and this demand is growing.[73] Understanding the central concepts of climate change law, and having the skills to design achievable solutions is a crucial learning outcome for MENA climate educators. Presently, legal educators in the region face unique challenges, including outdated teaching methods, a failure to teach relevant skills, financial constraints, the language of instruction, and conflict.

Slow uptake of innovative teaching methods at all levels of education

While there have been recent efforts spearheaded by the United Nations Environment Programme (UNEP) and the Association of Environmental Law Lecturers in the Middle East and North African Universities (ASSELLMU) to integrate climate change law into the curriculum in MENA universities, the region is still failing to achieve its education capacity, and greater action is required to realize the potential of its youth.[74] Students enter higher education in the MENA without crucial life skills.[75] Critical learning skills are missing from primary and secondary education in the region, particularly creativity, critical thinking, and problem-solving.[76] Despite decades of high investment in education, the quality of education remains low, attributed partly to an over-reliance on rote and passive learning at the expense of enquiry.[77] Anecdotal evidence suggests that often legal education in the region relies on rote learning of the relevant civil codes. Further, postgraduate skills training may be too formalistic and does not encourage innovation.[78] Lawyers trained this way may lack the critical faculties required to innovate, and will tend to follow rote methods of thinking, struggling to innovate and design creative solutions.

MENA educators, several of whom may have spent their careers operating in a passive learning environment, will have to make a paradigm shift to adapt to the demands of enquiry-based learning. Additionally, many MENA students' limited exposure to active learning will make for a steep learning curve as they take on enquiry-based learning, which requires them to be active learners. Students will likely experience discomfort stepping into this new learning experience as enquiry-based learning requires students to contextualise concepts through questions and experimentation.[79]

Weak focus on skills training in education in the MENA region

There is currently a weak tradition of tailoring course design to develop skills that match the needs of the labour market across the MENA region.[80] Educators have experienced little incentive to ensure credentials provide graduates with skills.[81] Historically, in many MENA countries, public sector employment was guaranteed for anyone with the required education credentials.[82] For example, in Egypt, there is a strong preference for working in the public sector, motivated by an imbalanced labour market and the prestige associated with a government job.[83] A legacy of "credentialist equilibrium" resulted, where education has been pursued merely as a step towards public sector employment with little or no link between credentials and skills.[84]

There is a large pool of unemployed graduates in the region, up to 40 per cent in some countries, partly attributed to a mismatch between graduates' skills and market needs.[85] Very high unemployment among university graduates in the Maghreb is one of the main reasons for the increasing migration of higher-skilled individuals.[86] Faculties in the region have the opportunity to adapt course design to develop skills that are relevant to the region's market needs.[87]

Financial constraints in the MENA region

Due to immense wealth and income disparities among MENA countries, some have described the MENA region as the most unequal in the world.[88] Economic disparities contribute to significant variations in quality across education in MENA higher education institutions.[89] Some MENA countries can neither afford to attract and retain the best academics that are skilled in EdTech, and nor can they match student preferences or fields of national relevance for economic development.[90] Moreover, some MENA countries, especially those outside of the Gulf Region, can hardly afford to invest in information and communications technology (ICT) which is expensive to purchase and maintain.[91]

Resource accessibility challenges

Teaching climate law courses requires teaching resources, and while repositories of climate law teaching materials are available, including dedicated journals, textbooks, and course syllabi,[92] the majority of materials are in English.[93] The principal language of instruction in the region is Modern Standard Arabic (MSA).[94] As a result, MENA academics often struggle to access the lion's share of available climate law teaching resources, raising the need for more translation of resources, as well as designing learning resources and platforms in Arabic, to provide greater access to learning resources on climate change law to MENA academics and students.

Access to education in countries affected by conflict

The region has been greatly affected by political instability and conflicts in its recent history. In recent years, armed conflicts have occurred in Egypt, Israel and Palestine/OPT, Iraq, occupations of Lebanon by Israel, Libya, Syria, and Yemen; several are ongoing.[95] Conflict and instability negatively impact education and pose a tremendous challenge to the educators and students affected.[96] In many cases, students are unable to access face-to-face education and educational resources safely. An obvious benefit of EdTech platforms is the elimination of geographical restrictions; regardless of location, anyone can take part. Online education platforms could offer a lifeline to students in developing countries, particularly those blighted by conflict, facing barriers to quality education. Technology requires a capable network to handle the demands, and if students in countries affected by conflict can get online, they can potentially access education.[97]

Preparing MENA climate actors through EdTech supported enquiry-based learning

Disruptive technology is changing the provision of legal services.[98] If MENA educators adopt new teaching methods using EdTech, it is possible to leapfrog ahead of more advanced educations systems elsewhere.[99] The following course design elements can address the unique challenges faced by climate law educators in the region.

Updating teaching methods

MENA educators can supplement doctrinal instruction on the narrow issue of climate law, with disruptive education methodologies, simultaneously developing students' skills and understanding. Introducing students to enquiry-based learning by directing them to find and read primary legal materials eases them into this new learning method; for example, by asking students to find and read the UNFCCC, Kyoto Protocol, and Paris Agreement, students get an uncomplicated introduction to structured enquiry-based learning while familiarizing themselves with crucial legal instruments. An excellent example of the successful implementation of this approach is the United Nations online course Introduction to Environmental Governance.[100] Building on this experience, students can move on to a guided enquiry exercise to find a variety of soft law.[101]

Holding online class discussions from the outset of the course allows the instructor to ensure that students grasp the essential concepts. Online, asynchronous class discussions by instructors could focus on relating the assigned legal instruments to students' existing knowledge of international law and administrative law; exploring the differences between the vital international treaties; ensuring students appreciate the practical impact of national commitments made by MENA countries in the Paris Agreement; or exploring the nature and role of soft law.

Having begun to contextualize the complex nature of climate law, within their current understanding, students could then complete an individual guided enquiry assignment, researching the existing domestic climate regime. Students can post their presentations to the online platform; these could be video presentations, audio presentations, or slides. Students can then critique each other's work. This feedback provides meaningful learning and allows the students to construct knowledge collaboratively.

Skills-based training to develop future climate actors

As addressing the climate threat in the region is a tangible necessity, rather than an ideological goal, evidence-based policymaking is the best basis for effective climate policies in the region. Policy processes tend to be centralized in developing countries.[102] Moreover, the process tends to be inherently political.[103] Students need to gain an understanding of the challenges of domestic policy implementation. To introduce students to the nature of evidence-based policymaking and domestic policy implementation, they could carry out some guided research. This research should be guided by the instructor to consider the value of the evidence as the basis of effective policymaking, how to research what evidence is available, and how to access and critically evaluate this evidence.[104] Based on their research, teams could upload their presentations to the online platform, for further peer critique and an online discussion.

For their capstone project, students can conduct an open enquiry into their domestic climate regime; first identifying a gap in the current regime and after that proposing a detailed legal solution. Students can work in teams to tackle this complex problem. Having the students focus on how to improve a real problem develops innovative thinking. Instructors can determine how wide a scope the students have in choosing their project. New instructors may be more comfortable assigning an area of study with which they are familiar. Engaging students in designing their learning goals in this way motivates the students, so ideally students would be allowed to pinpoint a narrow issue of climate law that appeals to them. Students should propose a defined work product that they will create, e.g., draft legislation or a draft policy. Each team should design the assessment rubric for their design project, based upon the course learning outcomes. Instructors should provide detailed feedback to the students about their proposal. Once agreed, teams should share their final assessment rubric on the online learning platform, so the other teams have access, enabling peer feedback.

As part of their capstone project, students can identify and recruit partners to work with; this could include regional and international stakeholders, for example, the Arab Youth Climate Movement (AYCM).[105] Students will develop their knowledge through these contacts, and this will help deepen their understanding of their area of focus. Additionally, these partnerships will help the students to build a valuable network of contacts.

Students could research a variety of climate law actors to understand the nature of their work; initially conducting research online. Students should be encouraged to organize interviews with local, domestic, or international climate actors. Understanding the realities of climate practice will give the students valuable insight into the challenging nature of the work. Some students will struggle with the real-world nature of this exercise if accustomed to a passive learning environment. Raising this discomfort in a class discussion and linking this to the importance of adopting a growth mindset can increase their confidence, sense of agency, and their willingness to step out from their cognitive comfort zones.[106]

These course design and facilitation suggestions would fulfil the five central tenets of enquiry-based learning. Namely, working on real problems, encouraging diverse ideas, providing opportunities for collaboration, constructively using authoritative sources, and performing a formative assessment.[107]

Regional collaboration and leveraging shared resources

Polycentric educational collaborations within the MENA region would enable educators to pool their economic and intellectual resources. This type of initiative was in mind when the ASSELMU formed in 2018. The ASSELMU aims to promote collaboration in the region, particularly regarding teaching climate change law in regional law schools.[108] ASSELLMU aims to bridge the significant gap in MENA legal education regarding climate change.[109] ASSELLMU plans to curate an interactive platform for climate scholars in the region to develop, integrate, and mainstream the teaching of climate change law.[110] This is precisely the type of vehicle which can help to revolutionize climate law education regionally.

For poorer countries suffering from financial constraints, cross-border education can bring significant benefits.[111] Cross-border collaborations facilitate academic mobility, allowing those with cutting-edge knowledge and resources to assist in training other regional faculties, enhancing knowledge flow and fostering innovation, towards a common goal.[112] Moreover, it enables students to gain skills aligned with the needs of their local labour market, even when such opportunities are not available locally.[113] Cooperating universities can collaborate in developing internationally relevant curricula and resources in MSA, and in line with their national standards and international best practice.[114] Collaborators could share the costs or labour, of translating resources available in English, or other languages, into MSA.

Cross-border tertiary education takes many forms, including joint educational programmes and courses supplied across-borders through e-learning and distance learning.[115] Theoretically, a law student in a country affected by conflict could enrol in an online climate law course offered in another MENA country. Collaboration on formal degree programmes comes with enormous challenges, including curricula incompatibility, diverse programme structures, and measures of qualification.[116] The challenge is having the course recognized by the student's home institution. The lack of any regional harmonization of accreditation and recognition of credentials hampers the viability of cross-national programme provision.[117] Additionally, disparate admission and progression requirements are a regional reality[118] and can challenge the viability of cross-institution collaborations. Successful models of international education collaborations regarding climate change exist between NYU campuses; as all participants belong to the same institution, there is no recognition problem.[119] This collaboration is an excellent model for how a climate-based collaboration could be carried out if participants can navigate the recognition obstacles.[120]

Climate change moot court competition

Moot court competitions offer an unparalleled opportunity to advance critical legal reasoning and advocacy skills; presently, no climate change competitions exist.[121] Over the last ten years, a successful initiative by the US Department of Commerce and the University of Pittsburgh Law School has provided legal skills training for 650 students from the Middle East, creating a burgeoning moot court movement in the region.[122] A regional climate change moot court competition could present an ideal vehicle to engage students in learning about climate law in the region. Due to travel restrictions precipitated by the COVID-19 outbreak in 2020, the majority of moot competitions moved online.[123] A hybrid moot court competition, where some teams compete online, or an entirely online moot court competition, presents a viable opportunity for MENA law schools to collaborate. An online moot court competition side-steps issues of having credentials recognized across borders, problems associated with conflict, and lack of finances. Therefore, this is a viable manner of facilitating a cross-border learning experience that develops student understanding of climate law and can develop students' critical thinking skills.

Conclusion

Disruptive education could revolutionize education provision in the MENA region, and realize the potential of the region's youth. MENA's youth are key stakeholders in regional climate adaption and mitigation. Securing climate mitigation and adaption measures in the region require adequate policies and legal frameworks. The causes and effects of climate change are complex and evolving quickly. Climate law actors need more than an understanding of current practices; they need skills to address challenges as they emerge and evolve. Adopting enquiry-based learning and regional educational collaborations can resolve many of the challenges facing MENA educators. A paradigm shift in how legal educators deliver learning is required to achieve this. By adopting the course design and facilitation methods recommended herein, MENA educators can empower students with the skills they need to become functional climate actors.

Notes

1 J. Peel, "Climate Change Law: The Emergence of a New Legal Discipline" (2012) 32 *Melbourne University Law Review* 922, 925.
2 D. S. Olawuyi, "Energy Poverty in the Middle East and North African (MENA) Region: Divergent Tales and Future Prospects" in I. Guayo, L. Godden, D. N. Zillman, M. F. Montoya, and J. J. Gonzalez (eds), *Energy Justice and Energy Law* (New York: Oxford University Press, 2020) 259.
3 S. S. Batie, "Wicked Problems and Applied Economics" (2008) 90 *American Journal of Agricultural Economics* 1176.
4 E. Fisher, E. Scotford, and E. Barritt, "The Legally Disruptive Nature of Climate Change" (2017) 80 *The Modern Law Review* 173.
5 Ibid, 174.
6 M. Mehling, H. V. Asselt, K. Kulovesi, and E. Morgera, "Teaching Climate Change Law: Trends, Methods and Outlook" (2020) *Journal of Environmental Law* 1.
7 Ibid, 2–3.
8 K. Bogenschneider and T. J. Corbett, *Evidence-Based Policymaking Insights from Policy-Minded Researchers and Research-Minded Policy* (Milton Park, Abingdon, Oxfordshire: Routledge, 2010), p. 2.
9 C. Wold, D. Hunter, and M. Powers, "Climate Change and the Law: Teacher's Manual", Lexis Nexis, 2013, p. 2, available at <www.lclark.edu/live/files/1822-teachers-manual-introduction-and-model-syllabi>.
10 S. Sutcliffe and J. Court, "Evidence-Based Policymaking: What is it? How Does it Work? What Relevance for Developing Countries?" Overseas Development Institute, November 2005, p. iii, available at <www.odi.org/publications/2804-evidence-based-policymaking-what-it-how-does-it-work-what-relevance-developing-countries>.
11 Ibid, 2.
12 P. Molthan-Hill, N. Worsfold, G. J. Nagy, W. L. Filho, and M. Mifsud, "Climate Change Education for Universities: A Conceptual Framework from an International Study" (2019) 226 *Journal of Cleaner Production* 1092 at 1093.
13 See Chapter 1 of this book.
14 Mehling and others, n 6, 20.
15 Ibid.
16 A. M. Salama and M. J. Crosbie, "Delivering Theory Courses in Architecture: Inquiry-Based, Active, and Experiential Learning Integrated" (2010) 4 (2) *International Journal of Architecture Research* 281; M. Prince and R. Felder, "The Many Faces of Inductive Teaching and Learning" (2007) 36 *Journal of College Science Teaching* 14.
17 D. C. Edelson, D. N. Gordin, and R. D. Pea, "Addressing the Challenges of Inquiry-Based Learning through Technology and Curriculum Design" (2011) *Journal of the Learning Sciences* at 392–393.

18 Ibid.
19 Salama and Crosbie, n 16.
20 Olawuyi, n 2; Atlantic Council, "Why the MENA Region Needs to Better Prepare for Climate Change", May 7, 2019 <www.atlanticcouncil.org/blogs/menasource/why-the-mena-region-needs-to-better-prepare-for-climate-change/>; D. Bloom, "How the Middle East is Suffering on the Front Lines of Climate Change", April 5, 2019 <www.weforum.org/agenda/2019/04/middle-east-front-lines-climate-change-mena/>.
21 Sutcliffe and Court, n 10.
22 S. T. El-Kogali and C. Krafft, *Expectations and Aspirations: A New Framework for Education in the Middle East and North Africa* (Washington, DC: World Bank 2020) 193.
23 S. Magana, *Disruptive Classroom Technologies: A Framework for Innovation in Education* (Thousand Oaks, CA: SAGE Publications, 2017), Preface xxiii.
24 M. Flavin, "Disruptive Technologies in Higher Education" (2012) 20 *Research in Learning Technology* 102.
25 "Disruptive" in this context means that technology has fundamentally changed the way we do things. The last seismic disruption to education was the invention of the printing press, which enabled exponentially higher dissemination of information and today, educational technology (EdTech) offers the same potential. See G. Conole, "MOOCs as Disrupting Technologies: Strategies for Enhancing the Learner Experience and Quality of MOOCs" (2016) 39 *RED-Revista de Educación a Distancia* 2.
26 Magana, n 23, 77.
27 Ibid.
28 Ibid, 6.
29 A. Aditomo, P. Goodyear, A. Bliuc, and R. A. Ellis, "Inquiry-Based Learning in Higher Education: Principal Forms, educational Objectives, and Disciplinary Variations" (2013) 38 *Studies in Higher Education* at 1239 at 1241.
30 J. Schwab, *The Teaching of Science* (Cambridge, MA: Harvard University Press 1962); J. Bruner, *The Process of Education* (Cambridge, MA: Harvard University Press 1960).
31 M. Pedaste, M. Maeots, L. Siiman, T. de Jong, S. A. N. van Riesen, E. T. Kamp, C. C. Manoli, Z. C. Zacharia, and E. Tsourlidaki, "Phases of Inquiry-Based Learning: Definitions and the Inquiry Cycle" (2015) 14 *Educational Research Review* at 49 and 54.
32 Hanover Research, "Impact of Student Choice and Learning Personalized Learning", November 2014, pp. 3 and 7.
33 Magana, n 23, 74.
34 Mehling and others, n 6, 2.
35 Mehling and others, n 6, 24.
36 Ibid, 16 and 24.
37 T. Bach, "Minding the Gap: Teaching International Climate Change Law Through Service Learning" (2016) 18 *Vermont Journal of Environmental Law* 173, 183, 185.
38 Mehling and others, n 6, 13.
39 Hanover Research, n 32, 7.
40 Lack of autonomy in legal studies is associated with the high levels of anxiety and depression in law students, see generally N. Skead and S. L. Rogers, "Stress, Anxiety and Depression in Law Students: How Student Behaviours Affect Student Wellbeing" (2014) 40 (2) *Monash University Law Review* 1–24.
41 C. Kong and Y. Song, "The Impact of a Principle-Based Pedagogical Design on Inquiry-Based Learning in a Seamless Learning Environment in Hong Kong" (2014) 17(2) *Educational Technology & Society* 127, 128.
42 B. Marr, "The Top 5 Tech Trends that Will Disrupt Education in 2020" *Forbes* (20 January 2020) <www.forbes.com/sites/bernardmarr/2020/01/20/the-top-5-tech-trends-that-will-disrupt-education-in-2020the-edtech-innovations-everyone-should-watch/#4978c37a2c5b>.
43 Flavin, n 24, 110.

44 El-Kogali and Krafft, n 22, 201.

45 Kong and Song, n 41, 138.

46 J. Hattie, *Visible Learning: A Synthesis of Over 800 Meta-Analyses Relating to Achievement* (New York: Routledge, 2009); S. Magana and R. J. Marzano, *Enhancing the Art and Science of Teaching with Technology* (Bloomington, IN: Marzano Research 2014).

47 M. Keppell, E. Au, A. Ma, and C. Chan, "Peer Learning and Learning Oriented Assessment in Technology-Enhanced Environments" (2006) 31 *Assessment & Evaluation in Higher Education* 459, citing J. D. Bransford, A. L. Brown, and R. R. Cocking, *How People Learn: Brain, Mind, Experience, and School* (Washington, DC: National Academic Press 2000).

48 Hanover Research, n 32, 3 and 7.

49 Keppell and others, n 47, 456–457.

50 X. Liu, C. J. Bonk, R. J. Magjuka, S. Lee, and B. Su, "Exploring Four Dimensions of Online Instructor Roles: A Program Level Case Study" (2019) 9 *Online Learning Journal* 29, citing G. Morine-Dershimer, "What's in a Case and What Comes Out?" in J. Colbert, K. Trimble, and P. Desberg (eds), *The Case for Education: Contemporary Approaches to Using the Case Methods* (Boston, MA: Allyn and Bacon, 1996).

51 How to implement structured self-reflection in law students is outside the scope of this work; see T. Casey, "Reflective Practice in Legal Education: The Stages of Reflection" (2014) 20 *Clinical Law Review* 317.

52 Kong and Song, n 41, 138.

53 Faculty Focus, "What Research Tells Us about Online Discussion", March 17, 2017 <www.facultyfocus.com/articles/online-education/research-tells-us-online-discussion/> citing H. Zhou, "A Systematic Review of Empirical Studies on Participants Interactions in Internet-Mediated Discussion Boards as a Course Component in Formal Higher Education Settings" (2015) 19 (3) *Online Learning Journal* 41.

54 Liu and others, n 50, 31 and 38.

55 Z. L. Berge, "Facilitating Computer Conferencing: Recommendations from the Field" (1995) 35 *Educational Technology* 22.

56 Liu and others, n 50, 30–31 and 38, citing G. Salmon, *E-moderating: The Key to Teaching and Learning Online* (London: Kogan Page 2002).

57 Ibid.

58 For further discussion on the essential role, introverts play in contributing to considered discourse, generally see H. K. Brown, *The Introverted Lawyer* (American Bar Association, 2017).

59 Liu and others, n 50, 38.

60 Keppell and others, n 47.

61 Kong and Song, n 41, 127.

62 Keppell and others, n 47, 456–457.

63 Kong and Song, n 41; Keppell and others, n 47, 456 and 462.

64 Bach, n 37, 182, citing J. K. Jones and A. L. Mereau, "Community Service and Service Learning", State University New York College of Environmental Science.

65 Ibid, citing M. F. Davis, "Access to Justice: The Transformative Potential of Pro Bono Work" (2004) 73 *Fordham Law Review* 904.

66 See generally Bach, n 37.

67 Bach, n 37, 184.

68 W. M. Sullivan, A. Colby, J. Welch Wagner, L. Bond, and L. S. Schulman, *Educating Lawyers: Preparation for the Profession of Law* (1st ed.) (San Francisco, CA: Jossey-Bass 2007), 12 and 197.

69 Olawuyi, n 2, 264.

70 Middle East Institute, "Implications of Climate Change on Energy and Security in the MENA Region", February 22, 2012 <www.mei.edu/publications/implications-climate-change-energy-and-security-mena-region>.

71 Olawuyi, n 2, 260.
72 See generally, T. Twining-Ward and K. Khoday, "Climate Change Adaption in the Arab States", United Nations Development Programme, July 2018.
73 Mehling and others, n 6, 2.
74 World Bank, "A New Education Approach is Needed to Prepare MENA Youth to Shape the Future", November 13, 2018 <www.worldbank.org/en/news/press-release/2018/11/11/a-new-education-approach-is-needed-to-prepare-mena-youth-to-shape-the-future>; El-Kogali and Krafft, n 22, 1.
75 Twelve core life skills are identified that need to be embedded in education in the MENA region, divided under four themes: learning, employability, personal development, and active citizenship. See United Nations Children's Fund, "Life Skills and Citizenship Education Initiative Middle East and North Africa Reimagining Life Skills and Citizenship Education in the Middle East and North Africa: A Four-Dimensional and Systems Approach to 21st Century Skills Conceptual and Programmatic Framework", UNICEF MENA Regional Office, 2017.
76 Ibid, 148.
77 El-Kogali and Krafft, n 22, 7 and 12.
78 International Legal Center New York, "Legal Education in a Changing World: Report of the Committee on Legal Education in the Developing Countries", Scandinavian Institute of African Studies 1975) 49.
79 The foundations of rote learning in the region stem from an oral tradition among Arabs that predates Islam, and has been used to preserve and spread Islamic teachings (El-Kogali, n 22, 7 and 12).
80 A. Jaramillo, A. Ruby, F. Henard, and H. Zaafrane, "Internationalisation of Higher Education in MENA: Policy Issues Associated with Skills Formation and Mobility", World Bank, 2011, 12, available at <https://openknowledge.worldbank.org/handle/10986/19461>.
81 El-Kogali and Krafft, n 22, 7.
82 Ibid.
83 El-Kogali and Krafft, n 22, 4, citing G. Barsoum, "Young People's Job Aspirations in Egypt and the Continued Preference for a Government Job" in R. Assaad and C. Krafft, *The Egyptian Labor Market in an Era of Revolution* (Oxford Scholarship Online 2015).
84 El-Kogali and Krafft, n 22, 6.
85 Jaramillo and others, n 80, 12.
86 J. A. Sabadie, J. Avato, U. Bardak, F. Panzica, and N. Popova, *Migration and Skills: The Experience of Migrant Workers from Albania, Egypt, Moldova, and Tunisia* (Washington, DC: World Bank European Training Foundation, 2010), 31, available at <https://documents.worldbank.org/pt/publication/documents-reports/documentdetail/415801468334877989/migration-and-skills-the-experience-of-migrant-workers-from-albania-egypt-moldova-and-tunisia>.
87 Jaramillo and others, n 80, 1.
88 L. Assoud, "Inequalities and Its Discontents in the Middle East", Carnegie Mellon Middle East Centre, March 12, 2020 <https://carnegie-mec.org/2020/03/12/inequality-and-its-discontents-in-middle-east-pub-81266>.
89 S. Al-Agtash and L. Khadra, "Internationalization Context of Arabia Higher Education" (2019) 8 *International Journal of Higher Education* 68, 69.
90 Jaramillo and others, n 80, 14–15.
91 Notably underserved in ICT are Iraq, Yemen, Libya, Djibouti, Comoros, and Somalia (Al-Agtash and Khadra, n 89, 71).
92 IUCN Academy of Environmental Law, "Climate Law Teaching Resources" <https://www.iucnael.org/en/online-resources/climate-law-teaching-resources>.
93 Mehling and others, n 6, 3.

94 MSA differs substantially from the language spoken daily in MENA countries, placing students at a linguistic disadvantage as they are effectively learning in a second language (El-Kogali and Krafft, n 22, 25).

95 BIIC and PEIC, "Protecting Education in the Middle East and North Africa Region", British Institute of International and Comparative Law and Protect Education in Insecurity and Conflict, June 2016.

96 Ibid, 13–14.

97 Marr, n 42.

98 Womble Boyd Dickinson, "Innovation and the Disruptive Impact of Technology on the Legal Sector", March 27, 2019 <https://womblebonddickinson.com/uk/insights/articles-and-briefings/innovation-and-disruptive-impact-tachnology-legal-sector>; Deloitte, "Objections Overruled: The Case for Disruptive Technology in the Legal Profession", 2017 < https://www2.deloitte.com/content/dam/Deloitte/uk/Documents/corporate-finance/deloitte-uk-technology-in-law-firms.pdf>.

99 El-Kogali and Krafft, n 22, 36.

100 See UNEP <https://web.unep.org/unepmap/introduction-environmental-governance-free-online-course-informea>.

101 Bach, n 37, 185.

102 Sutcliffe and Court, n 15, 11, citing J. W. Thomas and M. S. Grindle, "After the Decision: Implementing Policy Reforms in Developing Countries" (1990) 18 *World Development* 1163–1181.

103 Sutcliffe and Court, n 15, 12, citing S. Nutley, "Bridging the Policy/Research Divide: Reflections and Lessons from the UK" (2003) Keyn Paper at National Institute of Governance Conference; "Facing the Future: Engaging Stakeholders and Citizens in Developing Public Policy", Canberra, Australia, April 23–24, 2003.

104 Sutcliffe and Court, n 15, iv.

105 Climate Action Network International, "Arab Youth Climate Movement AYCM – Qatar" <www.climatenetwork.org/profile/member/arab-youth-climate-movement-aycm-qatar>.

106 See generally C. Dweck, *Mindset: The New Psychology of Success* (New York: Ballantine Books 2007).

107 These principles were designed for science-based inquiry, Kong and Song, n 41, 130.

108 See Chapter 1 of this book.

109 Ibid.

110 Ibid.

111 Jaramillo and others, n 80, 3.

112 Ibid, 13.

113 Ibid, 14.

114 Al-Agtash and Khadra, n 89, 74.

115 Jaramillo and others, n 80, 3.

116 Al-Agtash and Khadra, n 89 72.

117 Jaramillo and others, n 80, 2–3.

118 Ibid, 23.

119 New York University (NYU) facilitates collaboration between its 11 academic sites across the globe to study coastal urbanization and environmental change, J. A. Burt, M. E. Killilea, and S. Ciprut, "Coastal Urbanisation and Environmental Change: Opportunities for Collaborative Education across a Global Network University" (2019) 26 *Regional Studies in Marine Science* 100501.

120 Ibid.

121 Mehling and others, n 6, 21.

122 The programme has grown from four teams in 2011 to 28 teams in 2020, with students from 17 different countries; see Vis Middle East Pre-Moot Program <http://premoot.bcdr-aaa.org/home/>; Kluwer Arbitration Blog, "If You Want to Know the

Value of Two Weeks Ask the Middle East Vis Pre-Moot Community", (Kluwer Arbitration Blog, 5 April 2020) <http://arbitrationblog.kluwerarbitration.com/2020/04/05/if-you-want-to-know-the-value-of-two-weeks-ask-the-middle-east-vis-pre-moot-community/?doing_wp_cron=1595695613.6840510368347167968750>.
123 The Legal Writing Institute has produced "Guidance for Conducting Moot Court Competitions 2020–2021", for assisting those interested in hosting an online moot. For information contact the Legal Writing Institute <www.lwionline.org/>.

15 Financing climate resilient infrastructure in the MENA region

Potentials and challenges of blockchain technology

Roxana A. Mastor and Ioannis Papageorgiou

Introduction

This chapter examines the importance of blockchain technology as a tool for financing climate-resilient infrastructure in the MENA region. It describes the context and diversity of climate infrastructure needs in the region, and then explores legal imperatives for promoting the use of blockchain technology for meeting those needs.

Infrastructure decisions made today will also be important decades later, as infrastructure is the lifeline of current and future generations.[1] Infrastructure is both a culprit and a victim of climate change, as infrastructure, such as power plants, buildings, and transportation, contribute greatly to the increase of greenhouse gas emissions (GHGs), while also being incredibly vulnerable and at risk to changing climatic conditions. For example, infrastructure damaged or destroyed in a disaster can result in significant increases in carbon emissions, as rebuilding infrastructure is most often a carbon-intensive activity (e.g., rebuilding roads, power plants, etc.).[2] Also, historically, infrastructure systems have been designed, constructed, operated, and maintained according to contemporary climate and weather conditions, without taking into account future climate change and associated uncertainties that increase and intensify hazards.[3]

Every year, trillions of dollars are invested in infrastructure unable to withstand climate change risks, weather extremes, or natural disasters, as well as in infrastructure that perpetuates the climate crisis. It is estimated that from 2020 to 2030 global demand for new infrastructure could amount to $60 trillion,[4] while making infrastructure more climate-resilient would add three per cent to the upfront costs.[5] As infrastructure investments have an economic life expectancy of 30 years or more, it is crucial that climatic conditions and future variations are considered during their construction. Thus, building new infrastructure that is more climate-resilient makes sound economic sense; there is no resilient economy without a resilient infrastructure.

Nevertheless, financing climate-resilient infrastructure is not an easy endeavour. Infrastructure worldwide has suffered from chronic underinvestment for decades, and current global investments fall short on bridging this gap. Traditionally, governments have been the main source for infrastructure financing and management with limited private sector involvement.[6] However, the high level of investment necessary for infrastructure worldwide and the high cost of inadequate infrastructure, coupled with (for some countries) increased urbanization, political instability, conflicts, the refugee and internally displaced people crisis, and disasters have triggered many governments to mobilize private sector engagements and investments, through mechanisms such as public–private partnerships (PPPs), among others.[7]

Like the rest of the world, the MENA region is facing an infrastructure gap as well as large fiscal constraints. It is estimated that the infrastructure needs in the region are around $100 billion annually.[8] Unlike other regions of the world, the MENA region is exposed to extensive and diverse shocks and hazards that affect its growth and stability. The interplay of population density and growth, rapid urbanization, conflict, water scarcity, and climate change has intensified the need for resilient infrastructure able to withhold both conflicts and climatic disasters. The complex setting of the MENA region means that financing climate-resilient infrastructure requires strong partnerships, high investments, as well as diverse sources of capital.

The chapter is divided into five sections. After this introduction, the next section describes the context and diversity of infrastructure needs and investments in the MENA region and discusses the challenges in meeting these needs with traditional financing, focusing in particular on PPPs. The third section then makes the case for innovative financial instruments, supported by digitalization (i.e. blockchain technology) which can create the opportunity to address the barriers and challenges of old financing mechanisms such as PPPs. This section also showcases how the new technologies put forward by the Fourth Industrial Revolution (4IR) can play a significant role in MENA efforts to build climate resilient infrastructure and a sustainable economy. The fourth section examines legal and institutional challenges that will need to be addressed in order to effectively leverage blockchain technology solutions to advance climate infrastructure investments in the region. The final section concludes that the high infrastructure climate-resilient investment gap in the MENA region can be addressed by using blockchain technology as a technological enabler for upgrading old financing partnerships such as PPPs.

Promoting climate-resilient infrastructure investments in the MENA region

MENA infrastructure resilience challenges and needs

The rise in global population continues to place enormous pressure on infrastructure and cities worldwide. One of the regions that witnessed the highest rate of population growth in the world over the past century, increasing from around 100 million in 1950 to around 380 million in 2000, is the MENA region.

Currently, its population is at least 450 million or six per cent of the world population, thus more than doubling in size in the past 40 years.[9] According to the Food and Agriculture Organization of the United Nations, by 2050, the population is expected to reach 600 million.[10] Sixty-five per cent of the region's inhabitants live in cities (more than 280 million people), while 92 per cent are concentrated on three per cent of the land. City populations are expected to nearly double by 2040 as population growth in the rural areas determines potential rural-to-urban migrants, amplifying the residents' exposure and challenges when it comes to disasters, limited resources, inadequate provision of services and infrastructure.[11]

Furthermore, conflict and civil unrest have affected many countries in the MENA region over the years. This has inflicted great damage to human lives and the infrastructure systems they rely on, mostly felt in urban areas, as most of the forcibly displaced people from the region now live in cities and towns. This puts significant pressure on some cities where the sudden population increase leads to development of informal settlements.[12]

The MENA region is also water scarce, being home to only one per cent of freshwater resources.[13] For its domestic use, the region is increasingly reliant on desalinated water, which accounts for almost half of the world's desalination capacity. This also adds to its carbon footprint, as the thermal desalination plants are highly energy intensive.[14] Moreover, many of the region's largest cities and urban economies are at risk of rising waters, being located in fragile coastal zones which are prone to flooding. Both climate-related water scarcity and the threat of rising sea levels could cost their GDP significantly.[15] Additionally, some areas in the MENA region are prone to the occurrence of earthquakes, which can leave many human settlements severely impacted, as many substandard developed buildings cannot resist the strong seismic activities.[16]

Climate-resilient infrastructure in the MENA region could also help address some of the overconsumption and inefficiency when it comes to energy systems.[17] MENA countries are, on average, "50% more energy-intensive than the world average per unit of GDP".[18] This is due to several factors, such as heavily subsidized energy, limited competition in the state-owned electricity sector supply chain, lack of awareness of energy conservation and energy efficiency, as well as limited adoption of energy-efficient building codes and other regulations,[19] to name a few. The most energy-intense sectors in the MENA region are the building and transportation system, where practices such as lack of insulation, poor material and construction standards, lack of public transportation, and predominance of privately owned vehicles adds to the wasteful energy practices.[20] However, these two sectors, together with the power sector, have the highest potential to control their carbon dioxide (CO_2) emissions, helping the MENA region to reduce its GHGs to meet its National Determined Contributions (NDCs) targets in accordance with the Paris Agreement commitments.[21] In terms of climate change impacts, the MENA region is one of the most vulnerable in the world, registering the highest number of natural disasters since the 1980s, with "more than 370 natural disasters, affecting approximately 40 million people in MENA countries" and costing "their economies about US$20 billion".[22] The region is also expected to "face the greatest increase in temperatures worldwide, projected to rise by 2°C by 2030".[23]

MENA region infrastructure has been weakened by the interplay between population density, high levels of urbanization, conflict, water scarcity, flooding, earthquakes, energy inefficiency and overconsumption, and other related climate change events. These factors have led to infrastructure fragility and an increasing need for high quality and quantity of infrastructure.[24] The majority of the infrastructure needs are in "electricity generation and transportation, followed by water and sanitation, and information and communication technologies".[25] The infrastructure needs for the electricity and transport sector are each estimated at about "43 percent of total infrastructure needs in MENA, followed by [information and communication technologies] (9 percent) and water and sanitation (5 percent)".[26] It is estimated that the region will need around $100 billion annually in investments.[27] In order to achieve this annual spending, the MENA countries will need to use, on average, seven per cent of their GDP per annum.[28] For example the Arab Countries in the MENA region will need 10 per cent of their GDP per annum, compared with the current spending of only five per cent.[29] However, when it comes to adapting or addressing the financial challenges of infrastructure development, the MENA countries do not start off an equal footing. Being one of the most economically varied regions in the world, there is a wide disparity in wealth and income across the region. For example, the "GDP per capita range from $1000 in Yemen to more than $20000 in the Gulf Cooperation Council (GCC) countries". [30]

The following section builds upon the above assessment of the MENA region and analyzes the status of investments and investment actors in the region, focusing on PPPs.

Infrastructure investments in the MENA region: the limitations of public–private partnerships models

MENA countries are increasingly recognizing the importance of climate-smart infrastructure investments and their role in sustainable economic development of the region. Nevertheless, MENA countries still face significant fiscal constraints with regard to upgrading and building more climate-resilient infrastructure due to factors such as limited availability of public finances, low levels of foreign direct investments (FDIs), weak private investments, lack of coordination between international actors, and donors.

Traditionally, infrastructure projects in the MENA region have been financed and managed by the state with little private sector involvement.[31] For example, in the Gulf States the high revenues from hydrocarbons have been used to finance infrastructure projects without private sector involvement.[32] State-managed infrastructure investment and service delivery proves to be costly and highly inefficient. [33] That is why MENA region governments have been pursuing private sector involvement (both foreign and domestic private sector investment) in the delivery and financing of infrastructure services.

FDIs into the MENA economies have been far from linear, being affected by the "regional political instability, global financial shocks, high investment risks"[34] to name a few. Improvements to the business environment and investment climate, as well as putting in place relevant structural and institutional reforms, led to an increase in FDIs between 2000 and 2008. FDIs then dropped significantly after the financial crisis in 2008 and the regional turbulences (the "Arab Spring"),[35] decreasing by more than 65 per cent by 2011.[36] In 2015, "[FDIs] hit an all-time low along with the decline in oil prices which discouraged energy investment".[37] The FDIs in 2017 decreased by 40 per cent when compared with the 2007 levels. Between 2003 and 2017 the top FDI recipients were Egypt, Saudi Arabia, Jordan, and Algeria, while in 2017 the majority of the inflows were concentrated in the UAE and Egypt, followed by Morocco, Lebanon, and Oman.[38] Going forward, "Iraq, Libya and Syria are nations that will need reconstruction in the short- to medium-term and are likely to witness investment flows".[39] However, the MENA region only attracts "1.3% of the world's total FDI", while "intra-regional investment flows are weak", which is reflective of the lack of international and regional integration.[40] The sectors where 2017 intra-MENA FDI investments were focused were the "real estate (59% of the total value of projects), food and tobacco (8.2%), and renewables (7.5%)".[41]

The MENA region also attracts a network of regional and international organizations, many of them involved in financing infrastructure projects such as the World Bank and the Organization for Economic Cooperation and Development (OECD).[42] However, many of the multilateral development funds/banks face many constraints and limitations in terms of financing, such as sector-specific and geographic area criteria, making funding available only for specific outcomes, and requirements for sovereign guarantees from governments that cannot afford to do so.[43] Also, the number of regional and international banks involved in project finance and infrastructure deals has declined significantly, especially after the financial crisis, either "withdrawing from the market or scaling back their involvement in regional deals".[44] The remaining banks have reduced funding and technical capacity, especially for complex infrastructure projects and transactions in a high-risk market environment. Within this context of low bank liquidity, lack of or limited availability of risk mitigation tools, and lack of refinancing and credit enhancement mechanisms, traditional lenders have been discouraged to lend long-term on an unsecured basis. This has, nevertheless, influenced the investment appetite of lenders and made some infrastructure projects highly unattractive.[45]

Besides international organizations and multilateral development funds/banks, there are several foreign governments/donors (e.g., the European Union, the Netherlands, France through the *Agence Française de Développement* (AFD), Germany through *KfW Development Bank* and *KfW Group* -DEG, Japan, USA, Norway, Canada, etc.[46]) that are highly involved in financing projects in the region, as well as helping the much-needed capacity building. However, similar to the international organization and Multilateral Development Banks (MDBs), international donors face several challenges and limitations when deciding their funding priorities.[47]

Thus, in the context of constrained public finances, limited borrowing capacity, low bank liquidity, and lack of a coordinated effort from the international community, there has been a growing narrative around financing infrastructure projects through private sector participation.[48] In this regard, PPPs are being used worldwide to address the hurdles of the infrastructure investment gap and increase the uptake in private investment levels.[49] The involvement of the private sector in infrastructure development in the MENA region through the PPPs can prove to be a crucial vehicle to address the infrastructure needs in the region taking into account climate resilient considerations.[50] Nevertheless, in order to have successful PPPs, a few factors need to be taken into account, such as: transparency, "value for money", proper allocation of risks to parties best suited to manage them, and cost-effectiveness.[51] PPPs also require a "delicate balancing of government's objectives, investor's interest, financing sources and business opportunity".[52]

Although some countries in the MENA region have adopted policies and laws for PPPs (mainly in the Gulf) to support the finance diversification reforms and project development, the implementation of the PPPs in the region is still quite slow and has not yet produced the expected results.[53] For example, "PPP investment in the region amounted to a few billion US dollars as of 2016. These PPPs are mainly in a few countries – Jordan, Iraq and Morocco. They are concentrated in telecoms, electricity and seaports".[54] Some of the barriers and challenges for implementing PPPs in the region are "the unclear recognition of rights and obligations of public and private parties, insufficient regulatory framework, lack of transparency and accountability, and mistrust in the political environment".[55] Countries have been struggling to address these prerequisites in order to attract private capital and thus meet their infrastructure demands without much success, as currently the MENA region is showing the lowest level of PPPs globally.[56]

Going forward, in order to increase the transition to climate-resilient infrastructure in the MENA region, PPPs alone may not be able to adequately meet existing and emerging financing needs. It will, therefore, be necessary to reassess the traditional investment patterns, such as the PPPs, and embrace new technologies and business models that can instigate a successful transition to sustainable infrastructure in a low-cost and equitable way.[57] The distributed ledger technologies (DLT), such as blockchain, acting like a technological and financial enabler, can promote and improve the current processes for climate-resilient infrastructure investments in the MENA region—as explored in the section below. Although blockchain technology can unlock new sources of funding, the focus of this chapter is on how the technology can address the barriers and challenges of PPPs in the MENA region.

Blockchain technology as a technological and financial enabler for climate-resilient infrastructure in the MENA region

The infrastructure gap in the MENA region is widening, as many of the different investment actors are facing several barriers and challenges. As seen above, this is also the case for the PPPs, where the encountered challenges have prohibited their wider reach and implementation among MENA countries. However, in order to help close the global gap in infrastructure investment, an increasing number of United Nations agencies, multilateral, and national development banks, and the private sector have engaged in "the innovation turn".[58] In the current section, we analyze how new technologies put forth by the 4IR, such as blockchain technology, can play a significant role in MENA efforts to build climate-resilient infrastructure by trying to address some of the barriers and challenges of old financing mechanisms such as PPPs. We will focus primarily on how blockchain technology can address the barriers of PPPs implementation in the MENA region, namely lack of transparency, cost-effectiveness, proper allocation of risks between the relevant parties, unclear recognition of rights and obligations, and corruption.

Blockchain is one of the emerging technological innovations, which can advance the implementation and monitoring of Sustainable Development Goals and increase the level of investments in climate-resilient infrastructure. Blockchain is defined as a decentralized, continuously growing sequence of records, called "blocks", which track the overall record of transactions across a peer-to-peer network. The blocks are linked and secured using cryptography.[59] Widely known for bitcoin's success, blockchain is the architecture on which bitcoin was based and implemented. Unlike bitcoin—which used high amounts of energy and emitted large amounts of CO_2 "in order to validate transactions and sustain the network"—blockchain technology can be used in a more energy-efficient way.[60] Cryptocurrencies or digital currencies and, above all, the technology behind them, namely blockchain, are innovations with profound effects, even on the structure and operation of the internet itself. Blockchain technology goes beyond cryptocurrencies and financial transactions, and creates a trusted, decentralized, and secure network structure that can form the basis for a number of applications, such as smart contracts[61] and simplified supply chains. It enables the online exchange of value between two contracting parties in a safe and secure ecosystem without the need of trusted intermediaries. It thus "enables intangible or tangible assets like currencies, shares, infrastructure securities, data or obligations like contracts to be exchanged, without the need for intermediaries, via the trusted ledger".[62] Blockchain technology can be used to bring together all stakeholders involved in the value chains and those regulating them, such as producers, intermediates, consumers, and governments.

The benefits coming out from the blockchain technology functions add credibility to its suitability as a disruptive transformation tool within the PPPs for climate-resilient infrastructure, as well serving as an accelerator to achieve the United Nations 2030 Agenda[63] and move towards sustainable development for all. The blockchain technology can be used in climate infrastructure projects to be implemented through PPPs in a transparent, effective, and cost-efficient manner. Over the last years, blockchain technology has achieved the alignment between private and public sectors interest in infrastructure, while improving capital provision and lessen the cost of capital for the latter.[64]

Firstly, blockchain technology is about sharing information in an efficient and cost-effective manner. Once blockchain is used for a PPP project, smart contracts can eliminate many associated costs related to payment agreements and lengthy contracts by removing the banking and legal services fees, respectively. The smart contract is a self-executing contract between the parties within the blockchain network which can substitute the mentioned field services by embedding all relevant procedures.[65] Smart contracts can eliminate the need for bank guarantees and letters of credit, reducing the related financing costs. For example, according to the OECD, many governments are not managing their budgets and recordkeeping processes well, so an internal accounting system based on a blockchain technology could act as a great option to improve budget management and avoid additional expenses.[66] Worth mentioning is that the World Food Programme (WFP) in Jordan managed to save 98 per cent of administrative costs by replacing the need for a bank through a blockchain pilot, making reconciliation and accounting faster and cheaper.[67] Moreover, considering the involvement of multiple international firms during the implementation of infrastructure projects, the current technology reduces the foreign transaction fees associated with purchases through different foreign banks and currencies. While blockchain technology makes information readily available to all parties, it can also be designed to maintain privacy for important transactions and between selected parties.[68] Thus, blockchain technology can reduce reporting requirements significantly as the information is readily available on a single ledger, can reduce disputes, eliminate delays and any traditional fees and payments, as well as provide prompt transfer of payments upon the completion of work. All these features can make PPPs more efficient and cost effective.

Improving efficiency and cost-effectiveness is important, particularly for financing climate resilient and sustainable infrastructure projects. This is mainly due to their higher "upfront investments (capital expenditure) and have higher perceived technological risks compared to more traditional infrastructure".[69] These fundamental challenges (e.g., lack of liquidity, high transaction costs, and limited transparency) can be addressed through tokenization—"digitalization of real-world assets or financial instruments using blockchain technology".[70] Allowing physical and digital goods to be tokenized and exchanged could act as a tool to meet the financial demand. The use of such an accessible platform could act as an investment vehicle where everyone coming either from the public or the private sector can invest in PPP projects by purchasing tokens.

One of the most significant benefits of blockchain technology is eliminating the need for trust-based systems by eliminating counterparty risk. Due to its decentralized consensus relied on by all parties it removes the need for one central trusted authority. It does this through accountability, fair allocation of risks between the stakeholders, and transparency.[71] PPPs require that all parties involved are accountable to the work and their specific responsibilities according to the division of labour. Blockchain technology, by providing a shared consistent, immutable, and canonical data storage system, can keep all parties accountable to their specific responsibilities and roles. In a PPP, the private counterpart is expected to bear substantial risk, especially in a complex and dynamic setting such as the MENA region. Thus, the fair allocation of risks between stakeholders can influence the private sector involvement. Blockchain technology, through smart contracts, can maintain an agreed risk sharing and put in place a governance system to deal with such situations. Also, blockchain technology can build trust in PPPs by enhancing the transparency which benefits both the accountability and the fair allocation of risks. As mentioned above, the indisputable stored information in the blockchain system can be made available to all beneficiaries and the broader public. Smart contracts can also enhance transparency throughout the supply chain of a PPP project and therefore decrease the chance of corruption. Supply chain is identified as one of the key challenges in the construction industry[72] and luck of trust is the main reason behind that.[73] Data in the blockchain cannot be altered; hence the blockchain ledger remains unchanged. The immutability of the blockchain results in improved tracking of the product origin and overall logistics planning process. It can also "facilitate and improve the monitoring of financial, operational, social and environmental performance".[74]

Furthermore, the cumbersome process to create a PPP involves the unclear recognition of rights and obligations of PPP parties. It is common for PPP projects to be delayed because of poorly drafted agreements leading to cost overrun, delays, and even failure to deliver due to misinterpretation of the contract terms from each side. As a result, conflicts related to this process bring a lack of trust between the public and private counterparts.[75] To avoid this issue and to reduce disputes, smart contracts are mutually written into the blockchain network, including the agreed set of rules between the parties making clear the rights and obligations for both sides.

The adoption of blockchain technology has not been uniform when looking at the investment actors in the MENA region. However, many of the international organizations and some of the MDBs are exploring more and more the applicability of the technology in their everyday business.[76] This shift is quite promising for the MENA region where the infrastructure over the last years has been heavily dependent on development funding. For example, humanitarian technological innovation is evolving rapidly. Due to its emergency nature, humanitarian agencies have innovation high on their agenda in order to improve the alignment and coordination among established and new infrastructure initiatives. Thus, development actors, such as the United Nations Office for the Coordination of Humanitarian Affairs (UNOCHA), the United Nations Office for Project Services (UNOPS), and the United Nations Development Programme (UNDP) have already identified the advantages of the current technology and have utilized it to support their operations.[77] Identifying the need for change, humanitarian actors have been looking to change the business as usual concept within the sector and take advantage of innovative solutions for promoting changes.[78] According to the UNDP, blockchain's advantages including transparency, security verifiability, immutability, and resilience can act as an enabler of leveraging infrastructure investments offers, making the technology more efficient compared to the traditional funding mechanisms like PPPs.[79] Also, by reducing transaction costs, blockchain can improve liquidity on the market, providing possibilities for more economic activity. In addition, blockchain technology can guarantee to the donor or the implementing agency if the required funding has reached a project, avoiding overlapping with other funding sources.[80]

To boost information sharing, we mention as an example the Addis Ababa Action Agenda established by the Global Infrastructure Forum, led by the MDBs. This Agenda aims to improve coordination among established and new infrastructure initiatives between United Nations agencies, national institutions, development partners, and the private sector, by promoting capacity building on blockchain technology and digital currencies.[81] Also, the OECD, according to the G20's 2018 Roadmap to Infrastructure, identifies the need to improve standardization in infrastructure across project development and investment environment, and manage investment risks through a transparent framework, with mitigation strategies in place.[82]

Blockchain technology can not only address some of the barriers and challenges of PPPs, but can also fully transform this business model, by changing the way transactions are done as well as leveraging a range of stakeholders, which are crucial for making investments in climate-resilient infrastructure more attractive. Nevertheless, blockchain technology still operates in many countries in a legal and regulatory vacuum that could act as a deterrent for adoption and further development of the technology.[83] We focus on the legal and policy challenges in the following section, with an emphasis on smart contracts, providing as well some recommendations in addressing these challenges.

Regulating legal and policy challenges of blockchain: some recommendations

The emergence of blockchain in societies has changed the *status quo* by creating a new set of possibilities and bringing new players to the forefront. In order to further develop and be used as an effective tool, some of the fundamental legal and ethical challenges that blockchain poses need to be addressed. Thus, even though blockchain technology has the possibility to be used successfully for PPPs for climate-resilient infrastructure in the MENA region as seen in the previous section, there are still many legal and policy challenges that need to be overcome for this technology to be successfully used for PPPs. There is a gap and mismatch between the technological promise and the legal, policy, and institutional reality, as well as the lack of a broader change of the social, economic, and legal processes.

In many areas of the world, the current legal and policy systems have not been updated to be able to include and regulate groundbreaking technology, such as blockchain. There have been several attempts at regulating the technological field, such as with *lex cryptographica*,[84] as well as Lessing's work on "code as law".[85] According to Lessing's work, "technology architecture constitutes a form of regulation, together with state legislation, market forces, and social norms",[86] and these modes of regulation do not operate in isolation and their interaction is applicable in both the physical world and the cyberspace alike. All modalities create a direct and indirect effect, meaning the effect of the modality upon the individual liberties and their effect upon other modes of regulation. Lessing also highlights that the "interaction between law and architecture can be adversarial"[87] meaning that the law can either accept or reject if a value promoted by the architecture conflicts with the law, as well as the law and governmental reach will prove to be less effective the greater the degree of decentralization of the architecture.[88] Key questions that arose were in relation to "governance and government, private order and state authority, and the relationship between different 'calculative' spaces for assessing, monetizing and allocating value".[89] However, regulating blockchain seems to pose even more challenges, as "their architectural features are designed to enable the evasion of effective regulation and enforcement [...] to elude the rule of law".[90] Thus, the framework offered by *lex cryptographica* and Lessing can be used as a starting point for regulating blockchain, while also taking into account the specificity of distributed ledger technologies, such as blockchain. Going forward studies should focus not only on the impacts on the law of contracts and models of governance, but also on market and social forces and their interaction with the code. This would correctly address the weaknesses in the regulatory system for this technology, helping the relevant authorities adopt the appropriate regulatory actions.[91]

One significant legal development used under blockchain technology is the use of smart contracts, which have proved to provide several advantages in terms of clarity, precision, reduced monitoring costs, and more. Also, as seen in the previous section, the usage of smart contracts seems to address some of the challenges with using PPPs for climate-resilient infrastructure. However, when compared with legal contracts, the smart contracts have their share of limitations, related to execution and termination: the termination option needs to be properly coded into the software; privacy concerns make them rather unsuitable for transactions that require confidentiality; inadequacy to formalize certain types of legal obligations; pseudonymous nature of parties to the contract; prospective of contract standardization and limited possibility for customization; as well as their potential use for criminal activities.[92] Also, as the "effects of the contract are indelibly written in the relevant code, the parties can easily bypass [...] traditional contractual safeguards", such as "systems of guarantees prescribed by the (national and international) contract law system", as "principles of bona fide, or institutions, such as force majeure and the hardship clause, or vitiating factors".[93] Coming back to Lessing's theory of the relation between decentralization and decreased control power of the government, the "extreme fragmentation of the single nodes of the network, which is not controlled by a single well-defined entity",[94] renders regulating blockchain technology increasingly difficult for governments. This proves to be more and more important, as it has been assessed by some that even when the law deprives a smart contract of its legal validity, this would not affect *per se* the further use of smart contracts by its users, and thus its enforcement. Its enforcement would be ensured by the code by which it was enacted.[95] Also, as seen above, parties to a smart contract have contractual autonomy and can choose to forgo the legal guarantees. Thus, blockchain technology has the capacity to make changes to the law, but it has proven to be much more difficult for the law to change blockchain code properties.[96] This puts governments and regulatory authorities in a rather weak position.

Nevertheless, blockchain technology and other types of digital jurisdictions cannot operate in a legal vacuum, as legal jurisdictions will also apply and have a role within these technologies. It is the law that can and should decide upon whether blockchain and its self-executing rules are legally enforceable or not.[97] The legal recognition of smart contracts can be achieved through two different regulatory alternatives, which are up to the relevant legislator to choose between. The legislator can choose to enact specific provisions for blockchain-based smart contracts, or to refrain from introducing these types of provisions, leaving the legal recognition up to the judiciary, which will have to examine it on a case-by-case basis. The judiciary will need to assess whether the smart contract meets the general contract conditions imposed by the law for it to be legally enforceable. However, some specific features of smart contracts need to be considered when the judiciary will perform its assessment (e.g., acceptance of contract comes through performance; requirements that the contract needs to be in writing, etc.). Nonetheless, the judiciary's approach could give rise to certain difficulties for judges "who have to examine blockchain-based codes in each and every single case as precondition for their legal qualification".[98] However, this "approach adheres to the general principles of contract law, thereby safeguarding the consistency of contract law as well as the effectiveness of contractual consent".[99]

Also, in the absence of regulation and a strong judiciary system, another option to overcome some of the limitations imposed by smart contracts are the "hybrid agreements", especially for the ones that contain open-ended obligations (good faith, best efforts) or are very hard to code (representations and warranties). Smart contracts would be used to only "memorialize a limited set of promises as part of a larger, more complicated contractual relationship" to be incorporated in the legal agreements regulating the whole transaction.[100]

Ultimately the role of law and regulation for blockchain technology should be to establish fair and transparent rules between the relevant actors to facilitate new ways of civic and economic interaction as well as shielding against the use of power and appropriation by unregulated technological and financial platforms and entities.[101] Due to the rapid adoption and application of this technology to so many different mechanisms, markets, economic fields, and societal structures, there is a need for the legal and regulatory system to respond to these developments. Mostly it should start by answering key questions, such as what is the relationship between code and law? What jurisdiction should be applicable in a specific case taking into account the decentralized nature of the technology and what are the relevant institutions? What is the relevant regulatory regime? This could also offer the necessary input to determine whether new legislation, tailored to the blockchain technology, is necessary and appropriate, or if the current law of contracts and more general rule of law are sufficient.[102] Or it can lead us in a totally different direction, concluding as Tony Prosser proposes "whilst regulation may be needed, it is portrayed as a second-best choice for social organization; in principle free markets giving us economic freedom and consumer choices should be preferred whenever possible. Regulation is thus an always regrettable means of correcting market failure [...] regulation is part of economic management."[103]

Nevertheless, law should not be a deterrent for the further development of blockchain technology, and its application in sustainable groundbreaking initiatives. Innovative law-making is required to eliminate the regulatory barriers to the new technology, encouraging innovation for technological change.

Conclusion

In many MENA countries, blockchain technology usage, in general, and specifically for leveraging investments in climate-resilient infrastructure is still in its nascent stages. Nevertheless, there has been growing interest in utilizing blockchain in the public-private sector which is meant to bring more investments, hence increasing the technology's significance and utilization. However, in order to fully realize the benefits and expand the utilization of blockchain, governments, international development actors, and the private sector should cooperate and work together to leverage a broader use of blockchain and its application in the infrastructure sector as well. Although PPPs are supposed to result in economic growth improvement and enhance resilient infrastructure in a sustainable way there is still space for improvement. The prevailing PPP model is unsuitable to promote investments for climate-resilient infrastructure in the MENA region. Current infrastructure needs a more adaptive model able to evolve towards functionality and transparency. The public and private sectors should adapt and incorporate blockchain business cases for such PPP projects to maximize efforts in building resilient infrastructure. Thus, blockchain technology can truly change the traditional framework of PPPs by bringing additional private investments and improve the management of unreconciled public funds.

The utilization of blockchain technology for scaling up investments in resilient infrastructure is undoubtedly an approach to deal with the drawbacks of current traditional PPP frameworks and improve the efficiency of the business ecosystem. Leveraging blockchain as a digital infrastructure enabler will help the public sector to reach its required investment goals for building resilient communities, through a decentralized financing platform dedicated to infrastructure. While blockchain is certainly promoting collaboration, the establishment of a more dynamic MENA investor community (governments, private sector, international development actors, international and local banks, etc.) focusing on blockchain investments and infrastructure would enhance collaboration and fit the purpose for scaling up the volume of investments by engaging a variety of investors and incentivize further their engagement.

Implementation of any new system is a complex and cumbersome process, as it will be the case for the further adoption and implementation of blockchain technology for climate-resilient infrastructure in the MENA region. One of the key challenges will most certainly be the legal and regulatory uncertainty surrounding this technology. However, although the task at hand might be complex, we argue that it might be the right time to start testing the application of blockchain technology for PPPs projects to start with. This, however, should take into account the applicable legal framework in order to prevent any future hurdles in its application. The law ultimately should be an aid to the further deployment of the technology and not a deterrent. The blockchain technology could very well be the technological enabler for upgrading and transforming old financing partnerships, such as PPPs, for climate-resilient infrastructure, while the MENA region—due to its complexity—provides for the optimal environment, proving to be in dire need for innovative infrastructure investment processes.

Notes

1 S. E. Chang, "Socioeconomic Impacts of Infrastructure Disruptions" (Oxford Research Encyclopedia: Natural Hazard Science, 26 October 2016) 2.
2 Climate Bonds Initiative, Climate Resilience Consulting, World Resources Institute, "Climate Resilience Principles: A Framework for Assessing Climate Resilience Investments" (30 September 2019) <www.climatebonds.net/files/files/climate-resilienc e-principles-climate-bonds-initiative-20190917.pdf>.
3 B. M. Ayyub and A. C. Hill, "Climate-Resilient Infrastructure: Engineering and Policy Perspectives" (2019) 49 *The Bridge: National Academy of Sciences* 8.
4 Global Commission on Adaptation, "Adapt Now: A Global Call for Leadership on Climate Resilience" (13 September 2019) <https://cdn.gca.org/assets/2019-09/Globa lCommission_Report_FINAL.pdf>.
5 Ibid. However, there is still a benefit-cost ratio of 4:1, i.e. $1 of preventive measures is equal to $5 of repairs (R. Rahiman, "Addressing Critical Issues for Building Climate Resilient Infrastructure" The Energy and Resources Institute, G20 2019 Japan (31 March 2019) 3 <https://t20japan.org/wp-content/uploads/2019/03/t20-japan-tf3-8-c ritical-issues-climate-resilient-infrastructure.pdf>.
6 R. Arezki and F. Belhaj, "Developing Public-Private Partnership Initiatives in the Middle East and North Africa from Public Debt to Maximizing Finance for Development", The World Bank Group, Policy Research Working Paper 8863, May 2019, 5 <http://documents.vsemirnyjbank.org/curated/ru/860241558712523200/pdf/Develo ping-Public-Private-Partnership-Initiatives-in-the-Middle-East-and-North-Africa-From-Public-Debt-to-Maximizing-Finance-for-Development.pdf>.
7 Organisation for Economic Co-operation and Development (OECD), "Public-Private Partnerships in the Middle East and North Africa: A Handbook for Policy Makers", Private Sector Development Handbook, 2013, 2 <www.oecd.org/mena/ competitiveness/PPP Handbook_EN_with_covers.pdf>.
8 Arezki and Belhaj, n 6; D. S. Olawuyi, "Financing Low-Emission and Climate-Resilient Infrastructure in the Arab Region: Potentials and Limitations of Public-Private Partnership Contracts" in W. L. Filho (ed), *Climate Change Research at Universities: Addressing the Mitigation and Adaptation Challenges* (New York: Springer 2017).
9 The World Bank, "Population, Total - Middle East & North Africa" (The World Bank, 2019) <https://data.worldbank.org/indicator/SP.POP.TOTL?locations=ZQ>.
10 The Food and Agriculture Organization of the United Nations, "Annex 5: Trends, Challenges and Priorities – Near East and North Africa" (May 2016) 1 <www.fao.org /3/a-mq647e.pdf>.

11 Global Facility for Disaster Reduction and Recovery (GFDRR), "Middle East and North Africa Regional Urban Resilience Conference", April 2–4, 2019 <www.gfdrr.org/en/resilient-mena>.
12 J. Saghir, "Urban Planning Can Make the Middle East More Resilient to Outside Forces" (22 August 2019) <www.thenational.ae/opinion/comment/urban-planning-can-make-the-middle-east-more-resilient-to-outside-forces-1.901325>.
13 The World Bank, "Beyond Scarcity, Water Security in the Middle East and North Africa" (ReliefWeb, 2017) <https://reliefweb.int/sites/reliefweb.int/files/resources/9781464811449.pdf>.
14 "Climate Change Is an Existential Threat for the Middle East & the GCC" (22 September 2019) <https://nassersaidi.com/2019/09/22/climate-change-is-an-existential-threat-for-the-middle-east-the-gcc-article-for-aspenia-fall-2019/>.
15 Ibid.
16 A. O. El-Kholei, "Are Arab Cities Prepared to Face Disaster Risks? Challenges and Opportunities" (2019) 58 *Alexandria Engineering Journal* 479.
17 Climate resilient infrastructure refers to "buildings, structures and systems that reduce GHG emissions, and improve society's ability to adapt to, and cope with, the risks posed by climate change" , Olawuyi, n 8, 534.
18 A. Al Shamali, B. El-Issa, E. Elmaddah, I. H. Mansour, I. Rouabhia, K. Al Thobhani, N. El Saim, S. Fawaz, S. Al Harthey, S. A. Serriya, and S. Al-Zoghoul, "Energy and Climate in the MENA Region: Youth Perspective to a Sustainable Future," Friedrich Ebert Stifung, GermanWatch, Climate and Energy Project, A Youth Perspective position Paper, November 2019, 10 <http://library.fes.de/pdf-files/bueros/amman/15777.pdf>.
19 Ibid.
20 Ibid.
21 United Nations Framework Convention on Climate Change, *Adoption of the Paris Agreement*, 12 December 2015, UN Doc. FCCC/CP/2015/L.9; United Nations Framework Convention on Climate Change, "NDC Registry (Interim) for MENA Region Countries" <www4.unfccc.int/sites/ndcstaging/Pages/Home.aspx>.
22 Global Facility for Disaster Reduction and Recovery, n 16.
23 J. Saghir, "Climate Change and Conflicts in the Middle East and North Africa" Issam Fares Institute for Public Policy and International Affairs, American University of Beirut, Working Paper No. 50, June 2019, 4; D. Olawuyi, "Advancing Innovations in Renewable Energy Technologies as Alternatives to Fossil Fuel Use in the Middle East: Trends, Limitations, and Ways Forward" in D. Zillman, L. Godden, L. Paddock, and M. Roggenkamp (eds), *Innovation in Energy Law and Technology: Dynamic Solutions for Energy Transitions* (Oxford University Press 2018).
24 Olawuyi, n 8.
25 Arezki and Belhaj, n 6, 5; OECD, n 7, 14.
26 E. Göll and J. Zwiers, "Technological Trends in the MENA Region: The Cases of Digitalization and Information and Communications Technology (ICT)," Middle East and North Africa Regional Architecture: Mapping Geopolitical Shifts, Regional Order and Domestic Transformations, MENARA Working Papers, No. 23, November 2018, 209–210 <www.iai.it/sites/default/files/menara_wp_23.pdf>.
27 Arezki and Belhaj, n 6, 5.
28 OECD, n 7, 14.
29 S. D. Orsi, "Arab Spring Brings Winds of Change to the Maghreb and Mena Region: Does that Spell Opportunity for Infrastructure Development and Project Finance?" (2011) 11 *Richmond Journal of Global Law and Business* 77a at 89.0
30 M. Abouelnaga, "Why the MENA Region Needs to Better Prepare for Climate Change" (Atlantic Council Blogs, 7 May 2019) <www.atlanticcouncil.org/blogs/menasource/why-the-mena-region-needs-to-better-prepare-for-climate-change/>.

31 D. Ives, M. M. Hoscheidt, B. Jaeger, and J. S. Tocchetto, "Infrastructure Investments in the Middle East & North Africa" (2013) 1 *UFRGS Model United Nations Journal* 308, 332.

32 M. A. Ismail, *Public Private Partnership Contracts: The Middle East and North Africa*, 1st ed. (Abingdon, Oxfordshire: Routledge 2020) 2.

33 Organisation for Economic Co-operation and Development (OECD), "Fostering Investment in Infrastructure" (OECD, 9 January 2015) <www.oecd.org/daf/inv/investment-policy/Fostering-Investment-in-Infrastructure.pdf>.

34 N. Saidi and A. Prasad, "Background Note: Trends in Trade and Investment Policies in the MENA Region," (MENA-OECD Competitiveness Programme, MENA-OECD Working Group on Investment and Trade, November 27–28, 2018) 23 <www.oecd.org/mena/competitiveness/WGTI2018-Trends-Trade-Investment-Policies-MENA-Nasser-Saidi.pdf>.

35 Ibid.

36 OECD, n 33, 41.

37 Saidi and Prasad, n 34, 16.

38 Ibid, 23.

39 Ibid.

40 OECD, n 33, 41.

41 Saidi and Prasad, n 34, 23f.

42 Ives and others, n 31, 334.

43 Orsi, n 29, 88.

44 OECD, n 7, 16.

45 Ibid.

46 Enhancing Business Support Organisations and Business Networks (EBSOMED), "MEDA Finance 2019 – Survey: Grant, Technical Assistance and Lines of Credit for the Mediterranean Region", BUSINESSMED, the European Union, 2019 <http://ebsomed.eu/sites/default/files/MEDA Finance survey 2019.pdf>.

47 A. Jägerskog, M. Schulz, and A. Swain, *Routledge Handbook on Middle East Security* (1st edn, Abingdon, Oxfordshire: Routledge 2019).

48 Organisation for Economic Co-operation and Development (OECD), "Regional Conference: Investment and Inclusive Growth in the Midst of Crisis: Lessons Learned & Ways Forward," 2016, 1 <www.oecd.org/mena/competitiveness/FCS_Conf_Conclusions_final.pdf>.

49 OECD, n 7, 14; United Nations Department of Economic and Social Affairs, "Financing for Sustainable Development Report 2019: Inter-Agency Task Force on Financing for Development" 2019, 61 <https://developmentfinance.un.org/sites/developmentfinance.un.org/files/FSDR2019.pdf>.

50 Orsi, n 29, 89.

51 OECD, n 33, 74–75.

52 Ives and others, n 31, 333.

53 Saidi and Prasad, n 34, 39.

54 Arezki an Belhaj, n 6, 6.

55 Ives and others, n 31; Olawuyi, n 8.

56 G. Di Liddo, A. Rubino, and E. Somma, "Determinants of PPP in Infrastructure Investments in MENA Countries: A Focus on Energy" (2019) 46 *Journal of Industrial and Business Economics* 523, 525–526.

57 Organisation for Economic Co-operation and Development (OECD), "Blockchain Technologies as a Digital Enabler for Sustainable Infrastructure" OECD Environment Policy Paper No. 16, 2019, 9 <www.oecd-ilibrary.org/docserver/0ec26947-en.pdf?expires=1599820684&id=id&accname=guest&checksum=D34C5086D3F8806DDD6087200150483D>.

58 L. Bloom and A. Betts, "The Two Worlds of Humanitarian Innovation" (Refugee Studies Centre, University of Oxford, Working Paper Series No. 94, 3 August 2013) <www.rsc.ox.ac.uk/files/files-1/wp94-two-worlds-humanitarian-innovation-2013.pdf>.
59 D. L. K. Chuen, *Handbook of Digital Currency 1st Edition Bitcoin, Innovation, Financial Instruments, and Big Data* (Cambridge, MA: Academic Press 2015) 14.
60 OECD, n 57.
61 Institute of International Finance, "Getting Smart: Contracts on the Blockchain" (IIF, May 2016) <www.iif.com/Publications/ID/582/Getting-Smart-Contracts-on-the-Blockchain>.
62 OECD, n 57.
63 United Nations General Assembly (UNGA), *Transforming Our World: The 2030 Agenda for Sustainable Development*, 21 October 2015, UN Doc. A/RES/70/1, 29.
64 United Nations Development Programme (UNDP), "The Future Is Decentralised Block Chains, Distributed Ledgers, and the Future of Sustainable Development," 2018, 34 <www.undp.org/content/dam/undp/library/innovation/The-Future-is-Decentralised.pdf>.
65 S. Nanayakkara, S. Perera, and S. Senaratne, "Stakeholders Perspective on Blockchain and Smart Contracts Solutions for Construction Supply Chains" (The CiB World Building Congress Conference, Hong Kong, June 2019) 2.
66 OECD, n 57, 46.
67 K. Dodgson and D. Genc, "Blockchain for Humanity" (Humanitarian Practice Network, 29 November 2017) <https://odihpn.org/blog/blockchain-for-humanity/>.
68 OECD, n 65, 21.
69 D. Uzsoki, "Tokenization of Infrastructure: A Blockchain-Based Solution to Financing Sustainable Infrastructure" (International Institute for Sustainable Development, MAVA, January 2019) <www.iisd.org/sites/default/files/publications/tokenization-infrastructure-blockchain-solution.pdf>.
70 Ibid.
71 Institution of Civil Engineers (ICE), "Blockchain Technology in the Construction Industry", December 2018, 20–23 <https://www.ice.org.uk/ICEDevelopmentWebPortal/media/Documents/News/Blog/Blockchain-technology-in-Construction-2018-12-17.pdf>.
72 M. Thunberg, M. Rudberg, and T. Karrbom Gustavsson, "Categorising On-Site Problems: A Supply Chain Management Perspective on Construction Projects" (2017) 17 *Construction Innovation* 90–111 at 111.
73 G. Al-Werikat, "Supply Chain Management in Construction; Revealed" (2017) 6 *International Journal of Scientific and Technology Research* 106.
74 Uzsoki, n 69.
75 OECD, n 57, 40.
76 European Union, "Blockchain for Supply Chains and International Trade" (European Parliament, May 2020) 145 <www.europarl.europa.eu/RegData/etudes/STUD/2020/641544/EPRS_STU(2020)641544_EN.pdf>.
77 United Nations Office of Information and Communications Technology, "Usage of Blockchain in the UN System," 2017, 4 <https://unite.un.org/sites/unite.un.org/files/session_3_b_blockchain_un_initiatives_final.pdf>.
78 United Nations Office for the Coordination of Humanitarian Affairs, "World Humanitarian Summit
    Business Consultation on Innovation," May 2015, 6 <https://reliefweb.int/sites/reliefweb.int/files/resources/Business%20Consultation%20on%20Innovation.pdf>.
79 Ibid.
80 The United Nations Office for Project Services (UNOPS), "Could Bitcoin Technology Revolutionize Aid Distribution?" November 28, 2017 <www.unops.org/news-and-stories/insights/could-bitcoin-technology-revolutionize-aid-distribution>.

81 United Nations Economic and Social Commission for Asia and the Pacific, "Financing for Development in Asia and the Pacific: Highlights in the Context of the Addis Ababa Action Agenda" (2018) 21 <www.unescap.org/sites/default/files/FFD in Asia-Pacific - Highlights Action Agenda.pdf>.

82 Organisation for Economic Co-operation and Development (OECD), "Roadmap to Infrastructure as an Asset Class", 2018, 2 <www.oecd.org/g20/roadmap_to_infrast ructure_as_an_asset_class_argentina_presidency_1_0.pdf>.

83 Organisation for Economic Co-operation and Development (OECD), "Session 15A: Realising the Promise of Blockchain", 2018, 1–2 <www.oecd.org/going-digital/sum mit/summit-issues-n-session-15a.pdf>.

84 "*Lex cryptographica*" refers to the rules administered through self-executing smart contracts and decentralized (autonomous) organizations, i.e. rules coded in and enforced by quasi-autonomous technological systems. See generally A. Wright and P. De Filippi, "Decentralized Blockchain Technology and the Rise of Lex Cryptographica" (2015) *SSRN* <https://ssrn.com/abstract=2580664>.

85 See generally L. Lessig, *Code: And Other Laws of Cyberspace* (New York: Basic Books 1999).

86 P. Tasca and R. Piselli, "The Blockchain Paradox" in I. Lianos, P. Hacker, S. Eich, and G. Dimitropoulos (eds), *Regulating Blockchain: Techno-Social and Legal Challenges* (Oxford University Press 2019) 28.

87 Ibid.

88 Ibid.

89 See generally Lessig, n 85.

90 J. P. Quintais, B. Bodó, A. Giannopoulou, and V. Ferrari, "Blockchain and the Law: A Critical Evaluation" (2019) 2 *Stanford Journal of Blockchain Law and Policy* 86.

91 Tasca and Piselli, n 86, 29.

92 Quintais and others, n 90.

93 Tasca and Piselli, n 86, 30.

94 Tasca and Piselli, n 86, 31.

95 Ibid.

96 Ibid.

97 F. Möslein, "Conflicts of Laws and Codes: Defining the Boundaries of Digital Jurisdictions" in I. Lianos, P. Hacker, S. Eich, and G. Dimitropoulos (eds), *Regulating Blockchain: Techno-Social and Legal Challenges* (Oxford University Press 2019) 279.

98 Möslein, n 97, 280.

99 Ibid., 281.

100 Quintais and others, n 90.

101 I. Lianos, P. Hacker, S. Eich, and G. Dimitropoulos, *Regulating Blockchain: Techno-Social and Legal Challenges* (Oxford University Press 2019) 10.

102 Ibid, 11.

103 R. Herian, *Regulating Blockchain: Critical Perspectives in Law and Technology* (Abingdon, Oxfordshire: Routledge 2019) 3; T. Prosser, *The Regulatory Enterprise* (Oxford University Press 2010) 1.

16 Adopting an adaptive governance paradigm to advance climate law and policy in the MENA region

Andreas Rechkemmer

Introduction: Globalization, governance, and regional climate issues

This chapter discusses the need for a recalibration of law and policy efforts at the Middle East and North Africa (MENA) region level. This recalibration may be accomplished by adopting an adaptive governance paradigm that builds on local knowledge, participation, and social networks. This chapter examines the scope and content of an adaptive governance approach to climate change law and policy and some of the preconditions for its application, including the need to overcome various challenges to its implementation across the MENA region.[1]

After five decades of intense data collection, research and analysis, and multiple rigorous assessments, there is broad consensus within the contemporary natural and social sciences on the method of identifying the most pressing global environmental problems.[2] In his 2002 seminal essay, "The Global Environmental Agenda: Origins and Prospects", James Gustave Speth provided a key classification of major global environmental issues and sustainability threats that has become an important reference standard.[3] The two most important environmental challenges are climate change and atmospheric changes. Atmospheric and climatic issues range from the weakened and/or depleted stratospheric ozone layer due to chlorofluorocarbons and other substances, to the systemic effects of global warming—most significantly caused by increased industry, agriculture, and mobility-related CO_2 emissions and other greenhouse gases (GHGs), such as methane.[4] Increased GHGs produce a host of consequences, including sea level rise, changes in precipitation and weather patterns, floods and droughts, a stronger El Nino, and the statistical increase of natural hazards in coastal, high altitude, and/or dryland regions.[5] What is also significant about climate change is that its phenomena cut across a multitude of other global environmental changes through a plethora of causal interlinkages and correlations and feedback loops, which demonstrate that climate change and other changes and disturbances to the Earth system are indeed systemic and inextricably linked. Those "other" changes include the loss of biological and genetic diversity; desertification, land degradation, and drought; threats to global freshwater resources; and threats to the world's oceans.[6]

Almost all of today's global sustainability issues are interrelated with climate change, whether through correlation or causation. Furthermore, all of the contemporary environmental challenges, and their inherent adverse impacts, can be found in the MENA region, and are often predominant drivers of broad socio-ecological, socio-economic, and socio-political disturbance regimes and crises.

To date, an enormous body of literature has accumulated on globalization, global governance, and, more specifically, global (or international) environmental governance. Remarkably, most of this literature was produced within the past three decades. The terms *globalization* and *global governance* are relatively new: their use in literature, science, and political rhetoric started to become somewhat trendy only in the early 1990s. There are close interlinkages between the phenomenon called globalization and the phenomena of global environmental change that directly affect the sustainability of our planet, its natural systems, and its human livelihoods.[7]

Markets are commonly believed to promote efficiency through competition and the division of labour—through the kind of specialization that allows both people and economies to focus on what they do best.[8] A globalized economy offers greater opportunity for people and businesses to access more and larger markets around the globe, "[b]ut markets do not necessarily ensure that the benefits of increased efficiency are shared by all".[9] Thus, globalization cannot be simplified as a mere phenomenon of trade, supply chains, markets, or financial flows. It needs to be understood more systemically as a complex economic, political, social, cultural, geographic, and socio-ecological process reorganizing global economic, socio-political, and geopolitical landscapes. As a result of this globalization, the governance of public and private goods, as well as that of natural resources and other common goods, has experienced significant alterations and shifts, not least in the MENA region.

It is precisely this context in which a thorough analysis and evaluation of contemporary global and regional sustainability and climate governance should be embedded. Various agency forces (e.g., public, private, and civil society actors, and the power dynamics among them), structures (e.g., legal and policy frameworks, regimes, and institutions), and dynamic processes (e.g., science communication, markets, intergovernmental negotiations, or social movements) at play in sustainability governance have to be understood not only as interdependent but also as largely dependent on the reality that has been shaped by globalization. Since present-day globalization is essentially a neoliberal construct, the current predominant legal, political, and institutional architecture of global environmental governance is largely ineffective due to its path-dependent reliance on ideologies, resources, and power structures that continue to support mainstream economic interest rather than the common good. The irony is that the same matrix out of which globalization was once born, and through which it is perpetuated, has been allowed to dominate the discourse on global governance of the commons. In other words, the forces that were at the helm of the creation of the current global sustainability crisis are being called upon to produce its solution.[10]

Environmental problems can be categorized along three scales of appearance. *Local phenomena* are limited to the spatial dimension of states, e.g., emissions in industrial zones, air pollution in urban areas, or the locally restricted contamination of a river through chemical waste. *Regional phenomena* are of a transboundary, but geographically limited nature, e.g., toxic pollution of transboundary rivers, or drought periods in Northern Africa. *Global phenomena* affect world-wide shared resources and sinks, as do climate change and global warming, the pollution of the oceans, or the loss of biodiversity. Although conceptual considerations suggest that, following this classification, only global phenomena are of international or inter-regional concern, emphasis has to be laid on the fact that local or regional problems may, and often do, culminate to an extent of a global dimension. For example, a regional drought may trigger chain reactions, such as loss of agricultural productivity, famine and poverty, migration or social unrest affecting other world regions and the international community, and local water stress which may destabilize an entire region and a country's relationship with its neighbours. From a systems perspective, conceptual boundaries between scales appear arbitrary given today's highly interdependent and densely coupled environmental and socio-ecological challenges, amongst which climate change is the most severe.[11] Hence the regional scale, nested between local and global governance levels, ought to be the default unit of analysis of multi-level sustainability and policy problems.

With regard to the interplay between globalization and the environment, there are two distinct types of causal interaction. First, we know of grave environmental problems that are caused or magnified by globalization-related phenomena. These problems include land degradation caused by unsustainable land-use due to world market pressures and foreign direct investment (FDI) patterns; the global climate crisis due to industrialization processes and "exported" unsound technologies; or unsustainable energy consumption triggered by a dramatically increased global mobility.[12] Second, there are intermediate consequences, such as the erosion of environmental safety standards due to competition pressure leading to the deforestation of rain forests, or textile production patterns in Asian countries.[13]

This chapter consists of four sections. After the introduction, the second section examines the structural, procedural, and ontological differences between environmental policy and sustainability governance. The third section makes the case for the adoption of a regionally focused adaptive governance framework to advance climate law and policy innovation in MENA. It also proposes building blocks for adaptive regional climate governance. The fourth section provides a conclusion.

Reforming regional climate and sustainability governance

From environmental policy to sustainability governance

The term *sustainability governance* has been chosen deliberately, rather than *environmental governance*. For a variety of reasons, it appears to be more inclusive, accurate, and up to date. The term sustainability indeed includes aspects and dimensions that exceed the mere notion of environment. It connects issues of environmental protection, nature conservation, and resource efficiency with those of social justice, human rights, human health and well-being, or economic and social development. In the context of climate change, in particular, the socio-ecological and socio-political dimensions go far beyond the properties of governing the natural environment or natural resources. This is especially true for the MENA region where a plethora of environmental and natural resources issues interact with frequent disasters and a host of social, socio-economic, and political challenges.[14] Therefore, governing sustainability is more intricate and complex than governing the environment. The term *sustainability governance* may become the cutting edge of discourse in the age of global change, as it pertains to the management and collective steering of social-ecological systems, resting on but also going beyond legal and policy instruments and frameworks. Furthermore, while the term *sustainability policy* refers to a specialized thematic field and sub-category of public policy, it implies government and often nation states as principal actors. Therefore, it is important to ask the question: What then defines *sustainability governance*? Maria Ivanova has provided:

> Two traditional forms of governance have dominated world affairs until recently—national governance through governmental regulation and international governance through collective action facilitated by international organizations and international regimes. However, governing human relations has become a complicated endeavour that has transcended the national and interstate scale and moved to a global level involving multiple actors across national borders and multiple levels of regulatory authority - from subnational to supranational. In this context, institutional arrangements for cooperation are beginning to take shape more systematically and have now been recognized as critical to the effective tackling of any global problem. Public-private partnerships, multi-stakeholder processes, global public policy networks, and issue networks are regarded as important tools for global governance.[15]

Overcoming both nation-centered and overly "global" approaches

The Rio Earth Summit of 1992 became the first "playground" on which these concepts were brought in and tested, and subsequently reflected in UNCED's conference outcomes as well as in the entire Rio follow-up process.[16] Two points are crucial. First, the primary focus of the more traditional environmental policy had been on the nation state and its law-making, policymaking, and enforcement capacities. However, in the 1990s, the unfolding discourse on environmental governance emphasized the global level (international, inter-governmental, multilateral) as the place and means to address sustainability issues. Sustainability issues were addressed through problem formulation, agenda setting, negotiations, "deals" to bring about regimes, and an iterative process of institutionalization.

The hope was to get a critical mass of nation states committed to taking collective legal, political, and social action to curb GHG emissions, reduce waste and pollution, and protect biodiversity, freshwater and marine resources, and so forth.[17] Second, the 1992 Rio Earth Summit deliberately triggered a globalized process of multiple parallel negotiation streams aiming to address a wide array of environmental issues. The negotiation streams were poorly interrelated thematically and badly orchestrated institutionally, resulting in an eclectic collection of largely disjointed and topical legal agreements and fragmented policy frameworks.[18] Both trajectories were wrong.

The regional prerogative

First, global sustainability problems play out regionally. They typically occur across and beyond the borders of nation states but also show characteristics that correlate with the unique geographical, topographical, morphological, and sociocultural ontologies of the world's regions. Hence, the default level of governance interventions, through law and policy frameworks and processes, should be the regional. Second, climate change is a mega (or meta) phenomenon of completely unprecedented severity. Its phenomenon and consequences are present and cascade into literally every single sustainability issue of our time. Both ontologically and methodologically, it was a huge mistake to classify and address climate change as a subset of global environmental governance, allocating one of many conventions and regimes to it. Rather, it deserves and requires to be understood as the "mother of all sustainability challenges" and thus a far more comprehensive and robust governance approach is needed. In sum, a major shift in our thinking on legal and policy frameworks, instruments, and resources is required to increase the intensity and efficacy of our efforts if we want to even have a remote chance of success in dealing with climate change *and* other sustainability grand challenges of the 21st century.[19]

Diversifying governance modes

Meanwhile, multilateral cooperation has experienced a re-definition of its genuine connotation. Multilateral cooperation now incorporates non-state actors; the scientific community; and non-hierarchical and alternative regulatory and networking patterns. International cooperation has become a more complex and non-linear system of structures and agents that govern at multiple levels and through multiple means. This system includes, but is not limited to, formal and inter-governmental negotiation processes under the UN aegis, which are based on or produce legal agreements and regimes on single issues. Nation states are still important as principal actors, among many others. Schellnhuber and Biermann argue that the regulation of environmental problems cannot be based on decentralized mechanisms alone. Rather, they require effective and efficient international institutions and legislation.[20] Other scholars highlight the role of non-governmental organizations (NGOs) and the need for a more formalized legitimacy model for the same. In their article "The Role of NGOs and Civil Society in Global Environmental

Governance", Gemmill and Bamidele-Izu identify five major roles for civil society to play in sustainability governance: collecting, disseminating, and analyzing information; providing input to agenda-setting and policy development processes; performing operational functions; assessing environmental conditions and monitoring compliance with environmental agreements (a so-called "watchdog" function); and advocating environmental justice.[21]

All aforementioned aspects serve as a conceptual quilt to grasp the notions of sustainability governance beyond the nation state, keeping in mind that this field provides a perfect example for the properties and features of governance in general. Over the past decades, collaborative research has produced insights into the core problems and themes of environmental (or sustainability) governance, as a result of which we now understand this field and its intricacies much better. For instance, the Institutions and Global Environmental Change (IDGEC) research project focused on the role that institutions play both in causing and addressing sustainability problems, looking at institutions as sets of rights, rules, and decision-making procedures. IDGEC helped in our understanding of governance aspects, such as causality and the performance and design of institutions, as well as their ability to fix the problems they were designed for, their interplay, and the scales they are directed to.[22] The Earth System Governance project, building on IDGEC, identified five so-called analytical problems of global environmental (or sustainability) governance:

> The five analytical problems of Earth System Governance are the problem of the overall *architecture* of earth system governance, of *agency* beyond the state and of the state, of the *adaptiveness* of governance mechanisms and processes and of their *accountability and legitimacy*, and of modes of *allocation and access* in earth system governance—in short, the five As of earth system governance research. These five analytical problems are derived from an analysis of the current state of research and of theoretical developments as well as from societal demands on the academic community.[23]

Applied to the regional level and properly hierarchized and prioritized through the lens of climate change, the conceptual framework outlined in this section is a robust tool to inform reforms in law and policy. Such a framework for reform would suit and benefit the MENA region greatly, given its heterogeneous and diverse yet interconnected and correlated sustainability, economic, and governance issues aggravated and amplified by the regional dimensions of global climate change.

From traditional governance to regionally focused adaptive climate governance

Nature and scope of adaptive climate governnace

The common denominator of all sustainability governance efforts is the goal of reducing risk and building resilience. Leading governance frameworks suggest that changes can and have to take place with regard to reducing the vulnerability of social-ecological systems by limiting or reducing their exposure and sensitivity to an array of environmental risks, threats, and impacts, and by increasing their adaptive and coping capacity. These are the primary objectives and ultimately the key success criteria for sustainability governance in the face of global change. Therefore, *adaptability* is a prerequisite and crucial for sustainability governance and the most indispensable variable of the equation. Adaptive governance is based on the generic governance concept, adding a dynamic dimension and a resilience goal as variations to the ecosystem sciences process. The state-of-the-art definition of governance at the national level comes from the United Nations Development Programme (UNDP):

> The exercise of political, economic and administrative authority in the management of a country's affairs at all levels. It comprises the mechanisms, processes and institutions through which citizens and groups articulate their interests, exercise their legal rights, meet their obligations and mediate their differences.[24]

As UN-HABITAT notes:

> two aspects of this definition are relevant [...] First, governance is not government. Governance as a concept recognizes that power exists inside and outside the formal authority and institutions of government [...] Governance includes government, the private sector and civil society. Second, governance emphasizes "process." It recognizes that decisions are made based on complex relationships between many actors with different priorities.[25]

In addition to these properties, adaptive governance is characterized by the need for reorganization and renewal during crisis and change. Also, adaptive governance must be "location specific and tailored to local circumstances".[26]

However, even local action is subject to multilevel governance and cross-scale interactions. For example, urban systems are influenced by, and influence other, governance levels (e.g., national, regional, and supra-national) and scales (both temporal and spatial), a feature they share with climate change dynamics. Adaptive governance is embedded in a structure and flow of permanent level-scale complexity. "The real challenge is dealing with systems that are not only cross-scale but also dynamic, whereby the nature of cross-scale influences in the linked social-ecological system changes over time, creating fundamental problems for division of responsibility between centralized and decentralized agents."[27] With the focus of both researchers and decision-makers shifting from observing and assessing the bio-physical aspects of climate change to a more human and society centred understanding of the very nature of the problem, the social, behavioural, cultural, economic, and technological aspects have entered the centre stage of the sustainability governance discourse as shown in Figure 16.1.

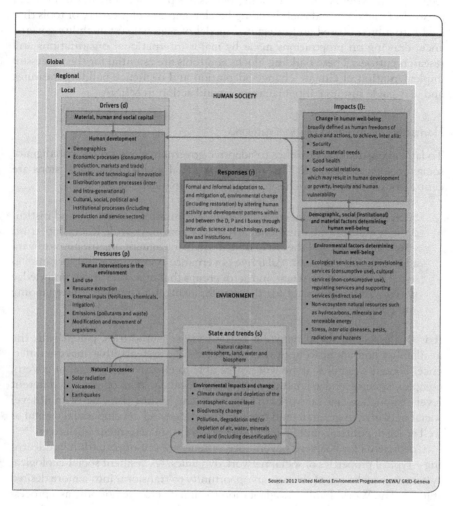

Figure 16.1 Drivers, pressures, state, impacts, and responses (DPSIR) analytical framework.
Source: United Nations Environment Programme (UNEP), Global
Environment Outlook (GEO) 5: Environment for the future we want, 2012,
available at www.unenvironment.org/resources/global-environment-outlook-5.

Overcoming challenges: building blocks for adaptive climate governance in MENA

The following are building blocks and tools for adaptive climate and sustainability governance at the regional level. The MENA region is diverse and heterogeneous, and yet universally and significantly affected by climate change. The building blocks offer a flexible and adaptible toolkit to inform policymaking and legislation in MENA. Seven building blocks are proposed, ranging from social networks to finance, and each building block comprises of a variety of tools that feed into a broad-based and integrated model of regional sustainability governance, drawing on propositions made by major international organizations and research entities. These building blocks and tools are essential for the successful implementation of climate change adaptation and resilience building measures and may guide sustainability management and action in MENA.

Harnessing the dynamics of social networks

Today's mainstream approaches to adaptive governance rely on the paradigm of social-ecological systems.[28] This means that social and societal dimensions are central for effective approaches to adaptive governance:

> Such governance connects individuals, organizations, agencies, and institutions at multiple organizational levels. Key persons provide leadership, trust, vision, meaning, and they help trans-form management organizations toward a learning environment. Adaptive governance systems often self-organize as social networks with teams and actor groups that draw on various knowledge systems and experiences for the development of a common understanding and policies.[29]

It is this aspect of social networks beyond formal authorities, paired with the availability and application of relevant knowledge systems, that make a difference. Formal structures alone will not be able to effectively manage, or govern, complex systems under conditions of crisis. "Stakeholders operate at different levels through social networks. This aspect emphasizes the role of multilevel social networks to generate and transfer knowledge and develop social capital as well as legal, political, and financial support to [...] management initiatives."[30]

Coefficients for adaptive governance systems are self-organization and learning—typical properties of social network dynamics. "A resilient social-ecological system may make use of crisis as an opportunity to transform into a more desired state."[31] The World Bank sees adaptation to climate change not only as a process of preparing for, and adjusting proactively to, the negative impacts of climate change but also as offering opportunities. Adapting, if placed under the overall umbrella of resilience building, will force us to work harder on the larger sustainability and social development agendas, which are of particular importance in MENA, and which may produce economic benefit. The World Bank concludes: "Building resilience requires not only robust decision making by those in positions of formal authority, but also a strong web of institutional and social relationships that can provide a safety net for vulnerable populations."[32] A central tool to harness the potential of social networks for adaptive sustainability governance is so-called bridging organizations. "An important factor in this context is organizations [...] that emerge to bridge local actors and communities with other scales of organizations. Such bridging organizations can serve as filters for external drivers and also provide opportunities by bringing in resources, knowledge, and other incentives."[33] They carry a central function for inter-organizational collaboration and can serve as platforms for "building trust, sense making, learning, vertical and/or horizontal collaboration, and conflict resolution. The bridging organization encompasses the function of [...] communicating, translating, and mediating scientific knowledge to make it relevant to policy and action."[34]

Finally, an important tool for enhanced governance is social learning. Given the magnitude and scale of sustainability problems, large-scale behavioural change is required from the individual to the societal level. It is a prerequisite to take into account that changes arising from new behaviours are multifaceted and need the right institutional arrangements and incentive schemes to become effective and sustainable. Sustainability is ultimately a deeply societal and behavioural issue, which means that any solution will have to fully imbibe factors of social learning and behavioural change, and their relation to livelihoods, settlements, energy, technology, and production and consumption. Research has shown that large-scale behavioural changes can be triggered and catalyzed through processes of social learning, a process of learning that occurs within a social context. Human beings learn from one another through observational learning, imitation, and modeling.[35]

Building a participatory knowledge base

"In recent years cooperative and collaborative efforts and participatory approaches have become increasingly popular in [...] management and governmental policy. Stakeholder meetings, engaging different actors in workshop settings, have been part of the process."[36] Through the described dynamics and effects of social networks, collective knowledge and understanding of risk factors and change dynamics is built to enhance the performance of governance systems. "All sources of understanding need to be mobilized, and management of complex adaptive systems may benefit from the combination of different knowledge systems."[37] These include formal as well as informal (local, traditional, indigenous) knowledge, which needs to be fed into management practices. "Successful management is characterized by continuous testing, monitoring, and reevaluation to enhance adaptive responses, acknowledging the inherent uncertainty in complex systems. It is increasingly proposed that knowledge generation [...] should be explicitly integrated with adaptive management practices rather than striving for optimization based on past records. This aspect emphasizes a learning environment that requires leadership and changes of social norms within management organizations."[38] Informal and traditional knowledge systems are particularly relevant for the MENA region and can be powerful and effective tools for local adaptation, especially in a developing and transitional country context, or for ecosystem-based approaches. They should be linked with formal systems to inform governance. As adaptation is motivated by the goal of managing risks and building long-term resilience, risk assessment is an indispensable building block for any action of adaptive governance. The World Bank proposes that we have to "understand its exposure and sensitivity to a given set of impacts and develop responsive policies and investments that address these vulnerabilities".[39] Such assessments are a primary building block for adaptive climate governance. They should be integrated into adaptive governance cycles that start with stakeholder-based identification of "what needs to be governed" and result in participatory and multi-level action, including but not limited to official government and technical agencies. It is essential that these elements are carried out as truly participatory processes involving not only "stakeholders" but all relevant social groups and networks. Figure 16.2 shows UN-HABITAT's model of participatory (urban) governance as a stepwise process that starts with stakeholder mobilization and risk assessment, followed by a thematic action prioritization that secures stakeholders' commitment. Step 3 consists of the actual formulation and implementation of strategies and actions, followed by monitoring and consolidation. Being a cycle, the process can enter a next loop with another round of stakeholder engagement. At each step of the process, participation is ensured. This framework may be applied at the regional level in MENA, such that stakeholder mobilization and participation schemes are organized at the sub-regional level, i.e. for North Africa, Levantine, and Gulf states separately.

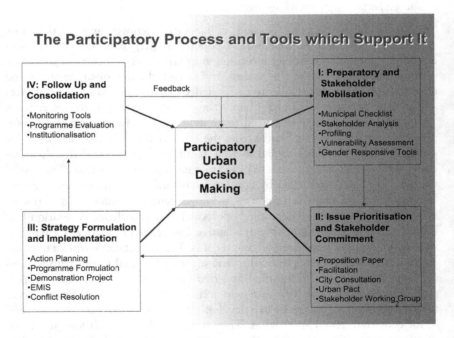

Figure 16.2 UN-HABITAT's participatory decision-making process. UN-HABITAT, note 36.

Integrative planning for enhanced institutional capacity

"Adaptation is [...] an ongoing cycle of preparation, response, and revision. It is a dynamic process, and one that should be revised over time based on new information."[40] Effective adaptive governance has to be structured and organized as a cyclical sequence of assessment, planning, intervention, and learning elements. This requires flexible institutions and multi-level governance systems. "The sharing of management power and responsibility may involve multiple and often polycentric institutional and organizational linkages among user groups or communities, government agencies, and non-governmental organizations, i.e., neither centralization nor decentralization but cross-level inter-actions. Nonresilient [...] systems are vulnerable to external change, whereas a resilient system may even make use of disturbances as opportunities to transform into more desired states."[41] This requires incorporating sustainability aspects into existing plans, policies, and projects. Governments play an important catalytic role in the adaptive governance process.

With the catalytical support of leadership teams, governments in MENA countries can develop specific roadmaps for adaptation that entail plans, policies, and actions as well as a blueprint for performance indicators and the prioritizing of activities. "Activities can include revising master plans (e.g. for land use, transportation, and other sectors) to reflect any significant changes in landscape (e.g. sea level, coastal zones, or floodplains) and natural resources (e.g. water supply), as well as related climate risks. Other related actions might include development of new strategies, high-level political initiatives, policy formulation, and new institutional structures [...]."[42] Plans and policies have to be complemented by another central tool, the development of performance indicators. "Measurement, reporting, and verification are important steps in evaluating the efficiency and effectiveness of a [...] adaptation effort. Demonstrating that an adaptation action or suite of actions has minimized vulnerability, reduced risk, and increased adaptive capacity helps to inform future decisions and satisfy taxpayers and external funders."[43]

Finally, it is important to apply a multilevel governance framework "to explore linkages between national, regional and local policies and to explore the strengthening of multilevel, regional and urban governance [...]. [It] calls for the narrowing or closing of the policy 'gaps' between levels of government via the adoption of tools for vertical and horizontal cooperation [...] [and] recognizes that national governments cannot effectively implement national climate strategies without working closely with regional and local governments as agents of change."[44]

Social resilience, critical sectors, and good governance

Adaptive climate and sustainability governance for MENA must place emphasis on social development aspects and involve addressing basic poverty reduction and the full spectrum of the UN 2030 Agenda and the Sustainable Development

Goals (SDGs).[45] Since adaptive governance is rooted in social networks and their dynamics and draws upon processes of social change, a particular emphasis on social development and the SDG agenda is a prerequisite. The social development dimension of adaptive governance must be linked with adaptation measures for specific sectors: "Climate change adaptation [...] requires collaborative problem solving and coordination across sectors [...] [such as] land use, housing, transportation, public health, water supply and sanitation, solid waste, food security, and energy. Adaptation efforts in any of these sectors will often involve multiple government agencies, as well as broad partnerships that include other governments, local communities, nonprofit organizations, academic institutions, and the private sector."[46] Adaptive climate and sustainability governance includes principles of good governance, as UN-HABITAT notes: "Good [...] governance is characterized by the interdependent principles of sustainability, equity, efficiency, transparency and accountability, security, civic engagement and citizenship."[47] This entails fighting crime and corruption as well as state and institutional fragility or an erosion of the rule of law, as these add to an overall erosion of resilience and coping capacity. Good governance also focuses on the poor, marginalized, and under-privileged, on women, children, families, and youth, and their proactive involvement in decision-making. Again, these factors are of particular relevance to the MENA region in the post Arab Spring context.

Conclusion: advancing adaptive governance in the MENA region

Within the UN system, global environmental governance over time has become a widely stretched and densely populated institutional framework consisting of a multitude of agencies, structures, and bodies—plus ongoing negotiation processes and conference series. Yet, this highly fragmented architecture of agencies and regimes working in the field of global sustainability reveals a number of organizational problems, such as an ineffective and certainly also inefficient multiplication of efforts due to a multitude of actors involved. Also, the principle of multilateralism at the global level has suffered serious setbacks since the events of 9/11 in 2001. National security issues have once more been allowed to dominate the global agenda, bringing forth a restoration of power politics based on national interests. This tendency not only challenges international law, but also undermines all efforts undertaken and already established towards the principles of collective action and regional governance.

There has been a withdrawal from global multilateral cooperation, as seen in strategic policy fields such as arms control; the Iran Nuclear Deal; the International Criminal Court and other courts under the UN aegis; the UNFCCC Paris Agreement (namely, the withdrawal of the USA). Additionally, there have been new bilateral trade treaties created to bypass and undermine the regulatory provisions of the World Trade Organization. Indeed, multilateralism has always been a difficult undertaking in the field of global climate governance due to the perceived "soft" nature of the theme of sustainability. It is precisely in this present vacuum—fragmented and ineffective global institutions on the one hand, and a general lack of interest in multilateralism on the other—that a new approach is necessary and proposed. Moreover, this vacuum is especially harmful for the MENA region. Being haunted by old and new wars, civil wars and other conflicts, fragile or failing statehood, abuse of power and autocratic rule, corruption, unemployment, social injustice and disparity, economic contractions, natural disasters, droughts and desertification, freshwater and food insecurity, a pandemic, climate change and a host of other environmental and sustainability issues, the MENA region suffers greatly from the absence of functional global institutions and multilateralism. What could make an important difference to climate change law and policy in MENA is a deliberate, orchestrated shift toward regional adaptive governance of climate and sustainability issues.

The establishment of public policy networks involving NGOs, community-based organizations (CBOs), and other civil society actors; transnational, regional, and local corporations; and the scientific community are vital in the goal to replace traditional development financing concepts. As are a bottom-up-approach; participatory aspects of policy formulation and implementation; a decentralized logic of intervention; and "partnership agreements". In addition, mainstreaming and integrating climate and ecological aspects and considerations into already existing regional legal and policy frameworks in other sectors, as well as re-clustering regional agreements as climate action frameworks, would catalyze and amplify regional adaptive governance of climate and sustainability issues in MENA.

Notes

1 This chapter draws from A. Rechkemmer, "Back to the Future: Learning from the Evolution of Global Sustainability Governance" in A. B. Brik and L. A. Pal (eds), *The Future of the Policy Sciences* (Cheltenham, UK: Edward Elgar 2021).
2 P. Ekins, J. Gupta, and P. Boileau (eds), *Global Environment Outlook – GEO-6: Healthy Planet, Healthy People* (Nairobi, Kenya: United Nations Environment and Cambridge, Cambridge University Press 2019).
3 J. G. Speth, "The Global Environmental Agenda: Origins and Prospects" in D. C. Esty and M. H. Ivanova (eds), *Global Environmental Governance: Options & Opportunities* (Yale School of Forestry 2002) *Forestry and Environmental Studies Publications Series* 8, 11.
4 See Chapter 1 of this book.
5 Ibid.
6 Ibid.
7 International Monetary Fund, "Globalization: A Brief Overview" (IMF, May 2008) <www.imf.org/external/np/exr/ib/2008/053008.htm>.
8 Ibid.
9 Ibid.
10 N. A. Ashford and R. P. Hall, *Technology, Globalization, and Sustainable Development: Transforming the Industrial State* (2nd ed.) (London: Routledge 2018).
11 Ekins and others, n 2; O. R. Young, L. A. King, and H. Schröder, *Institutions and Environmental Change: Principal Findings, Applications, and Research Frontiers* (Cambridge, MA: MIT Press 2008); A. Rechkemmer, "Environmental Refugees and Environmental Migration" in Gate 3/2000 <https://www.ssoar.info/ssoar/bitstream/h andle/document/11806/ssoar-2004-rechkemmer-global_environmental_governance.

pdf?sequence=1&isAllowed=y&lnkname=ssoar-2004-rechkemmer-global_environm
    ental_governance.pdf>.
12 Ibid.
13 E. Altvater and B. Mahnkopf, *Globalisierung der Unsicherheit* (Münster: Westphälisches
    Dampfboot 2002).
14 See Chapter 1 of this book.
15 M. H. Ivanova, "Partnerships, International, Organizations, and Global Environmental
    Governance" in J. M. Witte, C. Streck, and T. Benner (eds), *Progress or Peril? The
    Post-Johannesburg Agenda* (Washington, DC and Berlin: Global Public Policy Institute
    2003) 9.
16 See the Rio Declaration on Environment and Development, <http://www.unep.org/
    Documents. Multilingual/Default.asp?DocumentID=78&ArticleID=1163> (accessed
    July 12, 2015).
17 Young and others, n 11.
18 A. Rechkemmer, "International Cooperation and Environmental Politics after Rio
    and Johannesburg: Sychronicity of Realities in a Post-Postmodern World?" (2005)
    23 *Sicherheit und Frieden/Security and Peace* 40–44; A. Rechkemmer, *UNEO – Towards
    an International Environment Organization: Approaches to a Sustainable Reform of Global
    Environmental Governance* (Baden-Baden: Nomos 2005).
19 L. B. Andonova, M. M. Betsill, and H. Bulkeley, "Transnational Climate Governance"
    (2009) 9 *Global Environmental Politics* 52–73.
20 H. J. Schellnhuber, and F. Biermann, "Eine ökologische Weltordnungspolitik" (2000)
    55 (12) *Internationale Politik* 9–16.
21 B. Gemmill and B. Bamidele-Izu, "The Role of NGOs and Civil Society in Global
    Environmental Governance" in D. C. Esty and M. H. Ivanova (eds), *Global
    Environmental Governance: Options & Opportunities* (Yale School of Forestry 2002)
    Forestry and Environmental Studies Publications Series 8, 77.
22 Young and others, n 11.
23 F. Biermann and others, "Earth System Governance: People, Places and the Planet:
    Science and Implementation Plan of the Earth System Governance Project" (Earth
    System Governance Report 1, International Human Dimensions Programme on
    Global Environmental Change, Report 20, 2009).
24 UNDP, "Governance for Sustainable Human Development" (United Nations, 1997).
25 UN-HABITAT, "Tools to Support Participatory Urban Decision Making" (Urban
    Governance Toolkit Series, United Nations, Nairobi, Kenya, 2007).
26 The World Bank, "Guide to Climate Change Adaptation in Cities - Executive
    Summary" (The World Bank, 2011) 3–5.
27 C. Folke, T. Hahn, and P. Olsson, "Adaptive Governance of Social-Ecological
    Systems" (2005) 30 *Annual Review of Environmental Resources* 441–473.
28 F. Berkes and C. Folke, *Linking Social-Ecological Systems: Management Practice and
    Social Mechanisms for Building Resilience* (Cambridge University Press 2000); F. Berkes,
    J. Colding, and C. Folke (eds), *Navigating Social-Ecological Systems: Building Resilience
    for Complexity and Change* (Cambridge University Press 2002).
29 The World Bank, n 26, 19.
30 Ibid.
31 Ibid.
32 Ibid, 3–5.
33 Folke and others, n 27.
34 Ibid.
35 J. E. Ormrod, *Human Learning* (3rd ed.) (Upper Saddle River, NJ: Merrill Prentice
    Hall 1999).
36 Folke and others, n 27.
37 Ibid.

38 Ibid.
39 The World Bank, n 26, 3–5.
40 Ibid.
41 Folke and others, n 27.
42 J. Corfee-Morlot, L. Kamal-Chaoui, M. G. Donovan, I. Cochran, A. Robert, and P. J. Teasdale, "Cities, Climate Change and Multilevel Governance" (OECD Environmental Working Papers N° 14, 2009) 36–37.
43 Ibid, 39.
44 Ibid, 7.
45 UN General Assembly, *Transforming Our World: The 2030 Agenda for Sustainable Development*, 21 October 2015, UN Doc. A/RES/70/1.
46 The World Bank, n 26.
47 UN-HABITAT, n 25.

17 Advancing legal preparedness for climate change in the MENA region

Summary and options for policy makers

Damilola S. Olawuyi

Introduction

This book set out to examine the values, assumptions, and guiding principles that underpin climate change law and policy in the MENA region. With case studies from across the region, the book explored the applicable legislation, institutions, as well as lessons learned from emerging innovative and bottom-up approaches to climate regulation across the region. This final chapter reviews salient themes that were discussed and identifies directions for future action and research.

The MENA region, like many parts of the world, is facing a climate emergency.[1] Apart from climate-induced water scarcity, flooding, and fatal heat waves, climate change could have wide-ranging effects on social, economic, environmental, and infrastructure development in MENA countries. Consequently, responding to current and anticipated impacts of climate change is squarely at the forefront of the political and legislative agenda across the region. However, despite the rise in policy formulation on climate change, legal preparedness remains uneven across the region—in terms of having comprehensive laws, policies, training, and harmonized institutions on climate change. There is a need for a detailed and up-to-date assessment of emerging regulatory approaches to climate change mitigation and adaptation to effectively address the impacts of climate change on human development in the MENA region. This book is an attempt to meet this need.

The 17 chapters of this book explore the latest developments of climate change law and policy across the MENA region, with a focus on the key legal innovations across the region in response to the problem of climate change. Yet this book is not simply a stock-taking exercise. The chapters have also explored larger questions on legal and institutional frameworks that can help address broader issues of gender inequality, inadequate stakeholder engagement, lack of transparency, limited access to environmental information, and inadequate scope of available financing for climate change action, all of which stifle the design, approval, financing, and implementation of climate disaster response projects across the region. The case studies also explored innovative legal strategies to address these misalignments and inequities.

This final chapter of the book offers reflections on the case studies. It addresses how lessons from the diverse jurisdictions may inform thinking on how MENA countries can advance existing national strategies and visions on climate change, green economy, and low carbon future, amongst others, through clear and comprehensive legislation. Considering the wide-ranging impacts of climate change on water, energy, food, education, urban planning, infrastructure development, healthcare, aviation, tourism, sports, human rights, and other key sectors, comprehensive and wide-ranging responses are required. However, as shown in the case studies in this book, sectoral and piecemeal efforts on climate change have been prevalent in many MENA countries. This has not fostered a holistic consideration of the interplay and trade-offs between climate actions in key sectors and systems, especially the water, energy, and food (WEF) nexus.[2] Addressing climate change impacts in one domain, without addressing tradeoffs and impacts in other domains, may result in maladaptation and ineffectiveness.[3] Therefore, greater effort is required to mainstream climate change considerations into all aspects of development planning and decision making, in order to effectively anticipate and address short- and long-term impacts of climate change in a holistic manner.

This chapter discusses essential steps for mainstreaming climate change response into all aspects of governmental decision making, in order to accelerate legal preparedness for climate change across the MENA region. After this introduction, the second section discusses legal and institutional reforms that could advance legal preparedness for climate change in the region. The third section is the concluding section.

Mainstreaming climate change considerations into decision-making processes in the MENA region

There is an urgent need for all MENA countries to establish coherent and holistic governance frameworks that mainstream climate change considerations into all aspects of decision making and across key sectors to achieve decarbonization and low carbon transition in the MENA region. Climate change mainstreaming refers to the process of integrating climate change considerations into all aspects of development planning and policy making.[4] Under this governance model, climate change impacts are carefully considered in an integrated manner to ensure that climate change response in one sector does not create other adverse impacts in other sectors of the economy. This section discusses six important steps that could help stakeholders across different sectors to effectively consider and mainstream climate change mitigation and adaptation into all aspects of decision making.

Need for holistic law and policy responses

First, in order for climate change responses to be holistic, sustainable, and effective, there is a need to design policies that address the intersections of climate change with all other key sectors, ranging from human rights, to food, water, energy, education, aviation, and healthcare, amongst others. For example, as discussed in Chapter 4 of this book, there is a need to integrate a public health perspective into climate change planning and policy making to effectively address the short- and long-term dangers of climate change to public health institutions. It is equally important to integrate aviation perspectives into climate change policy and planning to ensure that decarbonization policies do not threaten the short- and long-term sustainability of the aviation sector.[5] Human rights considerations are important as well to avoid reactionary responses that may exacerbate human rights concerns or create other sectorial problems while tackling climate change.[6] Without an integrated policy response, projects designed to address climate change risk may exacerbate social exclusions, human rights violations, energy injustice, environmental impacts, and conflicts in already vulnerable communities.[7] Therefore, there is an urgent need to strengthen synergies and policy coherence in the design, approval, financing, and implementation of climate change mitigation and adaptation efforts across all sectors through an integrative approach. Climate change action can no longer be addressed from a narrow sector-by-sector lens.[8] Rather, climate change must be paramount in policy and decision-making at all levels of government. For example, Article 7(5) of the Paris Agreement specifically encourages countries to integrate climate adaptation into relevant socioeconomic and environmental policies and actions, where appropriate.[9] This will include climate-proofing extant energy infrastructure to enhance their resilience and adaptive capacity; upgrading infrastructure operating and maintenance practices to enhance climate resilience; upgrading building technologies to incorporate low-carbon and energy-efficient materials; and investing in new climate-smart infrastructure to replace aging and inefficient ones.[10] Similarly, the Paris Agreement in its preamble recognized that Parties should, "when taking action to address climate change, respect, promote and consider their respective obligations on human rights".[11] While Article 4(1) also emphasizes the need to achieve mitigation "on the basis of equity, and in the context of sustainable development and efforts to eradicate poverty".[12]

Clear and comprehensive climate change legislation

Second, mainstreaming climate change mitigation and adaptation into all aspects of governance and decision making will require clear regulatory frameworks on climate change. As seen from the case studies, one key limitation to climate change action across the region is the absence of climate change laws in many MENA countries. Some of the legal obligations to mitigate and adapt to climate change can be inferred or drawn from general environmental legislation, national visions, and other policy documents. However, such an approach is indeterminate and may not provide an opportunity for a clear and robust understanding of the critical interplay, tradeoffs, and synergies that exist in decision making and planning across the diverse domains impacted by climate change. For example, lack of comprehensive climate change legislation often means that several of the obligations relating to climate proofing and the use of market-based mechanisms to respond to climate change are found in different segregated and compartmentalized pieces of legislation or instruments—including some which do not specifically refer to important concepts such as climate proofing, carbon taxation, or the use of market-based mechanisms to respond to climate change. Consequently, while the need for climate change action is generally clear, the granularity of what it requires in practice remains unclear. This creates interpretation gaps on the legal basis and source of responsibility when it comes to supervising climate action.

Clear, comprehensive, and specific legislation on climate change is critical to effectively mainstream climate change into the policy apparatus of governments at all levels.[13] A clear legal framework on climate change could provide the legal basis and obligations for project planners and stakeholders to integrate climate resilience into the design, operation, and maintenance of public infrastructure. Under this model, climate change planning would not be considered in a reactionary manner or when there is a disaster. Rather, actions to reduce climate-related risks can become an integral part of, and harmonized with, overall national planning and development processes, as well as disaster and emergency response planning.[14] Climate legislation can also provide project developers and planners with clarity on the key design standards and measures to comply with at the project design and approval stage. Given the urgency and need to address climate change, clear and specific climate legislation could streamline or create exemptions for climate resiliency projects and provide incentives for private sector participation, especially in priority climate and sustainable development projects.[15] Climate legislation could also be the basis for clarifying the regulatory models that will underpin the climate change response in a country. As Dimitropoulos and Lokhandwala note, there are three major regulatory tools to help transition towards more sustainable development policy choices: command-and-control, market-based approaches, and nudges.[16] MENA countries will need to carefully consider complex socio-economic conditions in their countries and develop a responsive legal framework that provides an effective avenue to accelerating climate change action in a cost-effective manner.[17]

Institutional coherence and coordination

In addition to developing a clear and comprehensive legal framework on climate change, creating the right institutional set up for practical coordination and cooperation of the diverse stakeholders and institutions on climate change will also be essential.[18] While climate change may affect different sectors, the process of coordinating a climate response across different agencies and institutions that could be impacted has not been straightforward. Due to the different mandates, priorities, and financial resources between ministries, the attention and focus on mainstreaming climate change into their strategies, policies, and programmes varies significantly. Consequently, climate change mitigation and adaptation programmes continue to be implemented and articulated in a largely sectoral and fragmented manner.[19] This makes it complex and difficult to achieve institutional coherence and coordination of efforts across ministries and agencies in the design and implementation of climate change policies and responses.[20] Furthermore, the implementation of intersectoral coordination is often stifled by capacity questions. For example, mainstreaming climate change into the work of human rights agencies or food ministries would require expanding staff capacity or recruiting experts in climate science. Similarly, implementing climate smart infrastructure projects by public works authorities could require recruiting staff that can understand, analyze, and implement climate-related legislation and policies. Given that these respective institutions are currently not constituted or designed to analyze and implement climate change programmes, their ability to effectively analyze and implement different climate change-related information and data may be limited by differences in skill sets and expertise. Also, epistemic distinctions and lack of coordination by different actors and institutions is fueled by the divergent training, styles, and perspectives of the actors in the respective sectors and fields.

These problems raise the need for greater interoperability and standardization approaches that foster cooperation and minimize duplication amongst sectors and actors, in the design and implementation of climate change projects and programmes. National dialogue should be promoted and aimed at building shared and common understanding on climate change responses by institutional actors in various sectors. Such dialogue will examine to what extent the mandates of existing institutions are coherent, conflicting, and/or duplicative and also whether there are linked platforms in place to support knowledge and information sharing and intersectoral cooperation on climate change. For example, institutions can leverage on their respective expertise, facilities, and best practices by engaging with staff and experts across sectors to assist with reviewing and assessing the implication of multisector projects that respond to climate change. Inter-agency linkages and partnerships, through joint initiatives and knowledge sharing, could increase trust and enhance synergistic solutions that enhance climate change response. Furthermore, such dialogue should aim to address barriers to interoperability by promoting the information sharing and knowledge exchange on climate change by all relevant ministries and agencies in open and linked systems, as well as constituting cross-sectorial panels and committees that can provide an informed picture of climate mainstreaming efforts across the country.

Increased climate financing and investment

Linked to the question of capacity and institutional coordination, is the question of resources. Designing climate-smart infrastructure across all key sectors will come at a considerable cost. For example, the cost of upgrading existing infrastructure, expanding current institutions, staffing, training, field inspections, project review panels, and programme design will require MENA countries to unlock investment from private and public sectors.[21] MENA countries can leverage the significant revenue from extractive industries to drive investment in decarbonization and renewable energy production projects, especially solar power generation. As discussed in this book, the development of large-scale renewable energy systems has been specifically identified as a priority in the national visions of several MENA countries.[22] With infrastructure spending in the Gulf region alone projected to be up to US$200 billion over the next ten years, MENA countries have expressed the political will to invest income from extractive industries.[23] The investments hope to achieve long-term structural change in energy systems by integrating renewable and low carbon electricity into national grids. For example, Qatar is developing the 800 MW Al Kharsaah photovoltaic power project, one of the largest solar power plant projects in the region, which aims to accelerate low carbon energy transition in Qatar.[24] The Kingdom of Saudi Arabia has also launched a renewable energy programme aimed at investing about $50 billion in renewable energy infrastructure by the year 2023.[25] The ultimate target of the programme is to produce about 10 GW of power (30 per cent of the Kingdom's total electricity needs) from renewable energy by 2023; 25,000 MW will be from concentrating solar power plants and 16,000 MW will be from solar photovoltaics. Bahrain, Oman, and the UAE have also announced similar renewable investment plans.[26]

If these programmes and visions are effectively implemented and sustained, MENA countries could be well placed to achieve climate-smart infrastructure development, while also achieving social outcomes, such as improved energy access, and accelerated technology and employment opportunities in critical sectors. In addition to climate-smart infrastructure development, MENA countries can also leverage savings from the extractive industries to finance low carbon transition and decarbonization. For example, MENA countries are home to some of the largest sovereign wealth funds (SWFs) across the world.[27] Such SWF portfolios can provide opportunities for MENA countries to invest in assets and projects worldwide that contribute to decarbonization. For example, the Qatar Investment Authority, Kuwait Investment Authority, the Public Investment Fund of the Kingdom of Saudi Arabia, and the Abu Dhabi Investment Authority are founding members of the One Planet SWF Working Group, which aims to integrate climate change analysis and environmental considerations investment decisions.[28] Through this commitment, MENA countries aim to allocate SWF investments to finance "the smooth transition to a more sustainable, low-carbon economy" as envisaged by the Working Group.[29] By integrating climate change considerations into the design, financing, and implementation of SWF investments, MENA extractive industries can significantly advance global low carbon transition, reduce environmental risks, and promote the attainment of the SDGs domestically and abroad.[30]

Furthermore, investing in research and innovation can provide MENA countries with homegrown and state-of-the-art technologies and tools, needed to effectively address climate change and to transition from hydrocarbon-based economies to lower carbon and diversified economies. For example, Qatar's investment in cryogenic and chemical gas-to-liquid technologies and infrastructure has significantly addressed methane leakage from pipeline transportation. Similarly, the first commercial-scale carbon capture, utilization, and storage

(CCUS) facility in the region, the Al Reyadah Facility, was launched by the UAE in 2017. The US$122 million facility, largely financed by the Abu Dhabi National Oil Company (ADNOC), aims to capture up to 800,000 metric tonnes of CO_2 per year, as a way of addressing climate change and ocean acidification.[31] MENA countries will need to develop sustained research and innovation efforts that can accelerate the development and deployment of CCUS and other essential technologies that reduce environmental pollution from extractive industries and promote transition to renewable and low carbon energy.

Innovative pedagogies to train climate change lawyers and administrators

Fifth, the rise of climate change law and policy across the region raises the need for innovative pedagogies to train and prepare future lawyers and administrators for evidence-based policy making. A discussed by Bell in Chapter 14, addressing the disruptive problem of climate change will require innovative and disruptive teaching methods.[32] She notes that "addressing the particular challenges of climate law requires a departure from traditional teaching methods, and the benefits of enquiry-based learning supported by education technology (EdTech)".[33] EdTech emphasizes the use of online education platforms, such as online learning management systems and discussion fora, that will allow students to continue with their self-directed learning outside of the classroom.[34] This will require law schools across the region to design tailored courses that provide practical and skill-based learning using a wide range of online and in-class tools. Similarly, in developing climate change courses, it is particularly important to adopt enquiry-based learning methods that will integrate practical skills and knowledge, and prepare students to engage in real-world problem-solving.[35] For example, climate change courses should clearly emphasize practical aspects, such as development and implementation of local adaptation strategies, designing climate technologies and innovations, working with local efficiency practices, and collaborating with government in formulating nationally determined contributions and policies.[36]

Furthermore, MENA environmental law academics will need to pool their intellectual resources to cooperate in research and capacity development that advance climate action. Successful partnerships are necessary to achieve all aspects of sustainable development. This is well recognized by the United Nations Sustainable Development Goals (SDGs); Target 17.9 encourages countries to enhance international partnerships and support to implement all of the SDGs and develop North-South, South-South, and triangular cooperation.[37] Target 13.3 also encourages countries to "improve education, awareness-raising and human and institutional capacity on climate change mitigation, adaptation, impact reduction and early warning".[38] Similarly, under Target 4.7 countries are to "ensure that all learners acquire the knowledge and skills needed to promote sustainable development, including, among others, through education for sustainable development and sustainable lifestyles".[39] These goals and targets emphasize the need for environmental law academics to continue to develop collaborative research and teaching initiatives that could expand the development of innovative courses on climate change law and policy. This type of collaboration was in mind when the Association of Environmental Law Lecturers in Middle East and North African Universities (ASSELLMU) was established in 2018 to serve as a professional network for all MENA environmental law academics.[40] As stated in its constitution, ASSELLMU aims to provide a forum for the discussion and dissemination of environmental law in the MENA region and to identify and search for MENA solutions to the region's environmental problems. It also aims to "spearhead collaborative and multi-disciplinary projects to foster increased awareness for environmental problems, among students and people in MENA region".[41]

Since its establishment, ASSELLMU has organized a number of conferences and events that aim to enhance knowledge dissemination and information exchange amongst its members.[42] It is important for international organizations, private sector actors, as well as governments across the region, to provide technical and financial support to sustain such important efforts. ASSELLMU, and similar cooperative networks, will need to work with education stakeholders, such as the Ministry of Education, as well as private and public sector organizations, to identify essential executive courses for the development of professionals working on climate change-related projects. Educators can then develop a blend of practical and theoretical courses to provide lifelong learning opportunities for stakeholders tasked with designing and implementing climate change programmes and projects.

Need for climate change due diligence

The rise of climate change law and policy in the MENA region increases the need for business enterprises and private actors to anticipate and address climate-related legal risks in their investments and projects. Climate change due diligence is the process through which enterprises and public actors investigate the direct and indirect impacts of their investments and projects on climate change in order to avoid legal risks.[43] Legal risks refer to the risk of financial, reputational, or investment loss, legal liability, or dispute settlement costs to a company or institution that may arise from a failure to mitigate the direct and indirect impacts of climate change.[44] Corporate activity directly impacts climate change when it fails to mitigate and reduce greenhouse gases (GHGs) that cause climate change, thereby affecting the enjoyment of human rights. On the other hand, the indirect impacts of climate responses and actions on fundamental human rights and livelihoods may also trigger legal liability. For example investing in emission reduction schemes, such as clean development and REDD+ projects, may result in adverse impacts on human rights, especially through displacements of local communities from their lands and forests, exclusion of women and youths from decision-making processes, lack of transparency and accountability, and inadequacy of judicial remedies for victims.[45] A significant number of law suits are arising in different parts of the world that seek to establish legal and human rights liability of business enterprises for failing to mitigate climate change by reducing their levels of GHGs.[46]

As more MENA countries adopt climate change legislation and policies, similar arguments may be made before courts. Over time, failure to effectively anticipate and manage climate-related risks could carry significant financial, legal, and reputational risks for business enterprises operating in the MENA region. Such risks may manifest in the form of disruption of projects, suspensions, or closures of projects by supervisory bodies, hefty regulatory fines, director and shareholder liability, amongst others, and may impact a company's profitability or ability to maintain license to operate.[47] Businesses can avoid the increasing backlash associated with climate-related risks by integrating climate change and human rights considerations into investment analysis and management processes. There is a strong business case, in terms of cost, reputation, and effectiveness, for doing so, even if the domestic law is silent. Adopting sound internal screening processes, as part of corporate due diligence and investment risk management frameworks, could help enterprises to anticipate, mitigate, and address the implications of their investments and projects on climate change.

As for the role of lawyers representing companies and investors in transactions and projects, it is important to clarify and emphasize the need for climate change due diligence at early stages of such investments. International bodies and courts insist that business enterprises have a role in anticipating and addressing causes of climate change. As a result, companies, investors, lenders, and insurance underwriters must appropriately investigate, assess, and price climate-related risks that may arise in the transactional context. Such risks may be structural, transitional, and disclosures related. Structural risks relate to possible disruptions to a company's or third party's business or assets (e.g., facilities, infrastructure, land, or resources) due to physical impacts of climate change, such as rising sea levels, more extreme storms, floods, fires, and drought. On the other hand, transitional risks may arise from increased adoption of low carbon policies across the MENA region, which may reduce the value or worth of oil and gas-related assets or increase compliance costs. For example, ongoing transitions to low carbon energy systems may result in new standards that may fundamentally change business structures; limit the availability of funding for fossil-related projects; raise the cost of debts, capital, and long-dated securities; or ultimately result in stranded assets. For example, SWFs in New Zealand, Norway, and France have announced plans to divest from high carbon industries.[48] Norway's SWF has capped coal-related portfolios at 30 per cent, while a number of financial institutions are reducing lending for extractive projects.[49] These transitional risks will need to be carefully considered by lawyers when giving advice, especially given the rapid evolution of climate change standards and policies in the MENA region. Thirdly, as countries adopt climate change law and policy, business enterprises will have increased obligations to report on GHG mitigation efforts, as well as the financial impacts of climate-related risks and opportunities on the business and its shareholders.[50] For example in May 2017, 62 per cent of shareholders of ExxonMobil voted in favour of climate disclosures.[51] This decision shows how the rise of climate change law and policy in the region may come with increased expectation for climate disclosure by shareholders and regulatory agencies. Legal risks relating to non-disclosure, or release of misleading or inaccurate climate disclosures, will need to be carefully considered by lawyers when giving advice. Furthermore, the likely impact of climate-related risks and opportunities on the current and future financial position of a company as reflected in its income statement, cash flow statement, and balance sheet could play significant roles when negotiating asset value.[52] These and other obligations that may arise from emerging climate change law and policy must be carefully monitored and considered by lawyers as part of corporate due diligence and contract negotiation processes to prevent legal risks.

In essence, climate change mainstreaming should not be seen as a task for governments alone. Business enterprises, investors, lenders, insurance companies, and their lawyers alike, all have prominent roles to play in integrating climate change considerations into business decision-making processes and planning.

Conclusion

The MENA region faces a wide range of climate-related economic, social, and environmental challenges that require a dynamic legal innovation and response. This book has developed a profile of the multifarious regulatory and institutional gaps that limit the coherent development and application of climate change law and policy at national levels. While the manifestations of these challenges may be escalating in some countries, there is a consistent and urgent need for all MENA countries to effectively mainstream climate change considerations into all aspects of development planning and decision making. A starting point will be to develop clear and comprehensive legislation on climate change that clarifies the regulatory models and the roles of different agencies and institutions in the implementation of climate actions and projects. Furthermore, as clear climate change legislation and rules emerge, business enterprises, investors, lenders, insurance companies, and lawyers can reduce legal liability and risks arising from the direct and indirect impacts of their activities on climate change by integrating climate change considerations into business decision-making processes and planning.

The success, or otherwise, of integrating climate change considerations into decision making will, to a large extent, depend on local circumstances and contexts. Especially important are the regulatory frameworks, institutional capacity, and resources that are available to implement and monitor the design, approval, financing, and implementation of climate-smart infrastructure and decarbonization projects. Successful integration also depends on the political will of national authorities and industry stakeholders to support a comprehensive reform process that places climate change squarely at the heart of national planning and budgeting. Considering the extreme vulnerabilities of MENA countries to climate change, there is a need for continuous awareness and education on the part of governments to significantly increase investment in climate change research, education, and technology development programmes to boost legal preparedness to face climate change.

Notes

1 See Chapter 1 of this book.
2 See Chapter 8 of this book. See also D. Olawuyi, "Sustainable Development and Water-Energy-Food Nexus: Legal Challenges and Emerging Solutions" (2020) 103 *Journal of Environmental Science and Policy* 1.
3 See Chapters 9 and 10 of this book.
4 See United Nations Environment Programme (UNEP), "Mainstreaming Climate Change Adaptation into Development Planning: A Guide for Practitioners", Environment for the MDGs, UNDP-UNEP Poverty-Environment Initiative, 2011, available at <www.cbd.int/financial/climatechange/g-climatedapationguide-undp.pdf>.
5 D. Olawuyi, "Sustainable Aviation and the Transfer of Environmentally Sound Technologies to Africa: Paradoxes, Barriers and Prospects" in A. de Mestral, P. Fitzgerald, and T. Ahmad (eds), *Sustainable Development, International Aviation, and Treaty Implementation* (Cambridge University Press 2018) 154.
6 N. Roht-Arriaza, "Human Rights in the Climate Change Regime" (2010) 1 *Journal of Human Rights and the Environment* 211 (where the author identifies areas where current climate change regimes may cause human rights violations in local communities); D. Olawuyi, *The Human Rights Based Approach to Carbon Finance* (Cambridge University Press 2016) 1–25.
7 See D. Olawuyi, "Energy Poverty in the Middle East and North African (MENA) Region: Divergent Tales and Future Prospects" in I. Del Guayo, L. Godden, D. N. Zillman, M. F. Montoya, and J. J. Gonzalez (eds), *Energy Law and Energy Justice* (Oxford University Press 2020) 254–272.
8 See Chapter 9 of this book.
9 United Nations Framework Convention on Climate Change, *Adoption of the Paris Agreement*, 12 December 2015, UN Doc. FCCC/CP/2015/L.9, Article 7(5) [Paris Agreement].
10 Ibid, Article 9(e); D. Verner (ed) *Adaptation to A Changing Climate in the Arab Countries: A Case for Adaptation Governance and Leadership in Building Climate Resilience* (International Bank for Development and Reconstruction, The World Bank 2012) 197.
11 Paris Agreement, n 9, Preamble.
12 Ibid, Article 4(1).
13 Australian Government, "The Role of Regulation in Facilitating or Constraining Adaptation to Climate Change for Australian Infrastructure: Report for the Department of Climate Change and Energy Efficiency," Maddocks, 2011, stating that the resilience of infrastructure to the effects of climate change will depend upon the applicable regulatory framework that eliminates and responds to climate risks.
14 Olawuyi, n 7.
15 Ibid.
16 See Chapter 4 of this book.
17 Ibid. See also Chapter 11.
18 See Chapter 6 of this book.
19 See Chapter 8 of this book.
20 Olawuyi, n 2.
21 Ibid.
22 See Chapter 13. See also D. Olawuyi, "Advancing Innovations in Renewable Energy Technologies as Alternatives to Fossil Fuel Use in the Middle East: Trends, Limitations, and Ways Forward" in D. Zillman, M. Roggenkamp, L. Paddock, and L. Godden (eds), *Innovation in Energy Law and Technology: Dynamic Solutions for Energy Transitions* (Oxford University Press 2018) 354–370.
23 Olawuyi, ibid.

24 Reuters/The Qatar Peninsula, "Qatar to Build 800 MW Solar Power Plant on 10 sqkm Plot" *The Peninsula* (Doha, 19 January 2020) <https://thepeninsulaqatar.com/article /19/01/2020/Qatar-to-build-800-MW-solar-power-plant-on-10-sqkm-plot>.
25 Caline Malek, "Saudi Arabia Joins Club of Middle East's 'Green Energy' Leaders" (Arab News, January 19, 2020) <https://www.arabnews.com/node/1615406/saudi-a rabia>.
26 Olawuyi, n 22.
27 See D. Olawuyi, "Can MENA Extractive Industries Support the Global Energy Transition? Current Opportunities and Future Directions" (2020) *Extractive Industries and Society Journal* <https://doi.org/10.1016/j.exis.2020.02.003>
28 J. Beard, "Eight Asset Managers Unite for One Planet Initiative" (City Wire, 11 July 2019) <https://citywireselector.com/news/eight-asset-managers-unite-for-one-planet -initiative/a1249346>.
29 Ibid.
30 R. Sharma, "Sovereign Wealth Funds Investment in Sustainable Development Sectors" (UN 2017) <www.un.org/esa/ffd/high-level-conference-on-ffd-and-2030-age nda/wp-content/uploads/sites/4/2017/11/Background-Paper_Sovereign-Wealth-Fu nds.pdf>.
31 Power Engineering International (PEI), "MENA's First CCUS Project Now Operational" (PEI 6 January 2017) <www.powerengineeringint.com/2017/01/06/m ena-s-first-ccusproject-now-operational/>.
32 See Chapter 14.
33 Ibid.
34 M. Flavin, "Disruptive Technologies in Higher Education" (2012) 20 *Research in Learning Technology* 102.
35 D. C. Edelson, D. N. Gordin, and R. D. Pea, "Addressing the Challenges of Inquiry-Based Learning through Technology and Curriculum Design" (1999) *Journal of the Learning Sciences* 391, 392–393.
36 See Chapter 14.
37 United Nations, "Transforming Our World: The 2030 Agenda for Sustainable Development," 21 October 2015, UN Doc. A/RES/70/1, available at <https:// sustainabledevelopment.un.org/content/documents/21252030%20Agenda%20for %20Sustainable%20Development%20web.pdf>, p. 31, Target 17.9.
38 Ibid, p. 27, Target 13.3.
39 Ibid, p. 21, Target 4.7.
40 See Chapter 2.
41 ASSELLMU, *Constitution of the Association of Environmental Law Lecturers in Middle East and North Africa Universities* (Adopted in Casablanca, Morocco in 2019).
42 D. S. Olawuyi, "Spearding Environmental Legal Education in the MENA Region" (2019) 1 *MENA Business Law Review* 68.
43 D. Olawuyi, "Climate Justice and Corporate Responsibility: Taking Human Rights Seriously in Climate Actions and Projects" (2016) 34 *Journal of Energy and Natural Resources Law* 27.
44 Ibid.
45 Ibid.
46 According to a recent United Nations report, at least 1,550 climate change cases have been filed in 38 countries as of July 2020. See United Nations Enivronment Program (UNEP), *The UNEP Global Climate Litigation Report: 2020 Status Review* (UNEP, 2020) 8-10. See, for example, *Urgenda Foundation v The State of the Netherlands* (*Ministry of Infrastructure and the Environment*, 24 June 2015, Case Number: C/09/456689/ HA ZA 13-1396), available at <https://elaw.org/system/files/urgenda_0.pdf>; *Milieudefensie et al. v Royal Dutch Shell Plc., Urgenda Foundation (on behalf of 886 individuals) v The State of the Netherlands*, available at <http://climatecasechart.com/non-us-case/milieudefensie-

et-al-v-royal-dutch-shell-plc/>; The Greenpeace Southeast Asia & Philippine Rural Construction Movement Petition, "Petition to the Commission on Human Rights of the Philippines Requesting for Investigation of the Responsibility of the Carbon Majors for Human Rights Violations or Threats of Violations Resulting from the Impacts of Climate Change submitted by Greenpeace Southeast Asia and Philippine Rural Reconstruction Movement", available at <https://www.greenpeace.org/static/planet4 -philippines-stateless/2019/05/be889456-be889456-cc-hr-petition.pdf>.

47 Olawuyi, n 43; B. Ugochukwu, "Litigating the Impacts of Climate Change: The Challenge of Legal Polycentricity" (2018) 7 *Global Journal of Comparative Law* 91.

48 J. Capapé, "Financing Sustainable Development: The Role of Sovereign Wealth Funds for Green Investment" UN Environment, Green Economy, Policy Brief (2018).

49 Qatar Tribune, "Norway Sovereign Wealth Fund, World's Biggest, to Dump Oil and Gas" *Qatar Tribune* (Doha, 8 March 2019).

50 Task Force on Climate-related Financial Disclosures, "Final Report: Recommendations of the Task Force on Climate-Related Financial Disclosures" (June 2017) <https://assets.bbhub.io/company/sites/60/2020/10/FINAL-2017-TCFD-Report -11052018.pdf>, discussing the importance of climate-related financial disclosures in helping investors, lenders, and insurance underwriters to appropriately investigate, assess, and price climate-related risks.

51 M. McGrath, "Exxon Shareholders Back 'Historic' Vote on Climate" (BBC, 31 May 2017) <www.bbc.com/news/science-environment-40106278>.

52 Task Force on Climate-related Financial Disclosures, n 50.

Index

Page numbers in bold denote tables, those in *italic* denote figures.

Printed in the United States
by Baker & Taylor Publisher Services